Sustainable Agriculture Reviews

Volume 41

Series Editor

Eric Lichtfouse
Aix-Marseille Université, CNRS, IRD, INRA
Coll France, CEREGE
Aix-en-Provence, France

Other Publications by Dr. Eric Lichtfouse

Books
Scientific Writing for Impact Factor Journals
https://www.novapublishers.com/catalog/product_info.php?products_id=42242

Environmental Chemistry
http://www.springer.com/978-3-540-22860-8

Sustainable Agriculture
Volume 1: http://www.springer.com/978-90-481-2665-1
Volume 2: http://www.springer.com/978-94-007-0393-3

Book series
Environmental Chemistry for a Sustainable World
http://www.springer.com/series/11480

Sustainable Agriculture Reviews
http://www.springer.com/series/8380

Journal
Environmental Chemistry Letters
http://www.springer.com/10311

Sustainable agriculture is a rapidly growing field aiming at producing food and energy in a sustainable way for humans and their children. Sustainable agriculture is a discipline that addresses current issues such as climate change, increasing food and fuel prices, poor-nation starvation, rich-nation obesity, water pollution, soil erosion, fertility loss, pest control, and biodiversity depletion.

Novel, environmentally-friendly solutions are proposed based on integrated knowledge from sciences as diverse as agronomy, soil science, molecular biology, chemistry, toxicology, ecology, economy, and social sciences. Indeed, sustainable agriculture decipher mechanisms of processes that occur from the molecular level to the farming system to the global level at time scales ranging from seconds to centuries. For that, scientists use the system approach that involves studying components and interactions of a whole system to address scientific, economic and social issues. In that respect, sustainable agriculture is not a classical, narrow science. Instead of solving problems using the classical painkiller approach that treats only negative impacts, sustainable agriculture treats problem sources.

Because most actual society issues are now intertwined, global, and fast-developing, sustainable agriculture will bring solutions to build a safer world. This book series gathers review articles that analyze current agricultural issues and knowledge, then propose alternative solutions. It will therefore help all scientists, decision-makers, professors, farmers and politicians who wish to build a safe agriculture, energy and food system for future generations.

More information about this series at http://www.springer.com/series/8380

Shamsul Hayat • John Pichtel
Mohammad Faizan • Qazi Fariduddin
Editors

Sustainable Agriculture Reviews 41

Nanotechnology for Plant Growth and Development

Editors
Shamsul Hayat
Plant Physiology & Biochemistry Section,
Department of Botany, Faculty of Life
Sciences
Aligarh Muslim University
Aligarh, Uttar Pradesh, India

Mohammad Faizan
Tree Seed Center, College of Forest
Resources and Environment
Nanjing Forestry University
Nanjing, People's Republic of China

John Pichtel
Environment, Geology and Natural
Resources
Ball State University
Muncie, IN, USA

Qazi Fariduddin
Plant Physiology & Biochemistry Section,
Department of Botany, Faculty of
Life Sciences
Aligarh Muslim University
Aligarh, Uttar Pradesh, India

ISSN 2210-4410 ISSN 2210-4429 (electronic)
Sustainable Agriculture Reviews
ISBN 978-3-030-33998-2 ISBN 978-3-030-33996-8 (eBook)
https://doi.org/10.1007/978-3-030-33996-8

This Springer imprint is published by the registered company Springer Nature Switzerland AG.
The registered company address is: Gewerbestrasse 11, 6330 Cham, Switzerland

Preface

Nanotechnology is the branch of science which involves the study and application of nanoparticles (NPs), those entities which exist within the 1–100-nm-size range in at least one dimension. Nanotechnology applications include improvement of agricultural production using bioconjugated NPs (encapsulation), transfer of DNA into plants for the cultivation of pest- resistant varieties, nanoformulations of agrochemicals for pesticides and fertilizers for crop improvement, and nanosensors/nanobiosensors in crop protection for the identification of diseases and residues of agrochemicals. Preliminary results on the current agricultural use of nanotechnology by densely populated countries such as China and India indicate that this technology may offer a significant impact on reducing hunger, malnutrition, and child mortality. Different types of NPs (ZnO-NPs, Au-NPs, CuO-NPs, CNTs, $AgNO_3$-NPs, and TiO_2-NPs) have been studied for plant growth and development. Particle size, size distribution, shape, surface and core chemistry, crystallinity, agglomeration state, purity, redox potential, catalytic activity, surface charge, and porosity are all important for understanding the behavior of NPs. Thus, nanomaterials have witnessed increased scrutiny in the basic and applied sciences as well as in bionanotechnology. This book is dedicated to presenting the latest developments of the role of nanoparticles on plant growth and development.

This work is composed of 11 chapters. Chapter 1 provides the scope and challenges of nanoparticle applications in plants. Chapter 2 discusses phosphorus transformations and availability upon nanoparticle application, a topic not covered in any previous volume on nanoparticles. Recent progress in synthesis of metal/metal oxide nanoparticles by green methods and their applications is covered in Chap. 3. Chapter 4 summarizes the current understanding of the fate and behavior of ZnO-NPs in plants and their uptake, translocation, and impacts on mitigating several negative plant growth conditions. Latest knowledge of TiO_2 NPs including interactions, transport, and translocation within plants and future perspectives regarding their use is discussed in Chap. 5. Chapter 6 highlights the current understanding as well as future possibilities of Ag-NP research in plant systems. Chapter 7 addresses the role of silicon NPs in plant growth, photosynthesis, and stress tolerance. Chapter 8 presents the nature of copper oxide nanoparticles, their uptake and translocation mechanisms, and potential toxic effects on different plant species at both physiological and cellular levels. This chapter also addresses tolerance mechanisms generated by plants and a critical assessment of the necessity for further research. In Chap. 9, the effects of nanofertilizers with PGPR as an innovative method for improving crop productivity are described. Chapter 10 covers the interaction of engineered nanomaterials with the soil microbiome and plants and their impact on plant and soil health. In Chap. 11, a new threat to crop plants and soil rhizobia, arising from nanoparticles, is presented.

This book is not an encyclopedia of reviews but rather a compendium of newly composed, integrated, and illustrated contributions describing our knowledge of nanoparticles as they influence plant growth and development. The chapters incorporate both theoretical and practical aspects and may serve as baseline information for future research through which significant developments are possible. It is intended that this book be useful to the students, researchers, and instructors, both in universities and research institutes, especially in relation to biological and agricultural sciences.

With great pleasure, we extend our sincere thanks to all the contributors for their timely response, their excellent and up-to-date contributions, and their consistent support and cooperation. We are thankful to all who have helped us in any capacity during the preparation of this volume. We are extremely thankful to Springer Publishing for their expeditious acceptance of our proposal and completion of the review process. The subsequent cooperation and understanding by their staff are gratefully acknowledged. We express our sincere thanks to our family members for all the support they provided and the neglect and loss they suffered during the preparation of this book.

Finally, we are thankful to the Almighty who provided and guided all the channels to work in cohesion on the concept to the development of the final version of this treatise, *Nanotechnology for Plant Growth and Development*.

Aligarh, India — Shamsul Hayat
Muncie, IN, USA — John Pichtel
Nanjing, China — Mohammad Faizan
Aligarh, India — Qazi Fariduddin

Contents

About the Editors

Shamsul Hayat is Professor in the Department of Botany, Aligarh Muslim University, Aligarh, India, where he received his Ph.D. degree in Botany. Before joining the Department as Faculty, he has worked as Research Associate and Young Scientist in the same Department. He has also worked as Associate Professor in King Saud University, Riyadh, Saudi Arabia, and as a BOYSCAST Fellow at the National Institute of Agrobiological Sciences, Tsukuba, Japan. His major areas of research are plant hormone, nanoscience, and abiotic stress in plants. It has been reported from his group that phytohormones, such as brassinosteroids and salicylic acid, play an important role in increasing the photosynthetic efficiency of the plant and regulate the antioxidant system even under abiotic stress. He is also studying the protein profiling in hormone-treated plants under abiotic stress. He has been awarded Prof. Hira Lal Chakravarty Award by the Indian Science Congress Association, Kolkata, India; Associate of the National Academy of Agricultural Sciences, New Delhi, India; BOYSCAST Fellow by the Department of Science and Technology, Government of India, New Delhi; and Young Scientist by the Association of the Advancement of Science, Aligarh, India. He has been the Principal Investigator of the various projects sanctioned by the different agencies and guided five students for the award of Ph.D. degree and two students for the award of M.Phil. degree. He has published more than 160 research papers and chapters in leading journals of the world, such as *Environmental & Experimental Botany*, *Plant Physiology and Biochemistry*, *Environmental Pollution*, *Nitric Oxide*, *Protoplasma*, *Plant Signaling & Behavior*, and *Photosynthetica*, *Acta Physiologiae Plantarum* with a good citation records (citation: 10,400 and h index 48), and also published 10 books by Kluwer Academic, Springer, Wiley-VCH, Science Publisher, and Narosa Publishing House. He has presented his work at several national and international conferences in Japan, Brazil, Spain, China, and Saudi Arabia. He is a regular Reviewer and on the panel of editorial boards of the national and international journals. He is also member of important national and international scientific societies.

John Pichtel is Professor of Natural Resources and Environmental Management at Ball State University in Muncie, IN, USA, where he has been on the faculty since 1987. He received his Ph.D. degree in Environmental Science and his M.S. degree in Soil Chemistry/Agronomy both from The Ohio State University and his B.S. degree in Natural Resources Management from Rutgers University. His primary research and professional activities embrace management of hazardous materials, remediation of contaminated sites, and environmental chemistry. His published papers have addressed topics such as phytoremediation of metal-contaminated sites, microbial degradation of petroleum compounds, transformations of energetic compounds in soil, and reclamation of coal mine spoils. He has written two books addressing waste management and cleanup of contaminated sites and has authored or coauthored seven book chapters. He has been the Author or Coauthor of approximately 60 research articles which have been published in the *Journal of Environmental Quality*, *Environmental Engineering Science*, *Journal of Bioremediation and Biodegradation*, and other international journals. He has been a Visiting Scientist at the University of Stirling (UK), the University College Cork (Ireland), Tampere University of Technology (Finland), and the University of Siena (Italy). He was selected as a Fulbright Scholar in 1999 and again in 2005. He teaches courses in emergency response to hazmat incidents, management of solid and hazardous wastes, environmental site assessment, and site remediation. He is a Certified Hazardous Materials Manager. He holds memberships in the Institute of Hazardous Materials Managers, the Sigma Xi Scientific Society, the International Association of Arson Investigators, and the Indiana Academy of Science. He has received extensive training from the US Federal Emergency Management Agency for response to chemical, biological, radiological, and explosives hazards. He has served as a Consultant in hazardous waste management projects and has conducted environmental assessments and remediation research in the United States, the United Kingdom, Ireland, Finland, Italy, and Poland.

Mohammad Faizan is working as Postdoctoral Fellow at Nanjing Forestry University, Nanjing, China. He has received his Ph.D. degree in Botany from Aligarh Muslim University, Aligarh, India. His research interests include the manifestations of plant growth regulators especially brassinosteroids and nanoparticles under natural and unnatural conditions. The findings of his research works have been published in the international journal of high-impact factor such as *IET Nanobiotechnology*, *Photosynthetica*, *Plant Physiology and Biochemistry*, *Nitric Oxide*, and many more. He is also member of various societies.

Qazi Fariduddin is working as Professor of Botany at Aligarh Muslim University, Aligarh, India. He has been extensively working in the field of Agricultural Biotechnology to explore the abiotic stress tolerance mechanism in plants through physiological and molecular approaches. The findings of his work have revealed that exogenous application of brassinosteroids (BRs) improved the yield and quality of plants under low temperature, salt, water, and heavy metal stresses. He had visited Göttingen University, Göttingen, Germany, for 6 months under BOYSCAST

Fellowship. He has also visited Michigan State University, Michigan, USA, under Raman Fellowship. The findings of his research works have been published in the international journal of high-impact factor such as *Proceedings of the National Academy of Sciences of the United States of America*, *Food Chemistry*, *Plant Physiology and Biochemistry*, *Chemosphere*, *Journal of Integrative Plant Biology*, *Environmental and Experimental Botany*, *Ecotoxicology and Environmental Safety*, *Nature-Scientific Reports*, and many more. He has successfully completed various funded research projects from reputed funding agencies. He has also supervised four doctoral students and three M.Phil. students and a number of master students. He has been the Investigator of various projects and also attended several national and international conferences.

Contributors

Khan Bilal Mukhtar Ahmed Plant Physiology & Biochemistry Section, Department of Botany, Faculty of Life Sciences, Aligarh Muslim University, Aligarh, Uttar Pradesh, India

M. A. Ali Department of Soil Science, Faculty of Agricultural Sciences & Technology, Bahauddin Zakariya University, Multan, Pakistan

M. Arshad Institute of Environmental Sciences and Engineering (IESE), School of Civil and Environmental Engineering (SCEE), National University of Sciences and Technology (NUST), Islamabad, Pakistan

T. M. Salem Attia Soils, Water and Environment Research Institute, Agricultural Research Center, Giza, Egypt

H. K. Choi College of Pharmacy, Chung-Ang University, Seoul, Republic of Korea

N. I. Elsheery Agriculture Botany Department, Faculty of Agriculture, Tanta University, Tanta, Egypt

Mohammad Faizan Tree Seed Center, College of Forest Resources and Environment, Nanjing Forestry University, Nanjing, People's Republic of China

Ahmad Faraz Plant Physiology & Biochemistry Section, Department of Botany, Faculty of Life Sciences, Aligarh Muslim University, Aligarh, Uttar Pradesh, India

Qazi Fariduddin Plant Physiology & Biochemistry Section, Department of Botany, Faculty of Life Sciences, Aligarh Muslim University, Aligarh, Uttar Pradesh, India

M. Q. U. Farooqi Department of Soil Science, School of Agriculture and Environment, Faculty of Science, The University of Western Australia, Perth, Australia

A. Fatima Ranjan Plant Physiology and Biochemistry Laboratory, Department of Botany, University of Allahabad, Allahabad, India

B. Fazeli-Nasab Research Department of Agronomy and Plant Breeding, Agricultural Research Institute, University of Zabol, Zabol, Iran

Shamsul Hayat Plant Physiology & Biochemistry Section, Department of Botany, Faculty of Life Sciences, Aligarh Muslim University, Aligarh, Uttar Pradesh, India

M. Jafari Faculty of Chemistry, University of Mazandaran, Babolsar, Iran

A. Jose School of Biosciences, Mahatma Gandhi University, Kottayam, India

B. J. Joseph School of Biosciences, Mahatma Gandhi University, Kottayam, India

Shams Tabrez Khan Department of Agricultural Microbiology, Faculty of Agricultural Sciences, Aligarh Muslim University, Aligarh, Uttar Pradesh, India

S. Siavash Moghaddam Department of Agronomy, Faculty of Agriculture, Urmia University, Urmia, Iran

A. R. Nayana School of Biosciences, Mahatma Gandhi University, Kottayam, India

John Pichtel Environment, Geology and Natural Resources, Ball State University, Muncie, IN, USA

J. Popović-Djordjević Chair of Chemistry and Biochemistry, Faculty of Agriculture, University of Belgrade, Belgrade, Serbia

L. Pourakbar Department of Biology, Faculty of Science, Urmia University, Urmia, Iran

S. M. Prasad Ranjan Plant Physiology and Biochemistry Laboratory, Department of Botany, University of Allahabad, Allahabad, India

E. K. Radhakrishnan School of Biosciences, Mahatma Gandhi University, Kottayam, India

H. Rasouli Medical Biology Research Center (MBRC), Kermanshah University of Medical Science, Kermanshah, Iran

Fareen Sami Plant Physiology & Biochemistry Section, Department of Botany, Faculty of Life Sciences, Aligarh Muslim University, Aligarh, Uttar Pradesh, India

R. Z. Sayyed Department of Microbiology, PSGVP Mandal's Arts, Science & Commerce College, Shahada, Maharashtra, India

Husna Siddiqui Plant Physiology & Biochemistry Section, Department of Botany, Faculty of Life Sciences, Aligarh Muslim University, Aligarh, Uttar Pradesh, India

S. Singh Ranjan Plant Physiology and Biochemistry Laboratory, Department of Botany, University of Allahabad, Allahabad, India

Z. Zahra Institute of Environmental Sciences and Engineering (IESE), School of Civil and Environmental Engineering (SCEE), National University of Sciences and Technology (NUST), Islamabad, Pakistan

S. Zarayneh Department of Biology, Science and Research Branch, Islamic Azad University, Tehran, Iran

Chapter 1
Nanomaterials: Scope, Applications, and Challenges in Agriculture and Soil Reclamation

T. M. Salem Attia and N. I. Elsheery

Abstract Nanotechnology has attracted scientists for study and exploitation of the unique physical, chemical and biological characteristics of nanomaterials. Nanomaterials are being developed for applications to a wide range of fields including medicine, drug delivery, electronics, fuel cells, solar cells, food preparation, and space exploration. Nanomaterials have already provided numerous benefits to agriculture – nanotechnology possesses the capability to detect and treat plant diseases, enhance photosynthetic rate and nutrient absorption by plants, deliver active ingredients to specific sites and treat water to remove contaminants. The potential of nanotechnology in agriculture and its effect on the planet is vast. This chapter will address the benefits of nanomaterials to agriculture and also to reclamation of disturbed lands.

Keywords Nanotechnology · Plant germination · Pesticide detection · Nano-fertilizers · Land reclamation

1.1 Introduction

Nanotechnology has many definitions that embrace the properties and applications of nanomaterials. The most basic definition of nanomaterials is, "those materials that measure between 1 and 100 nm (NNI 2005). Material classification based simply on size does not, however, offer a satisfactory definition. Many nano-sized structures (e.g. weathered minerals) are present in the environment naturally (Masciangioli and Zhang 2003), and do not fall into the category of nanoparticles.

T. M. S. Attia
Soils, Water and Environment Research Institute, Agricultural Research Center, Giza, Egypt

N. I. Elsheery (✉)
Agriculture Botany Department, Faculty of Agriculture, Tanta University, Tanta, Egypt
e-mail: nshery@agr.tanta.edu.eg

S. Hayat et al. (eds.), *Sustainable Agriculture Reviews 41*, Sustainable Agriculture Reviews 41, https://doi.org/10.1007/978-3-030-33996-8_1

The requirements for classification as nanotechnology include the model that materials must possess unique physical, chemical, and/or biological characteristics, different from those found at bulk scale in the same material (NNI 2005). Compounds also must be formed on the principle of atomic scale control of assembly and structure.

The National Nanotechnology Initiative (NNI) divides the definition of nanotechnology into three requirements; in other words, any 'nanotechnology' must involve all three: (1) research and technology improvement at the atomic, molecular, or macromolecular levels, in the length scale of approximately 1–100 nanometers; (2) creating and using structures, devices, and systems that have novel properties and functions because of their small and/or intermediate sizes; and (3) ability to be controlled or manipulated at the atomic scale (NNI 2005). The key underlying concept in nanotechnology is that the properties and behavior of matter change markedly at the nanoscale. For example, nanoscale status observed size and structure dependent; consequently, optical, electrical, interfacial and tensional properties are changes.

In addition to increased reactivity, the catalytic activity of nanomaterials is often greater than that of the same elements at the macro scale. Enhancing the physical and chemical properties of nanomaterials can lead to the creation of novel functional materials with enormous benefits for addressing some of the great challenges of modern society, e.g., energy production and storage, water treatment, lighter and stronger vehicles, better health care, efficient computers, etc.

Huge investments in nanotechnology are resulting in production many types of nanoparticles (NPs) that are moving rapidly from laboratories to mass markets. Nanoparticles will continue to be more heavily used in consumer products and industrial and commercial scenarios.

There is some cause for concern regarding the hazard potential of NPs. In particular, the high reactivity may impart adverse biological effects. Nanoparticles are comparable in size to certain structural components of cells and are therefore small enough to penetrate biological barriers; additionally, most nanomaterials are persistent. Therefore, in order to encourage sustainable development of nanotechnologies and safeguard human health and ecosystems it is necessary to assess the risks side-by-side with research and development of nanomaterials.

Nanomaterials were developed for many applications in many fields such as Medicine, drug delivery, electronics, fuel cells, solar cells, food and space and etc. these application could be nominalized as follows: (1) Nanomaterials are developed to have many beneficial impacts in medicine according to the size of molecules that can deliver drugs directly to diseased cells in your body. When it's perfected, this method should significantly decrease the damage treatment such as chemotherapy does to a patient's healthy cells. (2) Nanotechnology holds some answers for how we enhance the capabilities of electronics strategies while we reduce their weight and power consumption. (3) Nanotechnology has many beneficial impacts on several aspects of food science, from how food is grown to how it is packaged. Nanomaterials developed will make a difference not only in the food taste, but also in food safety, and the health benefits that food provides. (4) Nanomaterials have a huge catalyst reactivity that reduce the cost of catalysts used in fuel cells to produce hydrogen ions from fuel such as methanol and to improve the efficiency of membranes used in fuel cells to isolate hydrogen ions from other gases such as oxygen. (5) Developed nanotech solar cells are lower cost than conventional solar cells. Currently, researchers developing batteries using nanomaterials. One such battery will be a noble as new after storage for decades. Also batteries manufactured from nanomaterials could be recharged significantly faster than conventional batteries. (6) Nanotechnology may hold the key to making space-flight more practical. Innovations in nanomaterials make lightweight spacecraft and a cable for the space elevator promising. By significantly reducing the amount of rocket fuel required, these advances could lower the cost of reaching orbit and traveling in space. (7) Nanotechnology has many benefits to solve the shortage of fossil fuels such as diesel and gasoline by making the production of fuels from low grade raw materials economical, increasing the mileage of engines, and making the production of fuels from common raw materials more effective. (8) Nanotechnology can increase the powerful of catalysts used to transform vapors released from cars or industrial plants into beneficial gasses. This was attributed to high surface area of catalysts made from nanoparticles to interact with the reacting chemicals than catalysts made from larger particles. The larger surface area allows more chemicals to react with the catalyst progressively, which makes the catalyst more efficient. (9) Nanotechnology has many applications to solve many problems in water purification. One of these problems is the removal of industrial wastes, such as a cleaning solvent called TCE, from groundwater. Nanoparticles have the ability for biodegradation of these chemicals and converted it to harmless components. Studies have shown that this method is more efficient and lower costs in ground water treatment than methods which require pumping the water out of the ground for treatment. (10) Nanotechnology can be used as sensors to detect very small amounts of chemical vapors. Various types of nanomaterials, such as carbon nanotubes, zinc oxide nanowires or palladium nanoparticles can be used as sensors. Because of the small size of nanotubes, nanowires, or nanoparticles, a few gas molecules are enough to change the electrical properties of the sensing elements. This allows detecting a very low concentration of chemical vapors. (11).

Among its many applications nanotechnology is proven to offer substantial benefits to agriculture. Utilization of nanomaterials in agriculture focuses in particular on reducing application of plant protection products, minimizing nutrient losses from fertilization, and increasing yields through optimized nutrient management. Nanotechnology has been proven with the ability to detect and treat certain plant diseases, enhance nutrient absorption by plants, deliver active ingredients to specific sites, and treatment of water supplies. Relevant materials include nanocapsules, nanoparticles and even viral capsids. The use of target-specific nanoparticles can decrease damage to non-target plant tissues and the amount of chemicals released to the environment. Nanotechnology can also be applied to plant breeding and genetic transformation.

Nanomaterials and nanostructures with unique chemical, physical, and mechanical properties (e.g. electrochemically active carbon nanotubes, nanofibers and fullerenes) have been applied in highly sensitive biochemical sensors. These nanosensors have related application in agriculture, in particular for soil analysis, biochemical sensing and control, water analysis, and pesticide and nutrient detection and quantification. Nanotechnology also plays an important role in the treatment of agricultural waste products.

1.2 Nanotechnology Applications in Agriculture

1.2.1 Plant Germination and Growth

The effects of nanomaterials on seed germination and growth of agricultural crops have been evaluated extensively. Zheng et al. (2005) studied the application of nano-sized and conventional TiO_2 on yield of naturally-aged spinach seeds. Seeds treated with nano TiO_2 increased in dry weight (73%), photosynthetic rate (threefold) and chlorophyll a formation (45%) compared with the control over 30 days. This effect may be attributed to photo-sterilization and photo-generation of superoxide and hydroxide anions by nano-TiO_2. These reactive oxygen species (ROS) can enhance seed stress resistance and support capsule penetration for water and oxygen absorption for rapid germination. Zheng et al. (2005) theorized that TiO_2 nanoparticles might have improved absorption of inorganic nutrients,

The key mechanism for increased speed of germination in the presence of NPs is the penetration of nanomaterials into the seed. Khodakovskaya et al. (2009) concluded that MWCNTs can penetrate tomato seeds and improve germination rate by increasing seed water absorption. The MWCNTs increased seed germination to 90% compared to 71% in the control in 20 days. MWCNTs likewise increased plant biomass. Shah and Belozerova (2009) showed that nanoparticles (Pd, Au at low concentrations; Si, Cu at high concentrations, and a combination of Au and Cu) had a positive effect on seed germination, enhanced shoot-to-root ratio and seedling growth. Phytotoxicity of [Application of (?)] nano-Al and Al_2O_3 significantly enhanced root extension of ryegrass and corn, respectively, whereas nano-Al supported radish and rape root growth.

The influence of NPs on plants can be positive or negative (Monica and Cremonini 2009). One of the most important concerns regarding application of nanomaterials for seed germination is potential phytotoxicity. The degree of phytotoxicity is a function of the type of nanomaterial and its prospective application. For example, the applicability of fluorescein isothiocyanate (FTIC)-labeled silica NPs and photostable cadmium-selenide (CdSe) quantum dots were tested for improving seed germination. It was concluded that FTIC-labeled silica nanoparticles enhanced seed germination in rice, while CdSe quantum dots prevented germination (Nair et al. 2011). Lin and Xing (2007) assessed phytotoxicity of nanomaterials (MWCNTs, Al_2O_3, ZnO, Al and Zn) and their effects on germination rates in radish, rape canola, ryegrass, lettuce, corn, and cucumber. They concluded that higher concentrations

(2000 mgL^{-1}) of nano-sized Zn (35 nm) and ZnO (~20 nm) prevented germination in ryegrass and corn, respectively. Root length of both species was influenced by 200 mgL^{-1} nano-Zn and ZnO.

The US EPA has agreed to the use of nano silver in agriculture in the United States (Bergeson 2010a, b); at present there are more than 100 pesticides which contain nano Ag by virtue of its anti-microbial characteristics. However, the impacts of nano-Ag on ecosystems and human health remains a concern. Lu et al. (2010) concluded that citrate-coated colloidal Ag nanoparticles were not genotoxic (genetic), cytotoxic (cell), or phototoxic (toxicity through photodegradation) to humans, but the same material was toxic in powder form. This was attributed to the "chemical change of spherical silver nanoparticles in the powder to form silver oxides or ions." Photoxicity of the powdered Ag nanoparticles was inhibited by coating with biocompatible polyvinylpyrrole (Lu et al. 2010).

Oancea et al. (2009) assumed that controlled release of active plant growth stimulators and other chemicals encapsulated in nanocomposites composed of layered double hydroxides (anionic clays) could be another possible opportunity for organic agriculture. However, important food organic certifiers (e.g. UK soil association, Biological Farmers of Australia) are opposed to using nanomaterials for organic agriculture (Scrinis and Lyons 2010). Recently, German-based organizations such as Naturland and the International Federation of Organic Agriculture Movements (IFOAM) considered food products grown with artificial nanomaterials as non-organic food (Naturland 2011 and IFOAM 2011).

1.2.2 *Plant Protection and Production*

Nanopesticides can be summarized as very small particles of pesticidal active components or other small engineered structures with useful pesticidal characteristics (Bergeson 2010b). Nanopesticides can enhance the dispersion and wettability of agricultural formulations (i.e., decreasing organic solvent runoff and harmful pesticide movement) (Bergeson 2010a). Nanomaterials and biocomposites exhibit useful characteristics such as rigidity, permeability, crystallinity, thermal stability, solubility, and biodegradability (Bouwmeester et al. 2009; Bordes et al. 2009), which are important for formulating nanopesticides. Nanopesticides possess a large specific surface area which increases affinity to the target (Jianhui et al. 2005). Subcategories of nanopesticides such as nanoemulsions, nanoencapsulates, nanocontainers and nanocages have recently been discussed (Bergeson 2010b; Bouwmeester et al. 2009; Lyons and Scrinis 2009) for plant protection (Table 1.1). Basically, nanomaterials degrade faster in the soil than plants with residue levels below the regulatory criteria in foodstuffs. Jianhui et al. (2005) reported the advance of sodium dodecyl sulfate (SDS)-modified photocatalytic TiO_2/Ag nanomaterial combined with dimethomorph (DMM), a commonly used pesticide in agricultural production. The modified formulation, 96 nm average granularity, improved dispersivity and

Table 1.1 Nanomaterials in agricultural plant protection and production

Purpose	Material	Findings	References
Smart agrochemical delivery system via plant roots of sunflower, tomato, pea and wheat	Magnetic carbon-coated nanoparticles	Nanoparticles moved through plant xylem and phloem within 24 h	Cifuentes et al. (2010)
Controlled release herbicide delivery system for atrazine	Polyhydroxybutyrate-co-hydroxyvalerate microspheres with atrazine (13 nm)	Good affinity of herbicide with polymer, decreased genotoxicity and increased biodegradability	Grillo et al. (2010)
Nanocomposite-based controlled release of herbicide, 2,4-dichlorophenoxyacetate (2,4-D)	Inorganic Zn-Al layered double hydroxide (ZAL) as release agent	Initial burst of 2,4-D followed by sustained release that depends on type of anions and their concentrations in release medium	Hussein et al. (2005)
Controlled delivery system for water-soluble pesticide (validamycin)	Porous hollow silica nanoparticles (PHSNs)	Pesticide was loaded into PHSNs (36 wt % loading capacity) and the release was in two stages: initial burst followed by sustained release	Liu et al. (2006)
Reduce bean rust disease severity	CNT conjugated with INF24 oligonucleotides	Treatment reduced rust severity	Corrêa et al. (2010)
Control of lentil pathogen and wilting	Silver nanoparticles-AgNPs (0.5–1000 ppm)	Faster plant growth compared to control; AgNPs did not reduce plant wilting	Ashrafi et al. (2010)
Physical and biological changes of *Brassica oleracea* in presence of nanomaterials	TiO_2 (5–8 nm) 0.05–2 mL of TiO_2 in 500 mL of Hoagland solution	Higher concentrations had negative impact on shoot length but positive impact on root length	Singh et al. (2010)
Carbon nanostructures on tomato germination	MWCNTs	Seed germination not related to MWNCTs (observed up to 7 days)	Lima et al. (2010)
Treatment of fungal pathogens in vitro and in chickpea and wheat plants	Amphotericin B nano disks (AMB-NDs) 0.1–2 mg/mL (in vitro), 0.1–10 mgL^{-1} (plants)	AMB-NDs inhibited fungi at 0.1 mg/mL (in vitro); effective for chickpea fusarium wilt control (preventive dosage of 0.1 mgL^{-1}) and wheat leaf rust controlled by foliar treatment	Perez-de-Luque et al. (2012)

breakdown of the pesticide in soil while enhancing its impact in vegetable seedlings (cabbage and cucumber). Modification of the nanomaterials using SDS significantly improved absorption of DMM. Guan et al. (2010) fabricated encapsulated nano-imidacloprid with above properties to be used for pest control in vegetable production. The SDS-modified Ag/TiO_2 imidacloprid nanoformulation was developed by a microencapsulation technique that used chitosan and alginate. It was applied to soybean plants that were transplanted to a soil of pH 6.2. The formulation residues in soil and plants degraded faster during the first 8 days, and concentrations were minimal to undetectable after 20 days. The SDS in the above applications was used to enhance photodegradation of the NPs in soil. Alternatively, Mohamed and Khairou (2011) developed highly photo-degradable Ag/TiO_2 particles (5–7 nm), manufactured using polyoxyethylene laurel ether (POL) and SDS, which was applied for 2,4-D herbicide degradation under visible and UV radiation. The POL-manufactured nanoparticles can photo-degraded faster during the same exposure period.

Toxicity or biosafety of pesticides is a major concern in agricultural production. With application of nanopesticides, there is uncertainty on their long-term impacts to human health and the environment. Xu et al. (2010) concluded that with better kinetic stability, smaller size, low viscosity and optical transparency, nanoemulsions can potentially be an improved pesticide delivery medium. The micro- or nanoemulsion as a carrier for pesticide delivery can increase solubility and bioavailability of nanopesticides. However, there is a need to evaluate possible uptake of nanopesticides by agricultural workers via inhalation. Shi et al. (2010) studied the toxicity of chlorfenapyr (nanopesticide) on mice. It was concluded that the chlorfenapyr nanoformulation at concentrations from 4.84 to 19.36 mg/kg was less toxic to mice than the common (non-NP) (?) formulation. Thus, nanopesticides may decrease adverse environmental and human risks as compared to common pesticides.

Formulation stability is an important practical issue at the nano level. Liu et al. (2008) successfully fabricated a stable nanopesticide (bifenthrin) using polymer stabilizers such as poly (acrylic acid)-b-poly (butylacrylate) (PAA-b-PBA), polyvinylpyrrolidone (PVP), and polyvinyl alcohol (PVOH). A flash nano-precipitation technique was used to prepare 60–200 nm bifenthrin nanoparticles.

Another important avenue of research is the development of nanomaterials as a protective layer to allow for slow release of conventional pesticides and fertilizers. For example, Corradini et al. (2010) discovered the possibility of using chitosan nanoparticles, a highly degradable antibacterial material for slow release of NPK fertilizer. Liu et al. (2006) fabricated kaolin clay-based nanolayers as a cementing and coating material for slow release fertilizers. Nano-clays possess interactive surfaces with high aspect ratio for encapsulating agrochemicals such as fertilizers, plant growth promoters, and pesticides (Ghormade et al. 2010).

1.2.3 Pesticide Residue Detection

The US Food and Drug Administration (FDA 2005) reported about 1045 chemicals as pesticide residues. Nanomaterial-based sensors can be used to detect many pesticide residues instead of traditional gas or liquid chromatography (GC/LC) -mass spectroscopy (-MS) techniques (Stan and Linkerhägner 1996; Sicbaldi et al. 1997; Balinova et al. 2007). Traditional techniques involve numerous steps including sampling, solid-phase extraction in the laboratory, sample analysis, and defining spectral peaks to determine pesticide residues. The U.S. Department of Agriculture (USDA) lists (?) single- and multi-residue methods based on GC/LC-MS to evaluate organophosphates, organochlorines, carbamates, triazines, triazoles, pyrethroids, neonicotinyls, and strobilurin residues in 85 agricultural commodities (USDA 2010).

Nanosensors for pesticide residue detection offer "high sensitivity, low detection limits, super selectivity, fast responses, and small sizes" (Liu et al. 2008). Table 1.2 presents nanosensors designed for the detection of pesticide residues such as methyl parathion (Kang et al. 2010; Parham and Rahbar 2010), parathion (Li et al. 2006; Wang and Li 2008), fenitrothion (Kumaravel and Chandrasekaran 2011), pirimicarb (Sun and Fung 2006), and dichlorvos and paraoxon (Vamvakaki and Chaniotakis 2007). Additionally, Dyk and Pletschke (2011) have reviewed enzyme-based biosensors for organochlorine, organophosphate, and carbamate residue detection. Some of these biosensors use C, Au, hybrid Ti, Au-platinum (Pt), and nanostructured lead dioxide (PbO_2)/TiO_2/Ti to immobilize enzymes on sensor substrate and to increase sensor sensitivity.

Application of nanomaterials as biosensors for pesticide residue detection is vast; nevertheless, some issues such as: (1) availability of nanomaterials sensitive to multiple pesticide residues; (2) simplicity of sensor manufacture techniques and instrumentation; (3) desired dependability and repeatability in trace level detection; (4) cost; and (5) concerns related to nanomaterial exposure to the surrounding environment must be considered. Also, the vast array of pesticides used in agricultural production might minimize using nanomaterial-based sensors for detection pesticide residue (Liu et al. 2008; Dyk and Pletschke 2011). Liu et al. (2008) reported that development of selective and stable sensing techniques to participate as biomolecules (enzymes, antibodies, etc.) with nanomaterials is needed. As a starting point, nanosensors can be used to detect major residuals that are highly hazardous to human health. Smart nanomaterials also can be used as sensors for pesticide detection. The smart nanomaterials and nanopesticides (Bergeson 2010b) that act as a source of pesticide as well as indicative sensor make no need of sensors for detecting pesticide residues in soil. Nanomaterials that rely upon slow, targeted release of the material but which can also indicate deficiency (e.g., via color change) of nutrients in soil could work as an advanced alert system for farmers to adjust dosage rate and frequency.

Table 1.2 Nanomaterials for pesticide residue detection

Pesticide/herbicide of interest	Sensing material	Detection limit	Sensor type	References
2,4,5-trichlorophenoxy acetic acid	Poly-*o*-toluidine zirconium(IV) phosphate nanocomposite	1 mM	Electrochemical	Khan and Akhtar (2011)
Fenitrothion in water	Nano TiO_2/nafion composite	0.13 mM	Electrochemical	Kumaravel and Chandrasekaran (2011)
Melamine in milk	18-crown-6 ether functionalized Au nanoparticles	6 ppm	Optical	Kuang et al. (2011)
Organochlorine and organophosphorus pesticides in cabbage	Amino-functionalized nanocomposite with tetraethylenepentamine	0.29 mg/kg	Chromatography	Zhao et al. (2011)
Methyl parathion in water	Nano-ZrO_2/graphite/paraffin	2 ng/mL	Electrochemical	Parham and Rahbar (2010)
Methyl parathion in vegetables (cabbage, spinach, lettuce)	Nano-Au/nafion composite	10^{-7} M	Electrochemical	Kang et al. (2010)
Methyl parathion, chlorpyrifos	Nano-size polyaniline matrix with SWCNT, single-stranded DNA and enzyme	1 ppm	Electrochemical biosensor	Viswanathan et al. (2009)
Fenamithion and acetamiprid in water	Cd-tellurium quantum dots with p-sulfonatocalix[4]arene	0.12 and 0.34 nM	Luminescence	Qu et al. (2009)
Parathion in water	Nano-ZrO_2/Au composite	3 ng/mL	Electrochemical	Wang and Li (2008)
Dichlorvos and paraoxon in drinking water	Acetyl cholinesterase and pyranine immobilized on nano-sized liposomes	10^{-10} M	Optical Nano-biosensor	Vamvakaki and Chaniotakis (2007)
Parathion in vegetables	Nano-TiO_2 on glassy carbon electrode	10^{-8} M	Electrochemical	Li et al. (2006)
Pirimicarb in vegetables	Molecular imprinted nano-polymers (methacrylic acid with carboxyl functional groups)	8×10^{-6} M	Piezoelectric	Sun and Fung (2006)

1.2.4 Plant Pathogen Detection

Bergeson (2010a) reported that application of pesticides and fertilizers [should (?)] occur after detection of pathogens and prior to onset of symptoms. Nanomaterials could be used for identification of bacterial, viral and fungal plant pathogens (Boonham et al. 2008; Yao et al. 2009; Chartuprayoon et al. 2010) in agriculture as a rapid analytical tool. Nanoparticles demonstrate high accuracy for detection of viral pathogens in plants (Baac et al. 2006). Nanoparticles also can be modified for use as a diagnostic tool to detect compounds related to a diseased condition. Nano-chips are kinds of microarrays which contain fluorescent oligo capture probes which the hybridization can be detected (López et al. 2009). These nano-chips are capable of detecting single nucleotide changes of bacteria and viruses due to their sensitivity and specificity (López et al. 2009). Yao et al. (2009) developed fluorescencet silica nanoparticles uploaded with an antibody to detect *Xanthomonas axonopodis* pv. Vesicatoria which causes bacterial spot disease in Solanaceae plants. Singh et al. (2010) used nano-gold-based immune sensors by using surface plasmon resonance (SPR) that could detect Karnal bunt (*Tilletia indica*) disease in wheat. The researchers attempted to detect the disease using an SPR sensor in wheat plots for seed certification and to formulate plant quarantines. Application of nanomaterials for detecting pathogens using nanosensors in field applications is highly valuable for rapid diagnosis and disease eradication. Plants affected by different stress disorders through physiological changes such, the induction of systemic defense, that regulated by plant hormones: jasmonic acid, methyl jasmonate and salicylic acid. Wang et al. (2010) merged this indirect stimulus to develop a sensitive electrochemical sensor by using a modified gold electrode with copper NPs to monitor salicylic acid levels in oil seeds for fungi (*Sclerotinia sclerotiorum*) detection. Further work in developing nanosensors to detect pathogens, their byproducts, and to monitor physiological changes in plants is needed.

1.2.5 Nanomaterials for (Mine?) Soil Reclamation and Environmental Remediation

Nanotechnology is a promising approach for reclamation of mine soils which involves removing or destroying contaminants and enhancing soil quality and fertility. Advantages of nanomaterials over traditional amendments for soil reclamation include higher reactivity due to smaller particle size, higher specific surface area and easier delivery of particles into porous media (i.e., soil). High reactivity leads to high efficiency and a rapid rate of soil reclamation, while easy delivery is advantageous for in situ application. Nanomaterials with great potential for mine soil reclamation include zeolites, zero-valent iron NPs, iron oxide NPs, phosphate-based NPs, iron sulfide NPs, and carbon nanotubes. This section places emphasis on the functions of these NPs in soil quality improvement; additionally, transport and mobility of NPs in the environment as well as their possible eco-toxicological effects are introduced.

1.2.5.1 Zeolites as a Soil Conditioner

Zeolites are crystalline hydrated aluminosilicates of alkali (Na^+, or K^+) and alkaline earth (Ca^{2+} or Mg^{2+}) cations characterized by an ability to hydrate/dehydrate reversibly and to exchange some of their constituent cations with those in aqueous solution without a major change in structure (Pabalan and Bertetti 2001). The unique feature of zeolites is that they possess an open, three-dimensional cage-like structure with a vast network of open channels throughout. The channels and pores, typically 0.3 to 0.7 nm in diameter, impart to the mineral large specific area (about 105 m^2 g^{-1}) for ion exchange and for selective capture of specific molecules (e.g., H_2O). Because of these structural features, zeolites generally are of low density compared with that of other minerals. Nearly 50 natural species of zeolites have been recognized, and more than 100 species have been synthesized in the laboratory (Mumpton 1985). Clinoptilolite is the most abundant zeolite species in sedimentary deposits and also the most mined zeolite mineral worldwide (Boettinger and Ming 2002). Zeolites occur in soils less than 5% (by weight) in content; again, clinoptilolite is the major zeolite species in soils.

By virtue of their ion exchange, adsorption, and molecular sieve properties, as well as their geographical abundance, zeolite minerals have generated worldwide interest for use in a broad range of applications. In agricultural enterprises, zeolites have been used as soil conditioners, slow-release fertilizers, and remediation agents for contaminated soil (Ming and Allen 2001). Published literature reveals that zeolite nanomaterials improve mine soil quality by increasing water holding capacity, increasing the clay-silt fractions, improving nutrient levels, and removing toxins (Ming and Allen 2001).

Reducing Soil Bulk Density and Improving Soil Water Holding Capacity

Natural zeolites possess several unique physical properties that make them attractive for improving soil physical properties. For example, the bulk density of zeolite minerals can be as low as 0.8 Mg m^{-3} as a result of its porous nature (Ming and Allen 2001). Mine soils often suffer from coarse texture which results in a high water infiltration rate, low water holding capacity and high bulk density (thus hindering root growth). Application of fine-grained zeolites (<0.05 mm) might increase effective silt and clay fractions, enhance water-holding capacity, and decrease bulk density which will ultimately improve crop growth. Githinji et al. (2011) studied the effect of zeolite (0.55–0.6 mm) at a rate of 15% (v v^{-1}) to sand (0.31 mm) media. They reported that bulk density decreased from 1.67 to 1.56 Mg m^{-3} and available water content increased two-fold. Wehtje et al. (2003) reported that Bermuda grass

(*Cynodon dactylon*) performance increased using zeolite-soil mixtures due to increased water holding capacity relative to control (non-amended soils).

Particle size distribution of zeolite minerals and application rate are important factors in improving soil physical properties. Petrovic (1990) reported that the optimum particle size of clinoptilolite added to golf course sand was between 0.1 and 1 mm for improving water infiltration, water availability, and aeration in soil. Huang and Petrovic (1990) reported that clinoptilolite particle size and amendment dosage are the main parameters for enhancing the water available to plants in a sand medium. They reported that available water in sand amended with 5 and 10% (g g^{-1}) clinoptilolite with a particle size of >1 mm was near 6 g kg^{-1}, whereas available water to plants in the same soil amended with same amount of <0.047 mm clinoptilolite was approximately 10 and 17 g kg^{-1}, respectively. Shoot-growth rate can increase by 26–60% on a sand-based putting green turf after using 10% clinoptilolite amendment (Huang and Petrovic 1995). Lopez et al. (2008) proposed solving drought-related problems by application of zeolite to soil which would act as a wicking (capillary) material to attract water from a shallow ground water table to the plant root zone. This would reduce dependence on precipitation or irrigation. They reported that grass planted in the zeolite-packed core structures survived, while the grass planted in the control area (no zeolite treatment) died.

Zeolites have also shown a substantial benefit for survival of vegetation in soils having poor structure, which have high bulk density, low water holding capacity, and where the available water depends mainly on precipitation (reference, year).

Improving Soil pH and Cation Exchange Capacity

Mine soils are usually acidic and infertile with low cation exchange capacity (CEC), resulting in poor nutrient status for plants. In contrast, pure zeolite materials usually have high CEC, ranging from 220 to 570 cmolc kg^{-1} (Boettinger and Ming 2002). Adding zeolites to a mine soil can increase overall CEC and pH in most cases (Ming and Allen 2001), which will improve soil nutrient holding capacity. Huang and Petrovic (1995) reported that application of 10% (g g^{-1}) zeolite to a sandy soil resulted in increased CEC from 0.08 to 15.59 cmolc kg^{-1} and pH from 5.4 to 6.6. Other studies reported that, with application of clinoptilolite to glacial till and marine clay at rates of 25% and 50% (g g^{-1}), CEC increased 2.6 ~ 3.3 times and pH increased from 4.2 to 6.5 (Katz et al. 1996).

It has also been reported that adding 0.2 ~ 2% zeolites to soil was beneficial to crop seed germination and crop production (Khan et al. 2009). As shown above, zeolites can increase the pH of acidic solution or soils due to its alkaline properties. The acid neutralization properties might be enhanced by virtue of the high CEC by which zeolites exchange solution protons (H^+) with Ca^{2+} ions. The acid neutralization capacity of zeolites is limited, however, compared with that of agricultural liming materials. Previous studies report that application of zeolites at 10% (g g^{-1}) to mine soils increased pH by 0.5–1 unit; in contrast, when using liming materials, pH increased by 2–3 units (Liu and Lal 2012).

It is not known whether zeolites can diminish acid generation in mine soils resulting from oxidation of sulfide minerals. But application of fine zeolites may block the pores in the coarse-textured mine soils and decrease oxygen dispersion to the underlying sulfide materials. Moreover, zeolites have the ability to adsorb gaseous molecules such as H_2S and SO_2 thereby decreasing harm to vegetation.

1.2.5.2 Nano-Enhanced Fertilizers

Zeolite-Enhanced Fertilizers

Mine soils usually lack nitrogen (N) and phosphorus (P); therefore, fertilizers are needed to ensure successful vegetative establishment (Burger and Zipper 2011). However, applying conventional N fertilizers often promotes growth of noxious weeds, suppressing the establishment and growth of crops and tree seedlings (Burger and Zipper 2011). Moreover, excessive use of mineral fertilizers has resulted in excessive nitrate leaching to groundwater, resulting in ground/surface water contamination. Therefore, using zeolites with nitrogen can provide a slow release fertilizer to meet crop needs while minimizing leaching losses (Ming and Allen 2001). In addition, improved fertilizer efficiency will decrease volatilization losses of gaseous N as NH_3 or N_2, especially when NH_4^+ fertilizers are exchanged onto zeolite exchange sites. In this way, the NH_4^+ ion is unavailable for conversion to the gaseous phase via microbial processes (Ming and Allen 2001).

Clinoptilolite is highly selective for K^+ and NH_4^+ relative to sodium (Na^+) or divalent cations such as Ca^{2+} and Mg^{2+} due to the location and density of negative charge in the structure and the dimensions of interior channels (Ming and Allen 2001). Hence, NH_4^{+-} and K^{+-} loaded zeolites are typically used as slow release fertilizers. Perrin et al. (1998) loaded clinoptilolite with NH_4^+ by soaking various size fractions in 1 M $(NH_4)_2SO_4$ for 10 days (d) and changing the soaking solution every 2–3 days. The solids were applied to 4-liter containers seeded with sweet corn (*Zea mays*). Soil fertilized with $(NH_4)_2SO_4$ leached 10–73% of added N (depending on application rate) whereas <5% of the added N leached from the $(NH_4)_2SO_4$-zeolite-amended soil, regardless of N application rate and zeolite particle size. Nitrogen use efficiency (NUE) ranged from72.0 to 95.2% using NH_4^+-clinoptilolite-amended soil after 42 days of plant growth, compared to 29.7–76.3% in soils fertilized with $(NH_4)_2SO_4$ only. Moreover, Lewis et al. (1984) reported that a NH_4-loaded clinoptilolite amendment could prevent injury by urea to radish (*Raphanus sativus*) plants while serving as an efficient slow-release N fertilizer. Barbarick and Pirela (1984) concluded that zeolites offer benefits to vegetation such as: preventing leaching losses of ammonium fertilizers; reducing ammonia toxicity to plants; and increasing crop yields.

Zeolites loaded with potassium have been used as a slow-release K-fertilizer (Williams and Nelson 1997; Carlino et al. 1998). Phosphorus (P) is also an important nutrient for vegetative establishment and reforestation in reclaimed mining areas. Rock phosphates such as apatites (e.g., Ca_{10} $(PO_4)6(OH)_2$) are commonly used sources of P in mine soil rehabilitation (Jacinthe and Lal 2007). However, availability of phosphorus depends upon the degree of apatite dissolution. Alkaline soil pH often impedes dissolution and decreases the quantity of soluble P. Jacinthe and Lal (2007) reported that rock phosphate has no effect on tree growth on reclaimed mine land, which was attributed to the relatively high pH of the soil ranging from 6.5 to 8.0. Zeolites have been used to address this problem: some researchers have used a combination of zeolite and ground apatite to enhance dissolution of apatite in order to deliver more available P even at high soil pH. This procedure creates exchange sites for Ca^{2+} in zeolites which decrease Ca^{2+} ions in the soil solution, thus supporting further apatite dissolution and phosphate release (Lai and Eberl 1986; Eberl et al. 1995 and He et al. 1999). Lai and Eberl (1986) combined rock phosphate with untreated and treated (NH_4^+, Na^+, and H^+) zeolite at a ratio of 1:5 and reported that the mixture contained 5–70 times more soluble P than that contained in the rock phosphate-only control. Using batch experiments, Allen et al. (1993) reported that the greater the zeolite to P rock ratio, the more P was released to solution, further confirming the role of zeolites in P rock dissolution.

Other Nano-Enhanced Fertilizers

Beyond zeolites, research has progressed on other types of nanomaterial-combined fertilizers. To achieve about 30–50% efficiency of the conventional fertilizers and no other management practices to increase the rate, Derosa et al. (2010) applied nanotechnology for fertilizer improvements. Lal (2008) suggested that applying nanotechnology to agriculture (including fertilizer development) is one of the best options for increasing crop production and supplying the world's increasing population with food. Suggestions that C nanotubes and zinc oxide nanoparticles are capable of penetrating tomato (*Lycopersicon esculentum*) root or seed tissues indicate that new nutrient delivery systems can be developed using the nanoscale porous domains on plant surfaces (2010). These phosphate-based nanoparticles have the potential to be used as P nanofertilizers for agricultural uses.

1.2.5.3 Nanomaterials for Remediating the Mine Soils Contaminated with Heavy Metals and Other Toxins

Zeolites

Natural and synthesized zeolites can immobilize heavy metals and radionuclides in contaminated soils and sediments, thus minimizing the risks of release to neighboring water bodies or by being taken up by plants/animals. Edwards et al. (1999) proved that mine soils polluted by Zn, Pb, Cu, and Cd and treated with synthesized zeolites (0.5–5% by weight) resulted in significant reductions (42–72%) of the labile and easily-available fractions of the metals. In addition to adsorption, zeolites raise soil pH which plays a role in metal immobilization (Edwards et al. 1999).

Other scientists, using leaching solutions such as 0.01 M $CaCl_2$ or dilute acetate to evaluate the stability of the heavy metals in the soil phase, have obtained similar results (Lin et al. 1998; Shanableh and Kharabsheh 1996; Moirou et al. 2001). The leachable fraction of the metals using these solutions was significantly reduced after soils were amended with 0.5–16% zeolites by weight (Lin et al. 1998; Shanableh and Kharabsheh 1996; Moirou et al. 2001).). Liu and Zhao (2007) and Liu (2011) discussed nanosized vivianite ($Fe_3(PO_4)_2 \cdot 8H_2O$) particles (~10 nm) and apatite ($Ca_5(PO_4)_3Cl$) particles (<200 nm) for heavy metal remediation.

Plants have also been used as indicators to evaluate metal toxicity and bioavailability in zeolite-amended soils. Using perennial ryegrass (*Lolium perenne L.*) and alfalfa (*Medicago sativa L.*) as indicator plants, Haidouti (1997) reported that application of zeolites at 1–5% ($g\ g^{-1}$) minimized Hg uptake by plants by up to 58 and 86% in roots and shoots, respectively. Chlopecka and Adriano (1996) found that addition of 1.5% ($g\ g^{-1}$) zeolite to a Zn-spiked soil was able to overcome the harmful effects of the metal and to increase growth and yield of maize and barley (*Hordeum vulgare*). The Zn concentration in plant tissue was also minimized by the amendment. Knox et al. (2003) reported that applying 2.5–5% zeolites to a metal-laden soil near a Zn-Pb smelter significantly increased growth of maize and oat (*Avena sativa*) and decreased Cd, Pb, and Zn accumulation in tissues. In contrast, neither plant could grow in the unamended soil. Mahmoodabadi (2010) reported that application of natural zeolites increased shoot dry weight, and number and dry weight of root nodules, and decreased Pb toxicity to soybean (*Glycine max*).

There are many reports, however, which indicate that application of zeolites for metal remediation reduced the growth of some crops and vegetables (Geebelen et al. 2002; Coppola et al. 2003; Stead 2002). It is generally believed that use of Na-type zeolites results in release of Na^+ to the soil solution and negatively affects plant growth even though the adverse effects of heavy metals were alleviated. Therefore, using Ca-type zeolites for heavy metal remediation is preferred at sites where revegetation is planned.

Zeolites possess unique selectivity for Cs^+ and Sr^{2+}; therefore, zeolites are effective agents for trapping radioactive ^{139}Cs and ^{90}Sr in soils contaminated from nuclear fallout, contact with water from reactor cooling reservoirs, or contamination from radioactive waste spills (Ming and Allen 2001). Similar to heavy metal remediation, the primary purpose of using natural zeolites is to immobilize radionuclides in the soil and to reduce or prevent uptake of those nuclides by plants (Ming and Allen 2001).

Iron Oxide Nanoparticles (nFeOs)

As an important constituent of soil and a necessary nutrient to plants and animals, iron (Fe) is classified as the 4th most abundant element in the earth. Fe oxides which commonly occur in soils and sediments usually occur as nano-crystals (5–100 nm in diameter). Reactive surfaces adsorb a wide range of both inorganic and organic substances through mechanisms such as surface complexation and surface precipitation (Bigham et al. 2002). By virtue of their notable absorption capacity for toxic substances and their environmentally friendly characteristics, many types of engineered iron oxide NPs have been synthesized and applied for in situ water/soil remediation processes. For example, nano-Fe oxide (nFeOx) solution can be pumped/spread directly onto polluted sites at low cost with insignificant risks of secondary contamination.

The nFeOs which have been intensively studied for heavy metal removal from water/wastewater include goethite (α-FeOOH, needle-like, 200 nm × 50 nm), hematite (α-Fe_2O_3, granular, 75 nm), amorphous hydrous Fe oxides (particles, 3.8 nm), maghemite (γ-Fe_2O_3, particle, 10 nm), and magnetite (Fe_3O_4, particles, approx. 10 nm) (Hua et al. 2012). These nFeOs have been widely researched for heavy metal removal from the aqueous phase via adsorption. Target contaminants have included Cu^{2+}, Cr^{6+}, Ni^{2+}, Pb^{2+}, Cr^{3+}, Zn^{2+}, As^{5+}, and As^{3+} (Hua et al. 2012). The use of nFeOs for polluted soil reclamation, however, has not been widely studied. Many researchers report that nanoparticles have the capacity for removal of heavy metals from the aqueous phase and can sequester the labile fractions of heavy metals from the soil solution by adsorption, thus decreasing their availability and mobility in soil.

Application of industrial wastes rich in iron oxides to contaminated soil resulted in substantial immobilization of heavy metals (Xenidis et al. 2010; Kumpiene et al. 2008; USEPA 2007), suggesting that application to mine soils could significantly immobilize soil-bound toxic substances. Shipley et al. (2011) used a column packed with soil mixed with 15% (g^{-1}) nano magnetite and leached it with mgL^{-1} arsenic. As concentrations were detected in the effluent for up to 132 days and the influent was containing 100 μg/L, and it was observed that solution injected through the column at a rate was 0.3 mL h^{-1}. Only 20% of the contaminant had leached after 208 days as compared with soil alone, that experienced negligible As adsorption. Shipley et al. (2011) reported that 12 heavy metals (V, Cr. Co, Mn, Se, Mo, Cd, Pb, Sb, Tl, Th, and U) could be simultaneously removed from soil by nFeOs. After 35 h of leaching, only Cr, Mo, Sb, and Co leached more than 20% of initial influent levels, revealing the high adsorption capacity of the nFeO nanoparticles, even for multiple toxins. Nano-hematite has similar adsorption capacity to the nano-magnetite (Shipley et al. 2011).

Remediation efficacy and deliverability of the NPs are controlled by chemical composition and NP stability and transport behaviors in the media (water, soil, and aquifer). Stability and transport of nFeOx depend on particle size, particle concentration, particle magnetism, solution chemistry.

For a given NP suspension, particle stability is largely governed by the electrostatic repulsion between particles (O'Carroll et al. 2013). The force is caused by particle surface charge; surface zeta potential is used to quantify the magnitude of charge or the electrostatic repulsion. When zeta potential is high, the repulsive force between particles is strong; thus, the nano suspension is more stable. Charged ions (e.g., H^+, OH^-, Na^+, or Cl^-) in the background solution can affect suspension stability by changing the particle surface charge (zeta potential). A pH value where the net surface charge becomes zero is termed point of zero charge (PZC), and the solution is least stable and most prone to form aggregates at pH values close to the PZC. Therefore, nanoparticle stability is influenced by solution pH, i.e., the extent to which it approximates particle PZC. The PZC for magnetite nanoparticles occurs at pH 7.1; the suspension will not be stable at pH values from 6 to 8 because the net particle surface charge decreases to about zero and rapid aggregation takes place due to minimal repulsion. In contrast, the NP suspension will be stable at pH 3–5 or 9–10, which are far from the PZC of magnetite NPs (Hu et al. 2010)

Nanoparticles in a concentrated solution are more likely collide with each other which makes them less stable than in a dilute solution where they form aggregates and eventually precipitate (He et al. 2008 and Baalousha 2009). He et al. (2008) reported that aggregation rates were higher for small hematite NPs due to changes in surface properties.

The force of magnetism among nFeO particles increases the probability of aggregation. Hong et al. (2009) observed that the stability and transport of magnetic NPs are adversely influenced by a combination of electrostatic and magnetic interactions. Hong et al. (2009) reported, during a column test with sand media, that the less-magnetic NPs were removed from the columns at a higher rate compared with the more-magnetic particles. The nonmagnetic nFeOs were readily transported: the majority of particles were retained at the column inlet for all transport experiments, the magnetic NPs were the greatest retained, indicating that magnetically changed the aggregation and subsequent straining cause a greater retention in the column.

Magnetic particles include maghemite (γ-Fe_2O_3), magnetite (Fe_3O_4), and zero valent iron (Fe^o), while a hematite (α-Fe_2O_3) nanoparticle is nonmagnetic. On the other hand, transport of magnetic NPs might be controlled by the magnificent of an external magnetic field to the system.

Natural organic matter can modify NP surfaces and change the particle PZC when absorbed. Changes of NP suspension stability by humic acids (HA) are likely due to the effect of the HA on particle PZC. Adsorption of HA often causes a decrease of magnetite PZC towards more acidic pH values. Hu et al. (2010) reported that PZC of magnetite NPs decreased from 7.1 (without HA) to 5.8 at 2 mg L^{-1} HA and 3.77 at 3 mg L^{-1} HA. When HA concentration was sufficiently high (e.g., 10 mg L^{-1}), the PZC decreased to values outside the range (pH 3–10) that it is commonly encountered in the natural environment. In this case, the suspension shows the highest stability under normal conditions (Hu et al. 2010). Similar results were obtained by other scientists (He et al. 2008; Baalousha 2009; Hong et al. 2009; Baalousha et al. 2008). An increase of solution ionic strength generally enhances NP aggregation such as nano $CaCl_2$ and TiO2 (Hu et al. 2010).

Iron oxides nanoparticles are generally assumed to have small or no toxicity to organisms according to limited reports. For example, Karlsson et al. (2009) assessed the ability of the nFeOs with varying sizes on cell death, mitochondrial damage, DNA damage and oxidative DNA lesions after exposure of the human cell line A549. They reported that the iron oxide (Fe_2O_3) NPs exhibited low toxicity with no clear difference between different particle sizes. Auffan et al. (2006) suggested that the organic coating on maghemite NPs served as a barrier to direct contact between particles and cells (human fibroblasts), further reducing possible toxic impacts. They found that the coated nFeOs produced weak cytotoxic and no genotoxic effects.

One key mechanism driving the toxicity of manufactured metal NPs is their ability to cause oxidative stress in cells by generating ROS. ROS damage proteins, lipids and DNA in addition to causing necrosis and apoptosis (Karlsson et al. 2009). However, Limbach et al. (2007) postulated that chemical composition rather than the nanoscale size is the most significant factor determining formation of ROS in exposed cells. Moreover, they observed that dissolved Fe ions promote a 20-times greater ROS production than exposure to the same amount of iron as Fe_2O_3 NPs, indicating that nano-sized Fe particles do not cause more toxicity than soluble Fe or Fe NPs having large particle sizes.

Nanoscale Zero-Valent Iron Particles (nZVI)

Nanoscale zero-valent iron (nZVI) technology, developed in the 1990s, was developed to degrade toxic halogenated hydrocarbon compounds and other petroleum-related products which pollute groundwater as a consequence of underground storage tank leakage, organic solvent spills, etc. (Zhang 2003). Metallic iron particles are highly effective reducing agents and are able to degrade many organic contaminants to benign compounds by reduction reactions. Relevant contaminants include chlorinated methane, chlorinated benzene, pesticides, polychlorinated biphenyls (PCBs), and nitro aromatic compounds (Zhang 2003). In addition to high degradation efficiency, this technology involves use of an eco-friendly material which is easily delivered to the subsurface due to its small particle size.

ZVI technology is also used to treat heavy metals in water and soil. Zero valent iron is a strong reductant with a reduction potential (E^0, Fe^{2+}/Fe^0) of −0.44 V (O'Carroll et al. 2013). Theoretically, some metals with E^0 much more positive than −0.44 V could be reductively immobilized by nZVI. Typical examples of such metals with environmental importance include CrO_4^{2-}/Cr^{3+} (E^0 = +1.56 V), $Cr_2O_7^{2-}/Cr^{3+}$ (E^0 = +1.36 V), and UO_2^{2+}/U^{4+} (E^0 = +0.27 V) (O'Carroll et al. 2013). The high-valent species (CrO_4^{2-}, $Cr_2O_7^{2-}$, and UO_2^{2+}) of those metals are usually more soluble and more toxic in the natural environment than their low-valent counterparts (Cr^{3+} and U^{4+}).

The nZVI converts the former to the latter through reduction reactions, thus reducing their solubility/mobility and toxicity (the overall process is termed reductive immobilization). For example, uranium (U) is the most common radionuclide pollutant found at many nuclear waste sites. It is detected in contaminated groundwater as highly soluble and mobile U^{6+} in the form of UO_2^{2+} (Cao et al. 2010). Fe oxyhydroxides adsorb UO_2^{2+} in soil and uranium mine tailings (Abdelouas 2006). However, acid mine drainage can dissolve and release the adsorbed uranium. These risks can be solved by converting uranium (U)to insoluble U^{4+} oxides using nZVI. Many reports conclude that, compared to other reductants (iron filings, galena (PbS), iron sulfide), nZVI is more efficient for reductively immobilizing U^{6+} from the aqueous phase, which may be attributed to its nano size, high reactivity, large surface area, and the reactive Fe(II) solubilized from nZVI (Yan et al. 2010; Fiedor et al. 1998; Crane et al. 2011; Dickinson and Scott 2010; Riba et al. 2008). The literature reveals that U^{6+} is removed by nZVI via reductive precipitation of UO_2^{2+} (U^{4+}) with minor precipitation of $UO_3 \cdot 2H_2O$ (U^{6+}) as confirmed by X-ray photoelectron spectroscopy (XPS) and X-ray diffraction (XRD) analyses (Yan et al. 2010)). Oxygen level, solution pH, and presence of bicarbonates and calcium ions all affect reductive immobilization processes (Yan et al. 2010; Fiedor et al. 1998). It has also been reported that nZVI is able to convert Cr^{+6} to Cr^{+3} in aqueous solution or soil media. Franco et al. (2009) reported that 97.5% of Cr^{+6} in a polluted soil was converted to Cr^{+3} by nZVI, which significantly decreased the chromium toxicity of the spoil. Similar results were obtained in soil using nZVI (Xu and Zhao 2007; Ponder et al. 2000).

Selenium (Se) is an important nutrient in animals, but high concentrations could be harmful when activities such as mining into shale for oil and phosphorus or irrigating arid and semiarid lands produce seleniferous soils (Lemly 1997). Plants can accumulate Se from impacted soils (Mackowiak and Amacher 2008). Plant accumulation and soil ingestion lead to Se bioaccumulation and Se poisoning in livestock and wildlife (Witte and Will 1993; Thomas et al. 2005). High-valence selenium species (SeO_4^{2-} or Se^{6+} and SeO_3^{2-} or Se^{4+}) are more soluble and mobile in the environment, and more toxic than are the low-valent species such as Se^{o} and Se^{2-}. nZVI has been applied to remove selenium from solution and reduce high-valent species to low-valent ones, thus reducing Se toxicity and solubility (O'Carroll et al. 2013). Olegario et al. (2010) reported that nZVI had high adsorption capacity for elimination of dissolved Se^{6+} (up to 0.1 mole Se/mole Fe). Using X-ray absorption near edge structure (XANES) spectroscopy and X-ray absorption fine structure (EXAFS) spectroscopy, they identified FeSe compounds in the solid phase as the reduced Se^{2-} species, transformed from Se^{6+}. They concluded that nZVI has the capability for reduction of soluble Se oxyanions to insoluble Se^{2-}.

The nZVI is able to treat other toxic elements in water or soil such as Hg^{2+}, Ni^{2+}, Ag^{+}, Cd^{2+}, As^{3+}, and As^{5+} (Li and Zhang 2006, 2007; Kanel et al. 2005, 2006). Decontamination mechanisms include reduction of metal ions to zero-valent metals on nZVI surfaces and/or adsorption of ions on nZVI. (The ZVI surfaces may contain a layer of iron oxidation products, e.g., iron oxides) (O'Carroll et al. 2013). Watanabe et al. (2009) reported that application of 0.01% nZVI ($g\ g^{-1}$) to a Cd-spiked soil decreased Cd accumulation in rice (*Oryza sativa*) seeds and leaves by 10% and 20%, respectively, compared to control.

Migration of bare nZVI NPs is estimated to be within a few cm in the subsurface soil (Saleh et al. 2008; Tratnyek and Johnson 2006). Due to a rapid NPs accumulation and interaction with surfaces of ambient porous media, substantial efforts have been made to enhance the stability and mobility of nZVI (e.g., using nanoparticle stabilizers), with the expectation that nZVI would diffuse throughout the entire contaminated aquifer and degrade pollutants in situ as soon as being injected into the subsurface.

Supported by laboratory column test results, several reports have claimed successful synthesis of nZVI with improved stability and mobility as well as reactivity

(He and Zhao 2005, 2007; Phenrat et al. 2008; Sakulchaicharoen et al. 2010). There is no solid evidence, however, on significantly increased mobility of such products in the field (O'Carroll et al. 2013). Stabilized nZVI has been visually confirmed to travel only as far as 1 m from an injection well, and evidence suggests that the maximum travel distance of up to 2–3 cm may be achieved in high permeability formations (O'Carroll et al. 2013). The differences between the laboratory and field data likely results from the fact that lab studies used lower Fe concentrations (<0.25 g L^{-1}), higher flow velocities (15–30 m day^{-1}), and simplified subsurface simulations in sand-packed columns. Field experiments have used higher Fe application rates (1–30 g L^{-1}), lower groundwater flow rates (0.1–10 m day^{-1}), and more complicated aquifer formations (O'Carroll et al. 2013), which result in greater aggregation and precipitation of nZVI. In addition, nZVI is oxidized more rapidly in column studies due to the presence of dissolved oxygen, creating maghemite and magnetite precipitates (Reinsch et al. 2010). These reports suggest that risk of nZVI dispersal in the environment and subsequent exposure of organisms does not occur extensively in current stages of nZVI technology.

No field experiments have been reported using nanoparticles for soil remediation. For mine soil recovery and vegetation establishment purposes, a thin soil surface layer (e.g., 50 cm depth) for plant root growth is typically required. In a method similar to the surface irrigation, a nanoparticle suspension could be applied over the targeted land surface. By exploiting nanoparticle size, the particles could engaged within the polluted surface layer only after the entire targeted soil column saturated and treated by the particles, thus it can reduce the risk of nanomaterial spread and subsequent secondary contamination can occur to neighboring soil and water bodies. With this approach, nZVI and other nanoparticles of high mobility may not be required for surface soil remediation.

There is a limited number of reports pertaining to the toxicological and ecotoxicological effects of nZVI use. Grieger et al. (2010) reported the possible effects of exposure to nZVI as follows: (a) low serious toxicity to aquatic organisms, with sublethal effects at minor concentrations (<1 mg L^{-1}); (b) histological and morphological changes in some species, during attach to organisms and cells; (c) some coatings decrease toxicity by reduced adherence; (d) release of Fe(II) from nZVI leads to ROS production as well as distruction of cell membranes causing cell death and lysis, and possible enhancement of biocidal effects of Fe(II); (e) aging of nZVI under aerobic conditions decreases nZVI toxicity, as Fe^{o} is rapidly oxidized.

Other metal-based SPs for environmental remediation include nanoscale manganese oxides and hydroxides, aluminum oxides, titanium oxides, zinc oxides, and magnesium oxides. All these can adsorb heavy metals from solution onto their surfaces (Bigham et al. 2002). Among metal oxide NPs, iron and manganese NPs are sensitive to the compact environment such as waterlogged soils or wetlands. Particles may be reduced to lower valence states thus losing some adsorption capacity. For manganese, zinc, and aluminum-based NPs, phytotoxicity might be useful under acidic soils conditions. Moreover, Limbach et al. (2007) reported that cobalt and manganese oxides (Co_3O_4 and Mn_3O_4) NPs produced more ROS (thus indicating more toxicity) than their respective salt solutions while titanium dioxide (TiO_2) and iron oxide (Fe_2O_3) nanoparticles were relatively inert.

Phosphate-Based Nanoparticles

Phosphate-based NPs are useful for removal of heavy metals from contaminated soil by creating highly insoluble and stable phosphate compounds. An example is treatment of lead-enriched soil. The solubility of common lead compounds in soils such as anglesite ($PbSO_4$), cerussite ($PbCO_3$), galena (PbS), and litharge (PbO) have been determined as $10^{-7.7}$, $10^{-12.8}$, $10^{-27.5}$, and $10^{+12.9}$, respectively (Ruby et al. 1994). In comparison, lead phosphate compounds such as pyromorphites ($Pb_5[PO_4]_3X$, where $X = F^-$, Cl^-, Br^-, and OH^-) have solubility products lower than 10^{-71} (Ruby et al. 1994). This indicates that lead phosphates are of significantly lower solubility than other Pb products in soils.

Conversion of less- to more stable Pb compounds using phosphate amendments is a thermodynamically preferred process which minimizes the leachability and availability of lead in the solid phase. Some phosphate amendments are more effective for *in situ* Pb precipitation and have been intensively studied (Ruby et al. 1994). Other metals having been investigated and effectively treated by phosphate (include Cu^{2+}, Zn^{2+}, Cd^{2+}, Co^{2+}, Cr^{3+}, Ba^{2+}, U^{6+}, and Eu^{3+} (Ma et al. 1995; Raicevic et al. 2005; Raicevic et al. 2006 and Basta and McGowen 2004).

Generally, soluble phosphate salts and particulate phosphate minerals are the commonly utilized phosphate forms for this purpose. The former includes phosphoric acid (Eighmy et al. 1997), NaH_2PO_4 (Stanforth and Qiu 2001), and $(NH_4)_2HPO_4$ (Basta and McGowen 2004), the latter various forms of apatite including synthetic apatites (Peld et al. 2004), natural rock phosphates (Ma et al. 1995; Raicevic et al. 2005; Raicevic et al. 2006; Basta and McGowen 2004), and biogenic apatites such as fishbone (Knox et al. 2006). Although both are extremely effective for in situ accumulation of heavy metals at the laboratory scale, problems in the field persist. For example, although soluble phosphates are mobile in the subsurface and thus more effective in heavy metal stabilization, excesses may result in eutrophication. Furthermore, excess quantities of phosphoric acid and ammonium phosphates may cause acidification of soil (Basta and McGowen 2004). Amendment dosage of 3% PO_4 (or 1% as P) by weight for soils has been studied by EPA and other scientists (USEPA 2001), and they suggest higher risk of a phosphate which can spill to water bodies and soil acidification following heavy metal remediation.

Solid phosphate application is hindered by large-sized particles which restrict phosphate mobility and delivery, which inhibits phosphate from contacting and reacting with heavy metals in subsurface layers. Also, finely ground solid phosphate particles are not mobile in soils, so mechanical mixing is required in the field for treatment. Considering the problems of phosphate application, Liu and Zhao [86] fabricated nanosized iron phosphate particles for heavy metal accumulation. Their NP formulation overcame the delivery problem and also the risk of secondary contamination related with the latter. The nanoparticle suspension, having the same mobility as the solution form due to the nanoscale particle size, is readily transported to the contaminated zone using conventional engineering methods (e.g., spray or well-injection). The NPs are also reported to be environmentally friendly because the phosphate in solid form is less bioavailable to algae than are those in soluble forms (Reynolds and Davies 2001). Algae-bioavailable P and N are primarily responsible for eutrophication in surface waters.

Liu and Zhao (2007) synthesized and applied a new class of iron phosphate (vivianite) NPs for in situ adsorption of Pb^{2+} in soils. Batch experiments revealed that the NPs significantly reduced leachability and bioaccessibility of Pb^{2+} in three soils (calcareous, neutral, and acidic), evaluated by the toxicity characteristic leaching procedure (TCLP) and physiologically based extraction test (PBET), respectively. When soils were treated for 56 d at rates ranging from 0.61 to 3.0 mg g^{-1} as PO_4^{3-}, the TCLP-leachable Pb^{2+} decreased by 85–95%, and the bioaccessible fraction decreased by 31–47%. Results from a sequential extraction technique indicated a 33–93% decrease in exchangeable Pb^{2+} and carbonate-bound fractions, and an increase in the residual Pb^{2+} fraction when Pb^{2+}-spiked soils were amended with the NPs. Additions of chloride to the treatment further lower TCLP-leachable Pb^{2+} in soils, suggesting the formation of chloropyromorphite minerals. Compared to soluble phosphate application for in situ metal immobilization, use of iron phosphate NPs resulted in an approximate 50% decrease in phosphate leaching.

Liu (2011) carried out remediation of Pb-contaminated soil (2647.9 mg Pb kg^{-1}) from a shooting range using manufactured apatite NPs. The apatite NP solution decreased the TCLP-leachable Pb fraction in the soil from 66.43% to 9.56% after 1-month amendment at a ratio of 2 mL solution to 1 g soil. The Pb concentration in the TCLP solution was decreased to 12.15 mg L^{-1} from 94.33 mg L^{-1}.When the amendment ratio was raised five-fold, leachable Pb decreased to 3.75 mg L^{-1} with only about 3% of the soil Pb leachable.

Phosphate-based NPs can also be used as P nano-fertilizers. In addition to supplying nutrient P to the plants, these NPs have the benefit of easy delivery (by spraying to the soil surface) with minimum P leaching to adjacent water bodies.

Iron Sulfide Nanoparticles

Sulfide-based NPs have been studied for removal of mercury (Hg) and arsenic (As) in water and soil/sediment by providing sulfide (S^{2-}) ligands and/or controlling surfaces. Reduced sulfur (S^{2-}) is considered a stabilizer/sink of heavy metals in reduced environments such as sediments or waterlogged soils, by forming highly insoluble metal sulfides (Moore et al. 1988). It has been estimated that a sediment sample is considered safe or nontoxic to aquatic organisms when the molar ratio of acid volatile sulfides (AVS) to total heavy metal concentrations (e.g., Cu + Ni + Zn) is >1 (Ankley et al. 1996). Ideally, all heavy metals are bound as insoluble metal-sulfide phases and thus the concentration of soluble (bioavailable) metals in pore water decreases (Ankley et al. 1996).

Sulfide (S^{-2}) has been proposed as an important inorganic compound for removal of Hg from the water column, thus halting formation of methylmercury (CH_3Hg). Methyl mercury is considered one of the most toxic Hg species, which can easily bioaccumulate and concentrate in fish and other aquatic organisms and become biomagnified through food chains. Consumption of fish and shellfish contaminated with CH_3Hg is the primary route of human exposure to mercury (Ankley et al. 1996). Dissolved, neutral mercury complexes (primarily HS^o and $Hg(HS)_2$ rather than Hg^{2+} or total dissolved Hg) are considered the main Hg(II) species controlling the extent of mercury methylation in contaminated sediments (Liu et al. 2009; Benoit et al. 1999). Iron sulfide amendments decrease concentrations of the neutral mercury complexes by formation of charged Hg(II)-polysulfides (e.g., $HgS_2{}^{2-}$, $HgSH^+$, HgS_2H^-) (Liu et al. 2009; Benoit et al. 1999; Drott et al. 2007; Xiong et al. 2009). In addition, formation of insoluble mercuric sulfide complexes also reduces conversion of the ionic Hg to volatile metal Hg in soil (Revis et al. 1989).

Liu et al. (2009) reported that synthesized mackinawite (FeS) was capable of reducing aqueous Hg approx. 0.75 mol Hg^{2+}/mole FeS. They proposed that 77% of the Hg removed was via precipitation of insoluble HgS species, and the residual 23% by adsorption to the FeS surface.

Under anoxic environments, iron sulfides have the ability to reduce the mobility and availability of As by adsorption and/or precipitation, depending on solution pH, iron sulfide type and As oxidation state (Renock et al. 2009; Wolthers et al. 2005; Gallegos et al. 2007, 2008). Wolthers et al. (2005) concluded that maximal As(V) adsorption by nano FeS occurred at pH 7.4 with an adsorption capacity of 0.044 mol As/mol FeS while the capacity was 0.012 As/mol FeS to As(III)..

The reduction capacity of iron sulfides is also practical for reductive immobilization of Tc^{+6} (Liu et al. 2008), Cr^{6+} (Patterson et al. 1997), and U^{6+} (Hua and Deng 2008), and also for reductive degradation of trichloroethylene (TCE) and tetrachloroethylene (PCE) (Butler and Hayes 1998, 1999, 2001). Again, the sulfide ion (S^{2-}) plays the major role in these reduction reactions and the decontamination mechanisms are similar to those of zero-valent iron NPs as discussed before.

Mackinawite is a widely reported iron sulfide synthesized for remediation studies in the laboratory. This compound is prepared by mixing Fe^{2+}-containing and S^{2-}-containing salts under anaerobic conditions, which produces micrometer-sized particles (Liu et al. 2008; Ankley et al. 1996; Xiong et al. 2009) which aggregate and precipitate within minutes (Xiong et al. 2009). Using carboxymethylcellulose (CMC) as a nanoparticle stabilizer, Xiong et al. (2009) fabricated a stable FeS spherical nanoparticle suspension with a particle size of 31.4 ± 4 nm diameter which remained suspended for at least 3 months. Shi et al. (2006) synthesized FeS NPs also using the CMC stabilizer, creating spherical particles with an average size of 4–6 nm. Xiong et al. (2009) reported that CMC- stabilized NPs enhanced adsorption of Hg in a sediment sample. When the molar ratio of FeS NP to Hg (sediment-bound) was set to 26.5, the Hg concentration in sediment pore water decreased by 97% and the TCLP leachability of sediment-bound Hg decreased by 99%, indicating that FeS NP amendment significantly decreased the labile Hg portion in the sample. In addition, FeS significantly decreased the availability of Hg species (HgS^o + Hg $(HS)_2{}^o$) by up to three orders of magnitude. The FeS spherical NP suspension was highly mobile in a clay loam sediment column, indicating high mobility for soil/sediment remediation. The authors observed that complete breakthrough of the NPs occurred at around 18 pore volumes (PVs), compared to 3 PVs for the inert tracer (Br^-).Morever, FeS spherical NPs were applied in the same tests, the majority (>99.7%) of particles were captured on top of the sediment column (Xiong et al. 2009). The work of Xiong et al. is probably the only one using FeS NPs to remediate soil-bound contaminants (Hg).

Other research suggests that FeS NPs can immobilize not only other heavy metals (especially As) but also some organic contaminant which exist in soils or in sediments. However, caution must be taken when using NPs in a mine soil reclamation plan; most iron sulfide (S^-) solids could oxidize to soluble sulfate species ($SO_4{}^{2-}$) due to their instability in aerobic environments (Liu et al. 2008; Ankley et al. 1996). Consequently, their adsorption capacity is lost and the contaminants previously retained on the FeS surface would be re-released to pore water and become remobilized (Ankley et al. 1996). Processes such as draining a pond or waterlogged land and dredging sediments are examples of how sediments may become exposed to air. In practical terms, it is difficult to maintain soil or sediment under anaerobic conditions for long periods, and a change in redox potential might result in a secondary contamination problem when using FeS NP amendments.

The generation of acidity in mine drainage and soils established from oxidation of the iron sulfide minerals (mostly pyrite, FeS_2) by oxygen (O_2) is inevitable after these buried minerals are exposed to the air through the mining process (Blodau 2006). Therefore, simple application of FeS minerals to the affected soils might exacerbate the AMD and soil acidity problems at a site. More stable adsorption materials such as iron oxide NPs (for As) or phosphate-based NPs (for heavy metals) could be better options.

Carbon Nanotubes

Carbon nanotubes (CNTs) are macromolecules consisting of sheets of C atoms covalently bonded in hexagonal lattices that seamlessly roll into a hollow, cylindrical shape with both ends commonly capped by fullerene-like tips (Niu and Cai 2012). According to their structures, CNTs could be categorized into single-walled C nanotubes (SWCNT) and multi-walled C nanotubes (MWCNT). The diameter of CNTs vary from hundreds of nanometers and micrometers to 0.2 and 2 nm for SWCNT, and from 2 to 100 nm for coaxial MWCNTs. CNTs are promising adsorbent materials for nonpolar organic contaminants such as trihalomethanes, polycyclic aromatic hydrocarbons, or naphthalene, dioxin, herbicides, DDT and its metabolites because of their large surface area, tubular structure and nonpolar properties (Niu and Cai 2012; Theron et al. 2008 and Mauter and Elimelech 2008.

CNTs have nonpolar characteristics, which leads to very low sorption of polar metal ions. Sorption is increased, however, after modification of the CNT surface by generating oxygen-containing polar functional groups (–COOH, –OH, or –C=O). These functional groups result in greatly increased negative charge on the CNT surface. The oxygen atoms in the functional groups provide a single pair of electrons to a metal ion, which increase the cation adsorption capacity of the CNTs (Rao et al. 2007). MWCNTs pretreated with nitric acid exhibited high adsorption for many heavy metal ions including Pb(II) (97.08 mg g^{-1}), Cu(II) (24.49 mg g^{-1}), and Cd(II) (10.86 mg g^{-1}) from aqueous solution. In addition, SWCNTs and MWCNTs, after oxidation with NaClO, demonstrated improved Ni(II) sorption properties. These treatments increased polarity of the CNT surface, making them more hydrophilic and, therefore, able to adsorb more charged metal ions from solution (Li et al. 2003 and Lu and Liu 2006).

Although CNTs may prove to be efficient adsorbents for many kinds of pollutants in both drinking and environmental waters, their practical application may be hindered by high cost (Theron et al. 2008). However, CNTs could be applied at the small-scale with sludge or to other solid wastes to remove contaminants which would render these wastes safe for land application.

In the aqueous phase pristine CNTs are prone to aggregation and precipitation due to their extreme hydrophobicity (Hyung et al. 2007; Jaisi and Elimelech. 2009). Dispersion of CNTs in the aqueous phase can be achieved either by modifying the surface structure and introducing hydrophilic (polar) functional groups (Jaisi and Elimelech. 2009; Jaisi et al. 2008) or by improving the interactions on the nanotube/water interface through addition of surfactants (Jiang et al. 2003), polymers (O'Connell et al. 2001a), or natural organic matter (Jaisi et al. 2008; Jiang et al. 2003; Zhou et al. 2012). The former method directly enhances the hydrophility of CNTs, while the latter option creates a thermodynamically suitable surface in water and also provides steric or electrostatic repulsion among dispersed CNTs, thus preventing aggregation (Hyung et al. 2007).

Natural organic matter may play a significant role in fate and transport of nanotubes in the environment. Hyung et al. (2007) stated that the water samples collected from the Suwannee River, USA, provided a MWCNT stabilizing capacity similar to that of fabricated solutions containing model natural organic matter (SR-NOM). For the same initial MWCNT concentrations, the concentrations of suspended MWCNTs in SR-NOM solution and Suwannee River water samples were significantly greater than those in a solution of 1% sodium dodecyl sulfate (a surfactant used to stabilize CNTs in the aqueous phase).

During study of the transport of carboxyl-functionalized SWCNTs in quartz sand-packed columns, Jaisi and Elimelech (2009) and Jaisi et al. (2008) reported that the performances of the nanotubes were generally similar to those obtained with colloidal particles and bacterial cells. For instance, ionic strength of the solution was increased due to increased SWCNT deposition in the column; additionally, divalent cations (e.g., Ca^{2+}) decreased SWCNT stability more than did monovalent cations (e.g., Na^{+}) at the same ionic strength. However, at very low ionic strengths, even in DI water, the SWCNT nature in sand media changed slightly, reflecting that simple physical constraints (i.e., straining) also played roles in nanotube mobility.

Jaisi and Elimelech (2009) reported that straining plays an important roles on nanotube mobility in soil. They compared the mobility of linear nanotubes and spherical fullerene nanoparticles in columns packed with the same soil. Fullerene removal rates were lower than were those of SWCNTs at the same ionic strength. Moreover, fullerene NPs were more affected by changes in ionic strength as compared with SWCNTs. The authors suggest that linear shape and structure, particularly the very large aspect ratio and its highly bundled (aggregated) form found in aqueous solution, were the main reasons for nanotube retention in soil columns. Furthermore, pore size distribution and pore geometry as well as heterogeneity in soil particle size, porosity, and permeability also participated in straining the flow through the soil by nanotubes. Thus, SWCNT mobility in soils is probably limited (Jaisi et al. 2008). The same results were found for MWCNTs (Xueying et al. 2009).

Natural soil environments are more heterogeneous and normally contain open soil structures (e.g., cracks, fissures, worm trails, and other open features) that can encourage mobility of SWNTs. Moreover, soil pore water is normally rich in dissolved organic molecules (e.g., humic and fulvic acids) that can improve the colloidal stability of nanomaterials (Jaisi et al. 2008). Due to limited study of NPs in soil, the discussions above reported significant suggestions on transport of all types of nanoparticles in the soil. On one hand, NP mobility may be lower and retention rate greater in soil media than reported using sand-packed columns in the laboratory due to the more complicated pore structures and pore distributions in soil. On the other hand, the existence of the preferential flow columns and natural organic matter in soil media would increase NP transport through soil and increase the risk of groundwater contamination.

1.2.5.4 Using Nanoenhanced Materials as Solid Waste Stabilizers/ Conditioners

Solid wastes possess a range of potential environmental contaminants (metals, salt, hydrocarbons, pathogens, noxious odors). Thus, to convert these wastes and impart benefits for landfill mining and soil reclamation, secondary environmental contaminations should be eliminated. Nano-enhanced materials have proved to enhance the environmental safety and public acceptance for landfill application of these wastes in mine or agricultural remediation. For example Li et al. (2007) reported that a low rate of nZVI (0.1% by weight) significantly eliminated noxious odors (caused by organic sulfur compounds), heavy metals, and organic contaminants in biosolids, indicating that nZVI could decrease the contamination of biosolids and increase their beneficial uses.

Turan (2008) concluded that co-composting of poultry litter mixed with 5% and 10% (g g^{-1}) natural zeolites removed 66% and 89% of salinity, respectively, in the end-product. Using 25%–30% (g g^{-1}) zeolites for biosolids remediation can remove many heavy metals (100% of Cd, 28–45% of Cu, 10–15% of Cr, 50–55% of Ni and Pb, and 40–46% of Zn) and decrease leaching of these metals (Zorpas et al. 2000). Zeolites also used at lower rates (0.5% and 1.0%) significantly lowered levels of labile Zn during experimental horticultural compost derived from sewage (Nissen et al. 2000). A rate of 1.0% zeolite caused significant reduction in total Zn and Cu transfer from soil to ryegrass plants over a 116 d growth period. The use of zeolites is a cost-effective amendment for compost to significantly reduce potential for soil metal mobility and soil to plant transfer (Villase˜nor et al. 2011). Villase˜nor et al. (2011) added three commercial natural zeolites to a pilot-scale rotary drum composting reactor where domestic sewage sludge and barley straw were co-composted. All three types of zeolites removed 100% of Ni, Cr, Pb, and significant amounts (>60%) of Cu, Zn, and Hg originating in the sludge (Villase˜nor et al. 2011). It was also reported that clinoptilolites reduced 50% of NH_3 emissions from the compost, thus avoiding N loss and unpleasant odors. Villase˜nor et al. (2011) claimed that addition of 10% zeolites produced composts compliant with Spanish regulations regarding heavy metal contamination. According to the authors, the zeolite-amended compost could either be applied directly to soil, or the metal-polluted zeolites could be separated from the compost prior to application to ensure the environmental safety.

Use of zeolites as heavy metal absorbents in compost is verified by other researchers (Zorpas and Loizidou 2008; Zorpas et al. 1999, 2002). Gadepalle et al. (2007) applied compost containing 5% zeolite to an As-contaminated soil and observed that zeolite addition reduced As uptake by ryegrass and that less than 0.01% of total As content in the soil may be absorbed by the plants. The literature above shows that amending solid wastes with relatively small quantities of nanomaterials could effectively reduce or eliminate the risk of secondary contamination associated with land applications of these wastes. This practice could expand the industrial or municipal waste lists which are safe for land application, thus reducing costs of waste disposal and ameliorating adverse environmental impacts. In addition, agricultural soils and drastically disturbed lands (e.g., mine soils) could benefit from these most cost-effective waste materials (soil amendments). Moreover, application of NPs to stabilize or condition conventional soil amendment materials (e.g., composts, biosolids, coal combustion by-products) could be a potential aspect of utilization of nanotechnology in agriculture at low cost.

Zeolites, nFeOs, phosphate-based NPs, and sulfide-based NPs are efficient in immobilizing inorganic contaminants in the solids, while C nanotubes have a high absorption capacity for organic pollutants and nZVI can destroy the organic wastewater contaminants (OWCs) present in wastes by reduction reactions. Finally, incubation of nanomaterials with solid wastes could stabilize the former and reduce the risks of nanomaterial spills and contamination resulting from direct application of NPs to the [to soil reclamation and environmental remediation].

1.2.5.5 Using Nanoenhanced Materials to Control Soil Erosion

Soil erosion caused by rainfall or wind at a closed mining site can result in loss of quality surface soil, exposure of buried sulfide minerals, and transport of sediments and pollutants to surface water bodies. Therefore, soil erosion management is of substantial importance in a mine soil reclamation plan. Nanoenhanced materials offer benefits to combat the harmful effects of soil erosion.

Andry et al. (2009) reported that surface runoff and soil loss can be significantly decreased by application of 10% of a Ca-type zeolite material when applied to an acidic soil under simulated rainfall. This was attributed to enhancement of wet aggregate stability and the large particle size of the sediment due to the amendments. Andry et al. (2009) suggested that zeolites can be more effective than lime in soil erosion management. Yamamoto et al. (2004) also applied a Ca-type of artificial zeolite at rates of 5–25% in sodic soils to control the rate of runoff and soil loss. They reported that the exchange of Ca^{2+} on zeolites with Na^+ in the sodic soil reduced clay dispersion, resulting in increased soil hydraulic conductivity and soil aggregation, which decreased runoff rate and soil loss. Zheng (2011) reported that using polyacrylamide (PAM, a polyelectrolyte used for soil erosion management) and magnetite NPs to an As-spiked soil subject to simulated rainfall decreased soil erosion while the NPs reduced As leaching. Wang et al. (2007) reported that using alumina nanoparticles (Al_2O_3, 140–330 nm) in conditioning a wastewater treatment sludge resulted in larger flocs and better dewatering effects than the single conditioning by polyelectrolyte only. Beneficial effects were more evident when finer nanoparticles (140 nm) were used. Wang et al. (2007) suggested that the NPs can increase the elongation of the chain-like structures of the polyelectrolyte, resulting in more effective bridging effects and better flocculation. PE (polyelectrolyte)-NP flocculation systems have been widely used for eliminating solid particles from solution (Ovenden and Xiao 2002; Yan and Deng 2000). Flocculation in such a system is induced by the sequential addition of a positively charged polyelectrolyte followed by negatively charged NPs such as bentonite and colloidal silica. The system produces better flocculation and drainage (dewatering) than conventional polymer-only flocculation systems (Ovenden and Xiao 2002). These results suggest that double application of polyelectrolyte and NPs could increase flocculation and improve soil particle size and particle stability, thus effectively managing soil erosion caused by wind or rain.

1.3 Conclusion

Nanotechnology has attracted many researchers as a result of the unique physical, chemical and biological characteristics of NPs that differ from those at the bulk scale for the same material. Nanomaterials are being developed for applications in many fields such as medicine, drug delivery, electronics, fuel cells, solar cells, food, and space exploration. Nanomaterials have provided many advances for agricultural purposes. Nanotechnology offers benefits for plant germination and growth. TiO_2 NPs increased dry weight (73%), photosynthetic rate (three-fold) and chlorophyll-a formation (45%) than control over a germination period of 30 days. Nanomaterials achieved better growth rate of spinach seeds than traditional TiO_2 indicating that nanomaterials have beneficial properties for plant germination.

Other nanoparticles have shown positive effects on seed germination and plant growth such as Pd and Au at low concentration; Si and Cu at high concentrations, and combinations of Au and Cu.

Nanomaterials can also be used to detect and treat plant diseases. Since nanomaterials could be used for resistance bacterial, viral and fungal plant pathogens in agriculture. Nano-chips are capable of detecting single nucleotide changes of bacteria and viruses due to their sensitivity and specificity. Nanomaterial-based nanosensors can be used to detect many pesticide residues instead of conventional gas or liquid chromatography (GC/LC)-mass spectroscopy (-MS) techniques. Some biosensors use C, Au, hybrid titanium (Ti), Au-platinum (Pt), and nanostructured lead dioxide (PbO_2)/TiO_2/Ti to immobilize enzymes on sensor substrate and to increase the sensor sensitivity.

Nanomaterials are also used for plant protection instead of manufactured pesticides. Nanopesticides can enhance the dispersion and wettability of agricultural formulations and result in reduced organic solvent runoff and pesticide migration off-site.

Nanomaterials and biocomposites exhibit useful characteristics such as rigidity, permeability, crystallinity, thermal stability, solubility, and biodegradability which are important for formulating nanopesticides. Nanopesticides possess a large specific surface area which increases the affinity to the target. Many kinds of nanopesticides are available such as nanoemulsions, nanoencapsulates, nanocontainers and nanocages. Nanomaterials have also been applied for enhancement of nutrient absorption by plants, delivery of active ingredients to specific sites and water treatment processes. The potential of nanotechnology in agriculture is huge; this exciting field needs greater scrutiny in order to understand and utilize all the benefits of nanomaterials.

References

Abdelouas A (2006) Uranium mill tailings: geochemistry, mineralogy, and environmental impact. Elements 2(6):335–341

Allen ER, Hossner LR, Ming DW, Henninger DL (1993) Solubility and cation exchange in phosphate rock and saturated clinoptilolite mixtures. Soil Sci Soc Am J 57(5):1368–1374

Andry H, Yamamoto T, Inoue M (2009) Influence of artificial zeolite and hydrated lime amendments on the erodibility of an acidic soil. Commun Soil Sci Plant Anal 40(7–8):1053–1072

Ankley GT, Di Toro DM, Hansen DJ, Berry WJ (1996) Technical basis and proposal for deriving sediment quality criteria for metals. Environ Toxicol Chem 15(12):2056–2066

Ashrafi SJ, Rastegar MF, Jafarpour B, Kumar SA (2010) Possibility use of silver nano particle for controlling Fusarium wilting in plant pathology. In: Riberio C, de-Assis OBG, Mattoso LHC, Mascarenas S (eds) Symposium of international conference on food and agricultural applications of nanotechnologies. São Pedro, SP, Brazil. ISBN 978-85-63274-02-4

Auffan M, Decome L, Rose J, Orsiere T, De Meo M, Briois V, Chaneac C, Olivi L, Berge-Lefranc JL, Botta A, Wiesner MR, Bottero JY (2006) In vitro interactions between DMSA-coated maghemite nanoparticles and human fibroblasts: a physicochemical and cyto-genotoxical study. Environ Sci Technol 40(14):4367–4373

Baac H, Hajós JP, Lee J, Kim D, Kim SJ, Shuler ML (2006) Antibody-based surface plasmon resonance detection of intact viral pathogen. Biotechnol BioengBiotechnol Bioeng 94(4):815–819
Baalousha M (2009) Aggregation and disaggregation of iron oxide nanoparticles: influence of particle concentration, pH and natural organic matter. Sci Total Environ 407(6):2093–2101
Baalousha M, Manciulea A, Cumberland S, Kendall K, Lead JR (2008) Aggregation and surface properties of iron oxide nanoparticles: influence of pH and natural organic matter. Environ Toxicol Chem 27(9):1875–1882
Balinova A, Mladenova R, Shtereva D (2007) Solid-phase extraction on sorbents of different retention mechanisms followed by determination by gas chromatographyemass spectrometric and gas chromatography-electron capture detection of pesticide residues in crops. J Chromatogr A 1150:136–144
Barbarick KA, Pirela HJ (1984) Agronomic and horticultural uses of zeolites: a review. In: Pond WG, Mumpton EA (eds) Zeo-agriculture. Use of natural zeolites in agriculture and aquaculture. Westview Press, Boulder, pp 93–104
Basta NT, McGowen SL (2004) Evaluation of chemical immobilization treatments for reducing heavy metal transport in a smelter-contaminated soil. Environ Pollut 127(1):73–82
Benoit JM, Gilmour CC, Mason RP, Heyes A (1999) Sulfide controls on mercury speciation and bioavailability to methylating bacteria in sediment pore waters. Environ Sci Technol 33(6):951–957
Bergeson LL (2010a) Nanosilver: US EPA's pesticide office considers how best to proceed. Environ Qual Manag 19(3):79–85
Bergeson LL (2010b) Nanosilver pesticide products: what does the future hold? Environ Qual Manag 19(4):73–82
Bigham JM, Fitzpatrick RW, Schulze DG (2002) Iron oxides. In: Dixon JB, Schulze DG (eds) Soil mineralogy with environmental applications. Soil Science Society of America, Madison, pp 323–366
Blodau C (2006) A review of acidity generation and consumption in acidic coal mine lakes and their watersheds. Sci Total Environ 369(1–3):307–332
Boettinger JL, Ming DW (2002) Zeolites. In: Dixon JB, Schulze DG (eds) Soil mineralogy with environmental applications, SSSA book series 7. Soil Science Society of America, Madison, pp 585–610
Boonham N, Glover R, Tomlinson J, Mumford R (2008) Exploiting generic platform technologies for the detection and identification of plant pathogens. Eur J Plant Pathol 121:355–363
Bordes P, Pollet E, Avérous L (2009) Nano-biocomposites: biodegradable polyester/nanoclay systems. Prog Polym Sci 34:125–155
Bouwmeester H, Dekkers S, Noordam MY, Hagens WI, Bulder AS, de Heer C, ten Voorde SECGS, Wijnhoven WP, Marvin HJP, Sips AJAM (2009) Review of health safety aspects of nanotechnologies in food production. Regul Toxicol Pharmacol 53:52–62
Burger JA, Zipper CE (2011) How to restore forests on surface-mined land Publication 460–123, Virginia Cooperative Extension (VCE), Stanardsville, Va, USA
Butler EC, Hayes KF (1998) Effects of solution composition and pH on the reductive dechlorination of hexachloroethane by iron sulfide. Environ Sci Technol 32(9):1276–1284
Butler EC, Hayes KF (1999) Kinetics of the transformation of trichloroethylene and tetrachloroethylene by iron sulfide. Environ Sci Technol 33(12):2021–2027
Butler EC, Hayes KF (2001) Factors influencing rates and products in the transformation of trichloroethylene by iron sulfide and iron metal. Environ Sci Technol 35(19):3884–3891
Cao B, Ahmed B, Beyenal H (2010) Immobilization of uranium in groundwater using biofilms. In: Shah V (ed) Emerging environmental technologies, vol 2. Springer, New York, pp 1–37
Carlino JL, Williams KA, Allen ER (1998) Evaluation of zeolite-based soilless root media for potted chrysanthemum production. HortTechnology 8(3):373–378
Chartuprayoon N, Rheem Y, Chen W, Myung NV (2010) Detection of plant pathogen using LPNE grown single conducting polymer nanoribbon. Abstract #2278, 218th ECS Meeting

Chlopecka A, Adriano DC (1996) Mimicked in-situ stabilization of metals in a cropped soil: bioavailability and chemical form of zinc. Environ Sci Technol 30(11):3294–3303

Cifuentes Z, Custardoy L, de la Fuente JM, Marquina C, Ibarra MR, Rubiales D, Pérez-de-Luque A (2010) Absorption and translocation to the aerial part of magnetic carbon-coated nanoparticles through the roots of different crop plants. J Nanobiotechnol 8(26):1–8

Coppola E, Battaglia G, Bucci M, Ceglie D, Colella A, Langella A, Buondonno A, Colella C (2003) Remediation of Cd- and Pb-polluted soil by treatment with organo-zeolite conditioner. Clays Clay Minerals 51(6):609–615

Corradini E, de Moura MR, Mattoso LHC (2010) A preliminary study of the incorparation of NPK fertilizer into chitosan nanoparticles. Express Polym Lett 4(8):509–515

Corrêa Jr JD, Rodrigues L, Lacerda RG, Ladeira LO (2010) Treatment of bean plants with carbon nanotubes conjugated INF24 antisense oligonucleotides reduce bean rust disease severity. In: Riberio C, de-Assis OBG, Mattoso LHC, Mascarenas S (eds) Symposium of international conference on food and agricultural applications of nanotechnologies. São Pedro, SP, Brazil. ISBN 978-85-63274-02-4

Crane RA, Dickinson M, Popescu IC, Scott TB (2011) Magnetite and zero-valent iron nanoparticles for the remediation of uranium contaminated environmental water. Water Res 45(9):2931–2942

Derosa MC, Monreal C, Schnitzer M, Walsh R, Sultan Y (2010) Nanotechnology in fertilizers. Nat Nanotechnol 5(2):91

Dickinson M, Scott TB (2010) The application of zero valent iron nanoparticles for the remediation of a uranium contaminated waste effluent. J Hazard Mater 178(1–3):171–179

Drott A, Lambertsson L, Bjorn E, Skyllberg U (2007) Importance of dissolved neutralmercury sulfides for methyl mercury production in contaminated sediments. Environ Sci Technol 41(7):2270–2276

Dyk JSV, Pletschke B (2011) Review on the use of enzymes for the detection of organochlorine, organophosphate and carbamate pesticides in the environment. Chemosphere 82:291–307

Eberl DD, Barbarick KA, Lai TM (1995) Influence of NH4-exchanged clinoptilolite on nutrient concentrations in sorghum-sudangrass. In: Ming DW, Mumpton FA (eds) Natural zeolites'93: occurrence, properties, use. Int'l Comm Natural Zeolites, Brockport, pp 491–504

Edwards R, Rebedea I, Lepp NW, Lovell AJ (1999) An investigation into the mechanism by which synthetic zeolites reduce labile metal concentrations in soils. Environ Geochem Health 21(2):157–173

Eighmy TT, Crannell BS, Butler LG, Cartledge FK, Emery EF, Oblas D, Krzanowski JE, Eusden JD, Shaw EL, Francis CA (1997) Heavy metal stabilization in municipal solid waste combustion dry scrubber residue using soluble phosphate. Environ Sci Technol 31(11):3330–3338

FDA (2005) Glossary of pesticide chemicals. Available at: http://www.fda.gov/downloads/Food/FoodSafety/FoodContaminantsAdulteration/Pesticides/ucm114655.pdf. Accessed 31 Jan 2011

Fiedor JN, Bostick WD, Jarabek RJ, Farrell J (1998) Understanding the mechanism of uranium removal from groundwater by zero- valent iron using X-ray photoelectron spectroscopy. Environ Sci Technol 32(10):1466–1473

Franco DV, Da Silva LM, Jardim WF (2009) Reduction of hexavalent chromium in soil and ground water using zerovalent iron under batch and semi-batch conditions. Water Air Soil Pollut 197(1–4):49–60

Gadepalle VP, Ouki SK, Van Herwijnen R, Hutchings T (2007) Immobilization of heavy metals in soil using natural and waste materials for vegetation establishment on contaminated sites. Soil Sediment Contam 16(2):233–251

Gallegos TJ, Sung PH, Hayes KF (2007) Spectroscopic investigation of the uptake of arsenite from solution by synthetic mackinawite. Environ Sci Technol 41(22):7781–7786

Gallegos TJ, Han YS, Hayes KF (2008) Model predictions of realgar precipitation by reaction of As (III) with synthetic mackinawite under anoxic conditions. Environ Sci Technol 42(24):9338–9343

Geebelen W, Vangronsveld J, Adriano DC, Carleer R, Clijsters H (2002) Amendment-induced immobilization of lead in a lead-spiked soil: evidence from phytotoxicity studies. Water Air Soil Pollut 140(1–4):261–277

Ghormade V, Deshpande MV, Paknikar KM (2010) Perspectives for nanobiotechnology enabled protection and nutrition of plants. Biotechnol Adv 29:792–803

Githinji LJM, Dane JH, Walker RH (2011) Physical and hydraulic properties of inorganic amendments andmodeling their effects on water movement in sand-based root zones. Irrig Sci 29(1):65–77

Grieger KD, Fjordbøge A, Hartmann NB, Eriksson E, Bjerg PL, Baun A (2010) Environmental benefits and risks of zero-valent iron nanoparticles (nZVI) for in situ remediation: risk mitigation or trade-off? J Contam Hydrol 118(3–4):165–183

Grillo R, Melo NFS, de Lima R, Lourenço RW, Rosa AH, Fraceto LF (2010) Characterization of atrazine-loaded biodegradable poly(hydroxybutyrate-cohydroxyvalerate) microspheres. J Polym Environ 18:26–32

Guan H, Chi D, Yu J, Li H (2010) Dynamics of residues from a novel nanoimidacloprid formulation in soyabean fields. Crop Prot 29:942–946

Haidouti C (1997) Inactivation of mercury in contaminated soils using natural zeolites. Sci Total Environ 208(1–2):105–109

He F, Zhao D (2005) Preparation and characterization of a new class of starch-stabilized bimetallic nanoparticles for degradation of chlorinated hydrocarbons in water. Environ Sci Technol 39(9):3314–3320

He F, Zhao D (2007) Manipulating the size and dispersibility of zerovalent iron nanoparticles by use of carboxymethyl cellulose stabilizers. Environ Sci Technol 41(17):6216–6221

He ZL, Baligar VC, Martens DC, Ritchey KD, Elrashidi M (1999) Effect of byproduct, nitrogen fertilizer, and zeolite on phosphate rock dissolution and extractable phosphorus in acid soil. Plant Soil 208(2):199–207

He YT, Wan J, Tokunaga T (2008) Kinetic stability of hematite nanoparticles: the effect of particle sizes. J Nanopart Res 10(2):321–332

Hong Y, Honda RJ, Myung NV, Walker SL (2009) Transport of iron-based nanoparticles: role of magnetic properties. Environ Sci Technol 43(23):8834–8839

Hu JD, Zevi Y, Kou XM, Xiao J, Wang XJ, Jin Y (2010) Effect of dissolved organic matter on the stability of magnetite nanoparticles under different pH and ionic strength conditions. Sci Total Environ 408(16):3477–3489

Hua B, Deng B (2008) Reductive immobilization of uranium(VI) by amorphous iron sulfide. Environ Sci Technol 42(23):8703–8708

Hua M, Zhang S, Pan B, Zhang W, Li L, Zhang Q (2012) Heavy metal removal from water/wastewater by nanosized metal oxides: a review. J Hazard Mater 211-212:317–331

Huang ZT, Petrovic AM (1995) Physical properties of sand as affected by clinoptilolite zeolite particle size and quantity. J Turfgrass Manag 1(1):1–15

Hussein MZb, Yahaya AH, Zainal Z, Kian LH (2005) Nanocomposite-based controlled release formulation of an herbicide, 2,4-dichlorophenoxyacetate incapsulated in zincealuminium-layered double hydroxide. Sci Technol Adv Mater 6:956–962

Hyung H, Fortner JD, Hughes JB, Kim JH (2007) Natural organic matter stabilizes carbon nanotubes in the aqueous phase. Environ Sci Technol 41(1):179–184

IFOAM (2011) IFOAM position paper on "The use of nanotechnologies and nanomaterials in organic agriculture". Available at: http://www.ifoam.org/press/positions/IFOAMPositionPaperNanotech2011_Approved.pdf. Accessed 3 Jan 2012

Jacinthe PA, Lal R (2007) Carbon storage and minesoil properties in relation to topsoil application techniques. Soil Sci Soc Am J 71(6):1788–1795

Jaisi DP, Elimelech M (2009) Single-walled carbon nanotubes exhibit limited transport in soil columns. Environ Sci Technol 43(24):9161–9166

Jaisi DP, Saleh NB, Blake RE, Elimelech M (2008) Transport of single-walled carbon nanotubes in porous media: filtration mechanisms and reversibility. Environ Sci Technol 42(22):8317–8323

Jiang L, Gao L, Sun J (2003) Production of aqueous colloidal dispersions of carbon nanotubes. J Colloid Interface Sci 260(1):89–94

Jianhui Y, Kelong H, Yuelong W, Suqin L (2005) Study on anti-pollution nanopreparation of dimethomorph and its performance. Chin Sci Bull 50(2):108–112

Kanel SR, Manning B, Charlet L, Choi H (2005) Removal of arsenic(III) from groundwater by nanoscale zero-valent iron. Environ Sci Technol 39(5):1291–1298

Kanel SR, Greneche JM, Choi H (2006) Arsenic (V) removal from groundwater using nano scale zero-valent iron as a colloidal reactive barrier material. Environ Sci Technol 40(6):2045–2050

Kang TF, Wang F, Lu LP, Zhang Y, Liu TS (2010) Methyl parathion sensors based on gold nanoparticles and Nafion film modified glassy carbon electrodes. Sensor Actuat B-Chem 145:104–109

Karlsson HL, Gustafsson J, Cronholm P, M¨oller L (2009) Size-dependent toxicity of metal oxide particles-A comparison between nano- and micrometer size. Toxicol Lett 188(2):112–118

Katz LE, Humphrey DN, Jankauskas PT, Demascio FA (1996) Engineered soils for low-level radioactive waste disposal facilities: effects of additives on the adsorptive behavior and hydraulic conductivity of natural soils. Hazard Waste Hazard Mater 13(2):283–306

Khan AA, Akhtar T (2011) Adsorption thermodynamics studies of 2,4,5-trichlorophenoxy acetic acid on poly-o-toluidine Zr(IV) phosphate, a nanocomposite used as pesticide sensitive membrane electrode. Desalination 272:259–264

Khan H, Khan AZ, Khan R, Matsue N, Henmi T (2009) Influence of zeolite application on germination and seedquality of soybean grown on allophanic soil. Res J Seed Sci 2(1):1–8

Khodakovskaya M, Dervishi E, Mahmood M, Xu Y, Li Z, Watanabe F, Biris AS (2009) Carbon nanotubes are able to penetrate plant seed coat and dramatically affect seed germination and plant growth. ACS Nano 3(10):3221–3227

Knox AS, Kaplan DI, Adriano DC, Hinton TG, Wilson MD (2003) Apatite and phillipsite as sequestering agents for metals and radionuclides. J Environ Qual 32(2):515–525

Knox AS, Kaplan DI, Paller MH (2006) Phosphate sources and their suitability for remediation of contaminated soils. Sci Total Environ 357(1–3):271–279

Kuang H, Chen W, Yan W, Xu L, Zhu Y, Liu L, Chu H, Peng C, Wang L, Kotov NA, Xua C (2011) Crown ether assembly of gold nanoparticles: melamine sensor. Biosens Bioelectron 26:2032–2037

Kumaravel A, Chandrasekaran M (2011) A biocompatible nano TiO_2/nafion composite modified glassy carbon electrode for the detection of fenitrothion. J Electroanal Chem 650:163–170

Kumpiene J, Lagerkvist A, Maurice C (2008) Stabilization of As, Cr, Cu, Pb and Zn in soil using amendments-a review. Waste Manag 28(1):215–225

Lai TM, Eberl DD (1986) Controlled and renewable release of phosphorous in soils from mixtures of phosphate rock and NH4-exchanged clinoptilolite. Zeolites 6(2):129 132

Lal R (2008) Promise and limitations of soils to minimize climate change. J Soil Water Conserv 63(4):113A–118A

Lemly AD (1997) Environmental implications of excessive selenium: a review. Biomed Environ Sci 10(4):415–435

Lewis D, Moore IFD, Goldsberry KL (1984) Ammonium- exchanged clinoptilolite and granulated clinoptilolite with urea as nitrogen fertilizers. In: Pond WG, Mumpton FA (eds) Zeo-agriculture: use of natural zeolites in agriculture and aquaculture. Westview Press, Boulder, pp 105–111

Li XQ, Zhang WX (2006) Iron nanoparticles: the core-shell structure and unique properties for Ni(II) sequestration. Langmuir 22(10):4638–4642

Li XQ, Zhang WX (2007) Sequestration of metal cations with zerovalent iron nanoparticles: a study with high resolution x-ray photoelectron spectroscopy (HR-XPS). J Phys Chem C 111(19):6939–6946

Li YH, Ding J, Luan Z, Di Z, Zhu Y, Xu C, Wu D, We B (2003) Competitive adsorption of Pb2+, Cu2+ and Cd 2+ ions from aqueous solutions by multi walled carbon nanotubes. Carbon 41(14):2787–2792

Li C, Wang C, Hua S (2006) Development of a parathion sensor based on molecularly imprinted nano-TiO_2 self-assembled film electrode. Sensor Actuat B-Chem 117:166–171

Li XQ, Brown DG, Zhang WX (2007) Stabilization of biosolids with nanoscale zero-valent iron (nZVI). J Nanopart Res 9(2):233–243

Lima AC, Ceragioli HJ, Cardoso KC, Peterlevitz AC, Zanin HG, Baranauskas V, Silva MJ (2010) Synthesis and application of carbon nanostructures on the germination of tomato seeds. In: Riberio C, de-Assis OBG, Mattoso LHC, Mascarenas S (eds) Symposium of international conference on food and agricultural applications of nanotechnologies. São Pedro, SP, Brazil. ISBN 978-85-63274-02-4

Limbach LK, Wick P, Manser P, Grass RN, Bruinink A, Stark WJ (2007) Exposure of engineered nanoparticles to human lung epithelial cells: influence of chemical composition and catalytic activity on oxidative stress. Environ Sci Technol 41(11):4158–4163

Lin D, Xing B (2007) Phytotoxicity of nanoparticles: inhibition of seed germination and root growth. Environ Pollut 150:243–250

Lin CF, Lo SS, Lin HY, Lee Y (1998) Stabilization of cadmium contaminated soils using synthesized zeolite. J Hazard Mater 60(3):217–226

Liu R (2011) In-situ lead remediation in a shoot-range soil using stabilized apatite nanoparticles. In: Proceedings of the 85th ACS colloid and surface science symposium. McGill University, Montreal

Liu R, Lal R (2012) A laboratory study on improvement of mine soil quality for re-vegetation through various amendments. In: Proceedings of the ASA-CSSA-SSSA international annual meetings. Cincinnati

Liu R, Zhao D (2007) Reducing leachability and bioaccessibility of lead in soils using a new class of stabilized iron phosphate nanoparticles. Water Res 41(12):2491–2502

Liu F, Wen L-X, Li Z-Z, Yu W, Sun H-Y, Chen JF (2006) Porous hollow silica nanoparticles as controlled delivery system for water-soluble pesticide. Mater Res Bull 41:2268–2275

Liu S, Yuan L, Yue X, Zheng Z, Tang Z (2008) Recent advances in nanosensors for organophosphate pesticide detection. Adv Powder Technol 19:419–441

Liu J, Valsaraj KT, Delaune RD (2009) Inhibition of mercury methylation by iron sulfides in an anoxic sediment. Environ Eng Sci 26(4):833–840

Lopez Z, Bawazir AS, Tanzy B, Adkins E (2008) Using St. Cloud clinoptilolite zeolite as a wicking material to sustain riparian vegetation. In: Proceedings of the 2008 joint meeting of the geological society of America, soil science society of America, American society of agronomy, crop science society of America, Gulf Coast association of geological societies with the Gulf Coast section of SEPM. Paper No. 54- 6

López MM, Llop P, Olmos A, Marco-Noales E, Cambra M, Bertolini E (2009) Are molecular tools solving the challenges posed by detection of plant pathogenic bacteria and viruses? Curr Issues Mol Biol 11:13–46

Lu C, Liu C (2006) Removal of nickel (II) from aqueous solution by carbon nanotubes. J Chem Technol Biotechnol 81(12):1932–1940

Lu W, Senapati D, Wang S, Tovmachenko O, Singh AK, Yu H, Ray PC (2010) Effect of surface coating on the toxicity of silver nanomaterials on human skin keratinocytes. Chem Phys Lett 487:92–96

Lyons K, Scrinis G (2009) Under the regulatory radar? Nanotechnologies and their impacts for rural Australia. In: Merlan F, Raftery D (eds) Tracking rural change: community, policy and technology in Australia, New Zealand and Europe. Australian National University E Press, Canberra, pp 151–171

Ma QY, Logan TJ, Traina SJ (1995) Lead immobilization from aqueous solutions and contaminated soils using phosphate rocks. Environ Sci Technol 29(4):1118–1126

Mackowiak CL, Amacher MC (2008) Soil sulfur amendments suppress selenium uptake by alfalfa and western wheatgrass. J Environ Qual 37(3):772–779

Mahmoodabadi MR (2010) Experimental study on the effects of natural zeolite on lead toxicity, growth, nodulation, and chemical composition of soybean. Commun Soil Sci Plant Anal 41(16):1896–1902
Masciangioli T, Zhang WX (2003) Environmental technologies at the nanoscale. Environ Sci Technol 37(5):102A–108A
Mauter MS, Elimelech M (2008) Environmental applications of carbon-based nanomaterials. Environ Sci Technol 42(16):5843–5859
Mohamed MM, Khairou KS (2011) Preparation and characterization of nanosilver/mesoporous titania photocatalysts for herbicide degradation. Microporous Mesoporous Mater 142:130–138
Moirou A, Xenidis A, Paspaliaris I (2001) Stabilization Pb, Zn, and Cd-contaminated soil by means of natural zeolite. Soil Sediment Contam 10(3):251–267
Monica RC, Cremonini R (2009) Nanoparticles and higher plants. Caryologia 62(2):161–165
Moore JN, Ficklin WH, Johns C (1988) Partitioning of arsenic and metals in reducing sulfidic sediments. Environ Sci Technol 22(4):432–437
Mumpton FA (1985) Using zeolites in agriculture. In: Innovative biological technologies for lesser developed countries, congress of the United States. Office of Technology Assessment, Washington, DC
Nair R, Poulose AC, Nagaoka Y, Yoshida Y, Maekawa T, Sakthi Kumar D (2011) Uptake of FITC labeled silica nanoparticles and quantum dots by rice seedlings: effects on seed germination and their potential as biolabels for plants. J Fluoresc 21:2057–2068
Naturland (2011) Naturland standards for organic aquaculture. Available at: http://www.naturland.de/fileadmin/MDB/documents/Richtlinien_englisch/Naturland-Standards_Aquaculture.pdf. Accessed 3 Jan 2012.
Nissen LR, Lepp NW, Edwards R (2000) Synthetic zeolites as amendments for sewage sludge-based compost. Chemosphere 41(1–2):265–269
Niu H, Cai Y (2012) Adsorption and concentration of organic contaminants by carbon nanotubes from environmental samples. In: Kim J (ed) Advances in Nanotechnology and the Environment. Pan Stanford Publishing, Singapore, pp 79–136
NNI (National Nanotechnology Initiative) (2005) What is nanotechnology? Accessed July (2005). http://www.nano.gov
O'Carroll D, Sleep B, Krol M, Boparai H, Kocur C (2013) Nanoscale zero valent iron and bimetallic particles for contaminated site remediation. Adv Water Resour 51:104–122
O'Connell MJ, Boul P, Ericson LM, Huffman C, Wang Y, Haroz E, Kuper C, Tour J, Ausman KD, Smalley RE (2001a) Reversible water-solubilization of single-walled carbon nanotubes by polymer wrapping. Chem Phys Lett 342(3–4):265–271
Ming DW, Allen ER (2001) Use of natural zeolites in agronomy, horticulture and environmental soil remediation. In: Ming DW, Bish DB (eds) Natural zeolites: occurrence, properties, applications. Mineralogical Society of America, Geochemical Society/Italian National Academy, Accademia Nationale dei Lincei (ANL), Saint Louis/Barcelonartd, pp 619–654
O'Connell MJ, Boul P, Ericson LM, Huffman C, Wang Y, Haroz E, Kuper C, Tour J, Ausman KD, Smalley RE (2001b) Reversible water-solubilization of single-walled carbon nanotubes by polymer wrapping. Chem Phys Lett 342(3–4):265–271
Oancea S, Padureanu S, Oancea AV (2009) Growth dynamics of corn plants during anionic clays action. In: Lucr_ari S¸ tiint¸ ifice, vol. 52. seria Agronomie
Olegario JT, Yee N, Miller M, Sczepaniak J, Manning B (2010) Reduction of Se (VI) to Se($^-$II) by zero-valent iron nanoparticle suspensions. J Nanopart Res 12(6):2057–2068
Ovenden C, Xiao H (2002) Flocculation behaviour and mechanisms of cationic inorganic microparticle/polymer systems. Colloids Surf A 197(1–3):225–234
Pabalan RT, Bertetti FP (2001) Cation-exchange properties of natural zeolites. In: Bish DL, Ming DW (eds) Natural zeolites: occurrence, properties, applications, vol 45. Mineralogical Society of America Reviews in Mineralogy and Geochemistry, Washington, DC, pp 453–518
Parham H, Rahbar N (2010) Square wave voltammetric determination of methyl parathion using ZrO_2-nanoparticles modified carbon paste electrode. J Hazard Mater 177:1077–1084

Patterson RR, Fendorf S, Fendorf M (1997) Reduction of hexavalent chromium by amorphous iron sulfide. Environ Sci Technol 31(7):2039–2044
Peld M, Kaia Tõnsuaadu K, Bend V (2004) Sorption and desorption of Cd2+ and Zn2+ ions in apatite-aqueous systems. Environ Sci Technol 38(21):5626–5631
Perez-de-Luque A, Cifuentes Z, Beckstead J, Sillero JC, Àvila C, Rubio J, Ryan RO (2012) Effect of amphotericin B nanodisks on plant fungal diseases. Pest Manag Sci 68:67–74
Perrin TS, Drost DT, Boettinger JL, Norton JM (1998) Ammonium-loaded clinoptilolite: a slow-release nitrogen ertilizer for sweet corn. J Plant Nutr 21(3):515–530
Petrovic AM (1990) The potential of natural zeolite as a soil amendment. Golf Course Manage 58(11):92–93
Phenrat T, Saleh N, Sirk K, Kim HJ, Tilton RD, Lowry GV (2008) Stabilization of aqueous nanoscale zerovalent iron dispersions by anionic polyelectrolytes: adsorbed anionic polyelectrolyte layer properties and their effect on aggregation and sedimentation. J Nanopart Res 10(5):795–814
Ponder SM, Darab JG, Mallouk TE (2000) Remediation of Cr(VI) and Pb(II) aqueous solutions using supported, nanoscale zero-valent iron. Environ Sci Technol 34(12):2564–2569
Qu F, Zhou X, Xu J, Li H, Xie G (2009) Luminescence switching of CdTe quantum dots in presence of p-sulfonatocalix[4]arene to detect pesticides in aqueous solution. Talanta 78:1359–1363
Raicevic S, Kaludjerovic-Radoicic T, Zouboulis AI (2005) In situ stabilization of toxic metals in polluted soils using phosphates: theoretical prediction and experimental verification. J Hazard Mater 117(1):41–53
Raicevic S, Wright JV, Veljkovic V, Conca JL (2006) Theoretical stability assessment of uranyl phosphates and apatites: selection of amendments for in situ remediation of uranium. Sci Total Environ 355(1–3):13–24
Rao GP, Lu C, Su F (2007) Sorption of divalent metal ions from aqueous solution by carbon nanotubes: a review. Sep Purif Technol 58(1):224–231
Reinsch BC, Forsberg B, Penn RL, Kim CS, Lowry GV (2010) Chemical transformations during aging of zero valent iron nanoparticles in the presence of common groundwater dissolved constituents. Environ Sci Technol 44(9):3455–3461
Renock D, Gallegos T, Utsunomiya S, Hayes K, Ewing RC, Becker U (2009) Chemical and structural characterization of As immobilization by nanoparticles of mackinaw wite (FeSm). Chem Geol 268(1–2):116–125
Revis NW, Osborne TR, Holdsworth G, Hadden C (1989) Distribution of mercury species in soil from a mercury contaminated site. Water Air Soil Pollut 45(1–2):105–113
Reynolds CS, Davies PS (2001) Sources and bioavailability of phosphorus fractions in freshwaters: a British perspective. Biol Rev Camb Philos Soc 76(1):27–64
Riba O, Scott TB, Vala Ragnarsdottir K, Allen GC (2008) Reaction mechanism of uranyl in the presence of zero-valent iron nanoparticles. Geochim Cosmochim Acta 72(16):4047–4057
Ruby MV, Davis A, Nicholson A (1994) In situ formation of lead phosphates in soils as a method to immobilize lead. Environ Sci Technol 28(4):646–654
Sakulchaicharoen N, O'Carroll DM, Herrera JE (2010) Enhanced stability and dechlorination activity of presynthesis stabilized nanoscale FePd particles. J Contam Hydrol 118(3–4):117–127
Saleh N, Kim H, Phenrat T, Matyjaszewski K, Tilton RD, Lowry GV (2008) Ionic strength and composition affect the mobility of surface-modified Fe0 nanoparticles in water-saturated sand columns. Environ Sci Technol 42(9):3349–3355
Scrinis G, Lyons K (2010) Nanotechnology and the techno-corporate agri-food paradigm. In: Lawrence G, Lyons K, Wallington T (eds) Food security, nutrition and sustainability. Earthscan, London (Chapter 16)
Shah V, Belozerova I (2009) Influence of metal nanoparticles on the soil microbial community and germination of lettuce seeds. Water Air Soil Pollut 197:143–148
Shanableh A, Kharabsheh A (1996) Stabilization of Cd, Ni and Pb in soil using natural zeolite. J Hazard Mater 45(2–3):207–217

Shi W-J, Shi W-W, Gao S-Y, Lu Y-T, Cao Y-S, Zhou P (2010) Effects of anopesticide chlorfenapyr on mice. Toxicol Environ Chem 92:1901–1907
Shi X, Sun K, Balogh LP, Baker JR (2006) Synthesis, characterization, and manipulation of dendrimer-stabilized iron sulfide nanoparticles. Nanotechnology 17:4554–4560
Shipley HJ, Engates KE, Guettner AM (2011) Study of iron oxide nanoparticles in soil for remediation of arsenic. J Nanopart Res 13(6):2387–2397
Sicbaldi F, Sarra A, Mutti D, Bo PF (1997) Use of gas-liquid chromatography with electron-capture and thermionic-sensitive detection for the quantitation and identification of pesticide residues. J Chromatogr AJ Chromatogr A 765:13–22
Singh D, Singh SC, Kumar S, Lal B, Singh NB (2010) Effect of titanium dioxide nanoparticles on the growth and biochemical parameters of Brassica oleracea. In: Riberio C, de-Assis OBG, Mattoso LHC, Mascarenas S (eds) Symposium of international conference on food and agricultural applications of nanotechnologies. São Pedro, SP, Brazil. ISBN 978-85-63274-02-4
Stan HJ, Linkerhägner M (1996) Pesticide residue analysis in foodstuffs applying capillary gas chromatography with atomic emission detection state-of-the-art use of modified multimethod S19 of the Deutsche Forschungsgemeinschaft and automated large-volume injection with programmed-temperature vaporization and solvent venting. J Chromatogr AJ Chromatogr A 750:369–390
Stanforth R, Qiu J (2001) Effect of phosphate treatment on the solubility of lead in contaminated soil. Environ Geol 41(1–2):1–10
Stead K (2002) Environmental implications of using the natural zeolite clinoptilolite for the remediation of sludge amended soils. PhD thesis, University of Surrey, Surrey, UK
Sun H, Fung Y (2006) Piezoelectric quartz crystal sensor for rapid analysis of pirimicarb residues using molecularly imprinted polymers as recognition elements. Anal Chim ActaAnal Chim Acta 576:67–76
Theron J, Walker JA, Cloete TE (2008) Nanotechnology and water treatment: applications and emerging opportunities. Crit Rev Microbiol 34(1):43–69
Thomas P, Irvine J, Lyster J, Beaulieu R (2005) Radionuclides and trace metals in Canadian moose near uranium mines: comparison of radiation doses and food chain transfer with cattle and caribou. Health Phys 88(5):423–438
Tratnyek PG, Johnson RL (2006) Nanotechnologies for environmental cleanup. Nano Today 1(2):44–48
Turan NG (2008) The effects of natural zeolite on salinity level of poultry litter compost. Bioresour Technol 99(7):2097–2101
USDA (2010) USDA pesticide data program analytical methods. Available at: http://www.ams.usda.gov/AMSv1.0/getfile?dDocName¼STELPRDC5049940. Accessed 29 Jan 2011
USEPA (2007) The use of soil amendments for remediation, revitalization and reuse. Solid Waste and Emergency Response (5203P) EPA 542-R-07-013, http://clu-in.org/download/remed/epa-542-r-07-013.pdf
USEPA, US Environmental Protection Agency Region 10 (2001) Consensus plan for soil and sediment studies: Coeurd'Alene river soils and sediments bioavailability studies (URS DCN: 4162500.06161.05.a. EPA:16.2), pp. 1–16. http://yosemite.epa.gov/R10/CLEANUP.NSF/fb6a4e3291f5d28388256-d140051048b/503bcd6aa1bd60a288256cce00070286/$FILE/soil amend consensus final 022801.PDF, 2012
Vamvakaki V, Chaniotakis NA (2007) Pesticide detection with a liposome-based nano-biosensor. Biosens Bioelectron 22:2848–2853
Villase˜nor J, Rodriguez L, Fernandez FJ (2011) Composting domestic sewage sludge with natural zeolites in a rotary drum reactor. Bioresour Technol 102(2):1447–1454
Viswanathan S, Radecka H, Radecki J (2009) Electrochemical biosensor for pesticides based on acetylcholinesterase immobilized on polyaniline deposited on vertically assembled carbon nanotubes wrapped with ssDNA. Biosens Bioelectron 24:2772–2777
Wang M, Li Z (2008) Nano-composite ZrO2/Au film electrode for voltammetric detection of parathion. Sensor Actuat B-Chem 133:607–612

Wang ZS, Hung MT, Liu JC (2007) Sludge conditioning by using alumina nanoparticles and polyelectrolyte. Water Sci Technol 56(8):125–132
Wang Z, Wei F, Liu SY, Xu Q, Huang J-Y, Dong XY, Yua JH, Yang Q, Zhao YD, Chen H (2010) Electrocatalytic oxidation of phytohormone salicylic acid at copper nanoparticles-modified gold electrode and its detection in oilseed rape infected with fungal pathogen Sclerotinia sclerotiorum. Talanta 80:1277–1281
Watanabe T, Murata Y, Nakamura T, Sakai Y, Osaki M (2009) Effect of zero-valent iron application on cadmium uptake in rice plants grown in cadmium-contaminated soils. J Plant Nutr 32(7):1164–1172
Wehtje GR, Shaw JN, Walker RH, Williams W (2003) Bermudagrass growth in soil supplemented with inorganic amendments. HortScience 38(4):613–617
Williams KA, Nelson PV (1997) Using precharged zeolite as a source of potassium and phosphate in a soilless container medium during potted chrysanthemumproduction. J Am Soc Hortic Sci 122(5):703–708
Witte ST, Will LA (1993) Investigation of selenium sourcesassociated with chronic selenosis in horses of western Iowa. J Vet Diagn Investig 5(1):28–131
Wolthers M, Charlet L, van Der Weijden CH, van der Linde PR, Rickard D (2005) Arsenic mobility in the ambient sulfidic environment: sorption of arsenic(V) and arsenic(III) onto disordered mackinawite. Geochim Cosmochim Acta 69(14):3483–3492
Xenidis A, Stouraiti C, Papassiopi N (2010) Stabilization of Pb and As in soils by applying combined treatment with phosphates and ferrous iron. J Hazard Mater 177(1–3):929–937
Xiong Z, He F, Zhao D, Barnett MO (2009) Immobilization of mercury in sediment using stabilized iron sulfide nanoparticles. Water Res 43(20):5171–5179
Xu L, Liu Y, Bai R, Chen C (2010) Applications and toxicological issues surrounding nanotechnology in the food industry. Pure Appl Chem 82:349–372
Xu Y, Zhao D (2007) Reductive immobilization of chromate in water and soil using stabilized iron nanoparticles. Water Res 41(10):2101–2108
Xueying L, O'Carroll DM, Petersen EJ, Qingguo H, Anderson CL (2009) Mobility of multiwalled carbon nanotubes in porous media. Environ Sci Technol 43(21):8153–8158
Yamamoto T, Yuya A, Satoh A, Takahasi H, Sumikoshi M, DeghaniSanij H, Agassi M (2004) Application of artificial zeolite to combat soil erosion. In: Proceedings of the American society of agricultural engineers, Canadian society for engineering of agricultural, food and biological system annual international meeting, Government Centre Ottawa, Ontario, Canada, August
Yan S, Hua B, Bao Z, Yang J, Liu C, Deng B (2010) Uranium (VI) removal by nanoscale zerovalent iron in anoxic batch systems. Environ Sci Technol 44(20):7783–7789
Yan Z, Deng Y (2000) Cationic microparticle based flocculation and retention systems. Chem Eng J 80(1–3):31–36
Yao KS, Li SJ, Tzeng KC, Cheng TC, Chang CY, Chiu CY, Liao CY, Hsu JJ, Lin ZP (2009) Fluorescence silica nanoprobe as a biomarker for rapid detection of plant pathogens. Adv Mater Res 79,82:513–516
Zhang WX (2003) Nanoscale iron particles for environmental remediation: an overview. J Nanopart Res 5(3–4):323–332
Zhao Y-G, Shen H-Y, Shi J-W, Chen X-H, Jin MC (2011) Preparation and characterization of amino functionalized nano-composite material and its application for multi-residue analysis of pesticides in cabbage by gas chromatographyetriple quadrupole mass spectrometry. J Chromatogr A 1218:5568–5580
Zheng L, Hong F, Lu S, Liu C (2005) Effect of nano-TiO_2 on strength of naturally aged seeds and growth of spinach. Biol Trace Elem Res 104:83–91
Zheng M (2011) A technology for enhanced control of erosion, sediment and metal leaching at disturbed land using polyacrylamide and magnetite nanoparticles [M.S. thesis], Auburn University, Auburn, Ala, USA
Zhou X, Shu L, Zhao H, Guo X, Wang X, Tao S, Xing B (2012) Suspending multi-walled carbon nanotubes by humic acids from a peat soil. Environ Sci Technol 46(7):3891–3897

Zorpas AA, Constantinides T, Vlyssides AG, Haralambous I, Loizidou M (2000) Heavy metal uptake by natural zeolite and metals partitioning in sewage sludge compost. Bioresour Technol 72(2):113–119

Zorpas AA, Loizidou M (2008) Sawdust and natural zeolite as a bulking agent for improving quality of a composting product from anaerobically stabilized sewage sludge. Bioresour Technol 99(16):7545–7552

Zorpas AA, Vassilis I, Loizidou M, Grigoropoulou H (2002) Particle size effects on uptake of heavy metals from sewage sludge compost using natural zeolite clinoptilolite. J Colloid Interface Sci 250(1):1–4

Zorpas AA, Vlyssides AG, Loizidou M (1999) Dewatered anaerobically-stabilized primary sewage sludge composting: metal leachability and uptake by natural clinoptilolite. Commun Soil Sci Plant Anal 30(11–12):1603–1613

Chapter 2
Phosphorus Phytoavailability upon Nanoparticle Application

Zahra Zahra, Muhammad Arshad, Muhammad Arif Ali, Muhammad Qudrat Ullah Farooqi, and Hyung Kyoon Choi

Abstract Nanotechnology has paved the way for overcoming numerous obstacles in agriculture by providing distinct improvements, beyond those of traditional methods. Engineered nanomaterials (ENMs) have been employed to improve plant growth and development, with specific goals of increasing crop production, suppressing plant disease and improving nutrient management. Nutrient deficiencies in soil limit crop yields worldwide. In recent decades, nanoparticles (NPs) have been considered as a mechanism for improving plant nutrition. This chapter draws attention to one of the crucial issues in agriculture management, i.e., the phytoavailability of naturally-bound nutrients in soil, particularly inorganic phosphorus (Pi), which has been an on-going concern for sustainable crop production. In most soils, substantial concentrations of inorganic and organic phosphates are present; however, approximately 88–99% of inorganic phosphorus is bound with calcium (Ca) and therefore unavailable to plants. About 30–40% of global crop yields are limited by low phosphorus availability. The content of this chapter focuses on the potential role of NPs for improving phytoavailability of soil phosphorus with special emphasis on soil properties affecting phosphorus availability; impacts of NPs on plant growth and development; potential phytotoxic effects of NPs and routes of entry in plants; and other mechanisms including biogeochemical processes.

Z. Zahra · M. Arshad (✉)
Institute of Environmental Sciences and Engineering (IESE), School of Civil and Environmental Engineering (SCEE), National University of Sciences and Technology (NUST), Islamabad, Pakistan
e-mail: marshad@iese.nust.edu.pk

M. A. Ali
Department of Soil Science, Faculty of Agricultural Sciences & Technology, Bahauddin Zakariya University, Multan, Pakistan

M. Q. U. Farooqi
Department of Soil Science School of Agriculture and Environment, Faculty of Science, The University of Western Australia, Perth, Australia

H. K. Choi
College of Pharmacy, Chung-Ang University, Seoul, Republic of Korea

S. Hayat et al. (eds.), *Sustainable Agriculture Reviews 41*, Sustainable Agriculture Reviews 41, https://doi.org/10.1007/978-3-030-33996-8_2

Keywords Engineered nanomaterials · Nutrients · Phosphorus · Phytoavailability · Soil-plant interactions

2.1 Introduction

Phosphorus (P) is an important element in the environment and its availability has major implications on ecosystem function and structure. Phosphorus is an essential nutrient both for plants and animals. Phosphorus deficiency is regarded as a main factor in limiting global crop production (Marschner and Rengel 2012). It was reported that crop yields are ~ 30–40% below desired levels because of low phosphorus availability (Vance et al. 2003). In Pakistan, ~ 90% soils were reported with low available phosphorus concentrations, with consequent deficiency experienced by crop plants (Akhtar et al. 2016). Agriculture is considered the backbone of a nation's economy, especially for the developing countries, that enables the provision of food to a huge population. World population is predicted to increase to 8 billion in 2025 and increase further to 9.6 billion in 2050 (FAO 2016). Therefore, it is necessary to increase agricultural production to meet the needs of the growing population.

2.2 Phosphorous as a Macronutrient

Phosphorus (P) is considered a macronutrient due to its requirement in large quantities by plants (Pagliari et al. 2018). A sustainable supply is of utmost importance for optimal plant growth and development. Phosphorus is taken up by plant root as the dihydrogen phosphate (H_2PO^{4-}), hydrogen phosphate (HPO_4^{2-}) and orthophosphate (PO_4^{3-}) ions. Phosphorus is an essential constituent of adenosine triphosphate (ATP), deoxyribonucleic acid (DNA) and ribonucleic acid (RNA). Phosphorus deficiency can affect different plant functions, seed development, root structure and ultimately crop yield.

2.3 Soil Status

In most soils, high concentrations of inorganic and organic phosphates are present; however, approximately 88–99% of the inorganic phosphorus is naturally bound with calcium (Ca) and therefore unavailable to plants. At the global scale, numerous soils suffer phosphorus deficiency due to fixation (Gyaneshwar et al. 2002). Phosphorus availability is also a function of soil pH (Vance et al. 2003). In alkaline and acidic soils, phosphate (PO_4^{-3}) ions are adsorbed onto positively charged minerals like Ca, aluminium (Al) and iron (Fe) oxides (Hinsinger 2001). Moreover, phosphorus content is influenced by soil type and its inherent physical and chemical characteristics (Karaman et al. 2006), for example, content of certain types of clay minerals.

As a consequence of intensive crop cultivation, soil fertility declines due to the uptake of vital nutrients by plants from soil. About 90% of soils in Pakistan suffer from phosphorus deficiency (GoP 2017). To meet phosphorus requirements, phosphate fertilizers have been applied which become immobilized, and only a small proportion, e.g., about 10–30%, is taken up by plants (Holford 1997). The situation will become more critical in the future as global population increases, while global phosphorus reserves decline. It is estimated that readily available phosphorus will be depleted by 2050; estimated phosporous reserves and their sustainability have been shown in (Fig. 2.1) (Cordell et al. 2009, 2011).

In agriculture, approximately 90% of phosphorus is derived from phosphate rock (Brunner 2010), a non-renewable resource which occurs in only limited locations. Most phosphate reserves i.e., 75%, are present in Morocco which exports the most phosphate ore. Although China and USA also possess significant phosphate reserves, they do not export to global markets, thereby further limiting the supply chain.

Global future demands for phosphate rock related to crop and food production are high than the reserves. It is estimated that total demand for food production will increase by 40% in 2030 and 70% in 2050 (Foresight 2011). Geopolitical instability could induce inflation in the prices of phosphate fertilizers with a consequent decrease in food production as had occurred in 2008 (Cordell and White 2011). In 2016, global total phosphate rock consumption was estimated at 60,973,000 tonnes

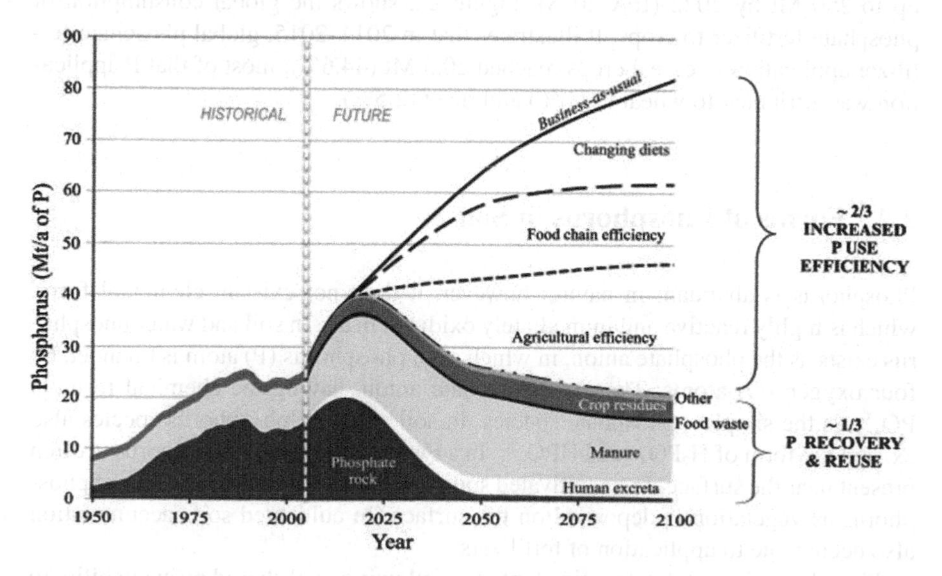

Fig. 2.1 Phosphorus reserves over the years and sustainability options. (Source: Cordell et al. 2011)

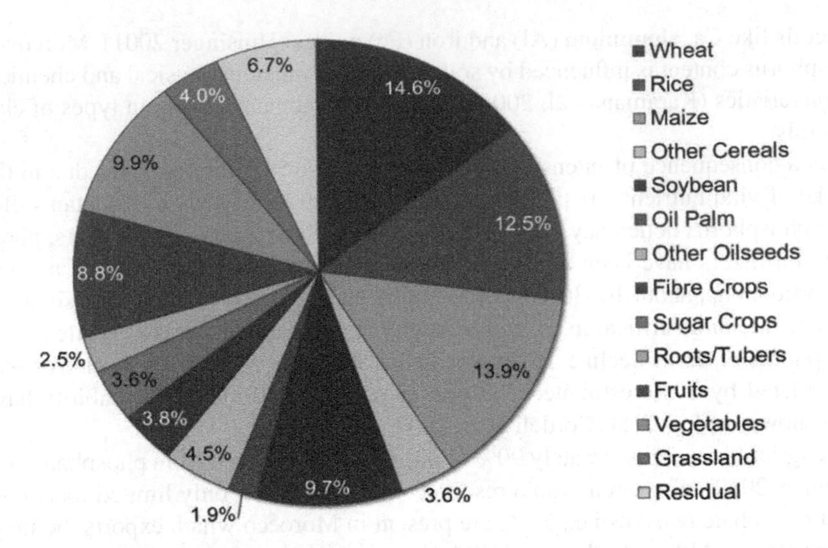

Fig. 2.2 Global usage of phosphate fertilizer with respect to crop (2014–2015). (Source: International Fertilizer Association (IFA) and International Plant Nutrition Institute (IPNI), 2017 (Heffer et al. 2017))

(IFA 2018). In Pakistan, the phosphate fertilizer used for agriculture purposes reached 1,209,145 tonnes for year 2016 (FAO 2016). According to IFA statistics, global demand for phosphate fertilizer increased by 2.0% in 2016–2017 (IFA 2017). The global phosphate rock supply is predicted to increase by 9% relative to 2017, up to 250 Mt by 2022 (IFA 2018). Figure 2.2 shows the global consumption of phosphate fertilizer to crops. It illustrates that in 2014–2015, global phosphate fertilizer application to cereal crops reached 20.5 Mt (44.6%); most of that P application was attributed to wheat (14.6%) and rice (12.5%).

2.4 Forms of Phosphorus in Soil

Phosphorus is abundant in nature; however, it does not exist in elemental form, which is highly reactive and immediately oxidized in air. In soil and water phosphorus exists as the phosphate anion, in which each phosphorus (P) atom is bounded by four oxygen (O) atoms. The orthophosphate anion, having the chemical formula PO_4^{3-}, is the simplest phosphate species. In soil solution, phosphorus species also exist in the form of $H_2PO_4^-$ and HPO_4^{2-}. In a typical soil profile, phosphorus is often present near the surface. In uncultivated soils, this occurs due to the cycling of phosphorus as vegetation is deposited on the surface. In cultivated soils accumulation also occurs due to application of fertilizers.

Phosphorus in soil is categorized into the following pools based on availability to plants: readily-available phosphorus; the inorganic labile pool that is rapidly

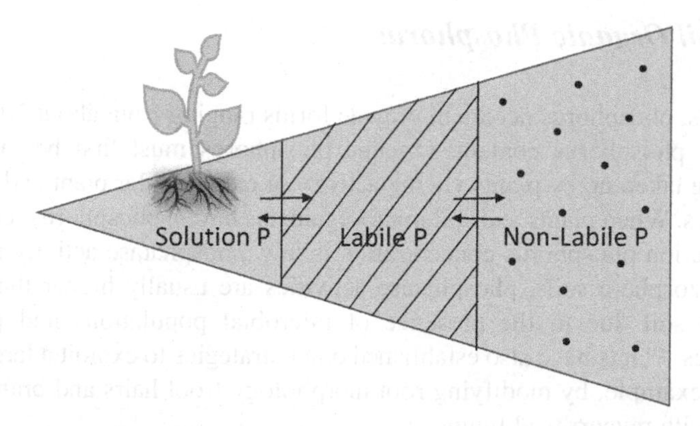

Fig. 2.3 Schematic illustration of the forms of phosphorus in soil. (Menzies 2009)

released from soil to refill the soil solution; and organic non-labile P, which is slowly available (Fig. 2.3) (Menzies 2009).

2.4.1 *Soil Solution Phosphorus*

Plants acquire phosphorus in the form of $H_2PO_4^-$ and HPO_4^{2-} from the soil solution; however, uptake of HPO_4^{2-} by plants is slower than is uptake of $H_2PO_4^-$. The proportion of these two forms is governed by pH; in acid solutions (pH < 7), $H_2PO_4^-$ is dominant, and in alkaline solutions (pH > 7), HPO_4^{2-} is dominant. In soil, the concentration of soluble phosphorus is relatively low and constitutes only a small portion of the annual ecosystem or crop requirement (Menzies 2009).

2.4.2 *Inorganic Labile Phosphorus*

Soil inorganic phosphorus consists of a solid phase, also termed the active or labile phosphorus pool, that readily replenishes P in soil solution. When phosphates are taken up by plants, the concentration in solution decreases and is subsequently replenished by phosphate released from the inorganic phosphorus pool. Soil inorganic phosphorus in the active pool is essentially the main source of phytoavailable phosphorus (Pagliari et al. 2018).

Inorganic phosphates that make up the labile pool are attached to Al and Ca or other elements and form dissolvable solids. The phosphorus in this pool is mostly present as specifically adsorbed orthophosphate. Soil particles either act as a sink (adsorption) or source (desorption) of phosphate to soil water; therefore, phosphorus desorbs into solution for possible uptake by plants; alternatively, the phosphorus concentration increases due to mineralization or fertilizer application, thus maintaining an equilibrium within the soil solution (Menzies 2009).

2.4.3 Soil Organic Phosphorus

In most soils, phosphorus occurs in organic forms ranging from about 20% to 90% of the total phosphorus content. Organic phosphorus must first be mineralized before being taken up by plants via the activity of extracellular plant and microbial phosphatases. When plants and soil microorganisms have a phosphorus requirement but soil solution phosphorus concentration is low, phosphatase activity is accelerated. In rhizosphere soils, phosphatase activities are usually higher than in non-rhizosphere soil due to the presence of microbial populations and plant root phosphatases. Plants have also established other strategies to exploit a large volume of soil, for example, by modifying root morphology (root hairs and branches) and symbiosis with mycorrhizal fungi.

The availability of organic phosphorus to plants depends on mineralization rate rather than the total amount present in soil. Mineralization and immobilization balance is governed by microbial population variability and activities; these processes are affected by environmental factors (Menzies 2009) as discussed below.

2.5 Abiotic Factors Affecting Phosphorus Availability

Myriad factors contribute to phosphorus availability in soils, for example soil properties (texture, pH, EC, organic matter percentage, etc.). Temperature and pH are discussed here in brief.

2.5.1 Temperature

Temperature strongly influences the equilibrium between mineralization and immobilization of soil phosphorus. Above 30 °C, which is considered as an optimum temperature for bacterial growth, higher organic phosphorus mineralization rates are observed. At 45 °C, soil phosphatases increased at their optimum level which could further contribute to mineralization processes. In contrast, immobilization is increasingly preferred at temperatures below 30 °C (Menzies 2009).

2.5.2 pH

In alkaline soil the dominant cation is usually Ca^{2+} which reacts with phosphate causing immobilization and thus unavailability to plants. In soils with pH less than 5.5, Al and Fe are dominant cations and react readily with phosphate. These insolu-

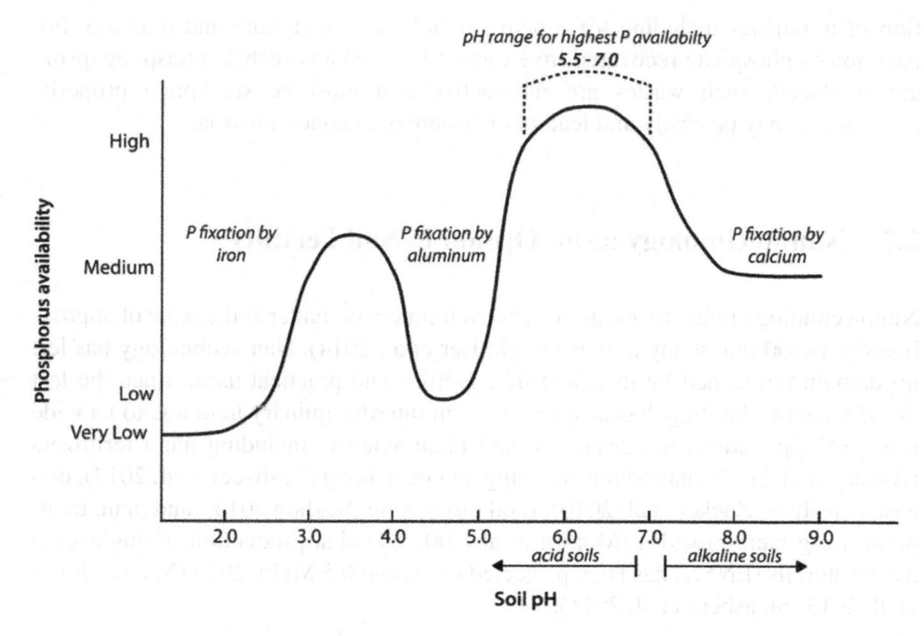

Fig. 2.4 Soil pH range as it affects phosphorus availability. (Source: Ramírez Avila et al. 2011)

ble compounds of Al and Fe phosphates are usually not phytoavailable. The range of phosphorus availability in soils as a function of pH appears in Fig. 2.4. Soil pH ranges between 5.5 and 7.0 usually contribute to optimum phytoavailability of phosphate (Menzies 2009).

In addition to the above factors, the presence of organic matter in soil can decrease phosphate adsorption on to goethite. The surface charge density decreases upon pH increase which consequently decreases the organic matter adsorption capacity (Vindedahl et al. 2016).

2.6 Environmental Concerns

Phosphorus plays dual role in the environment; on one hand, it is a major macronutrient. Phosphorus is an important component of nucleic acids, phospholipids, coenzymes and high-energy phosphate bonds, i.e., adenosine triphosphate (ATP) and adenosine diphosphate (ADP). Phosphorus also aids phospholipids in the formation of cell membranes (Daneshgar et al. 2018). On the other hand, synthetic fertilizers applied to agricultural soil may lead to increased P concentrations in water bodies through runoff, which could cause eutrophication, consequently resulting in death of aquatic life and deterioration of aquatic environments (Carpenter and Bennett 2011). Moreover, the processing of low-grade phosphate rock can result in production of impurities including toxic heavy metals, e.g., cadmium and uranium. For each ton of phosphate recovered, five tons of by-products such as phosphogypsum are produced; such wastes are radioactive and must be stockpiled properly. Stockpiling may be costly and lead to environmental contamination.

2.7 Nanotechnology as an Option in Soil Fertility

Nanotechnology refers to research and development of matter at the scale of approximately 1–100 nm in any dimension (Kaiser et al. 2014). Nanotechnology has left no domain untouched by its scientific novelties and practical uses. Since the last decade, nanotechnology has progressed as an interdisciplinary field due to its wide range of applications in agriculture and plant science, including nano fertilizers (Askary et al. 2017), nano-bioprocessing and packaging (Yashveer et al. 2014), disease detection (Zucker et al. 2010), food processing (Sekhon 2014), and plant treatment using nano-biocides (Moraru et al. 2003). Global production of engineered nanomaterials (ENMs) has been projected to exceed 0.5 Mt by 2020 (Maurer-Jones et al. 2013; Stensberg et al. 2011).

2.8 Applications of NPs in Agriculture

Nanotechnology has paved the way for addressing several obstacles in the agriculture sector by providing unique and marked improvements over traditional methods. ENMs have been used to improve plant growth and development with specific goals to increasing crop production and combatting plant diseases. The global scientific community has been inspired to explore the potential impacts of ENMs at various plant stages (Fig. 2.5).

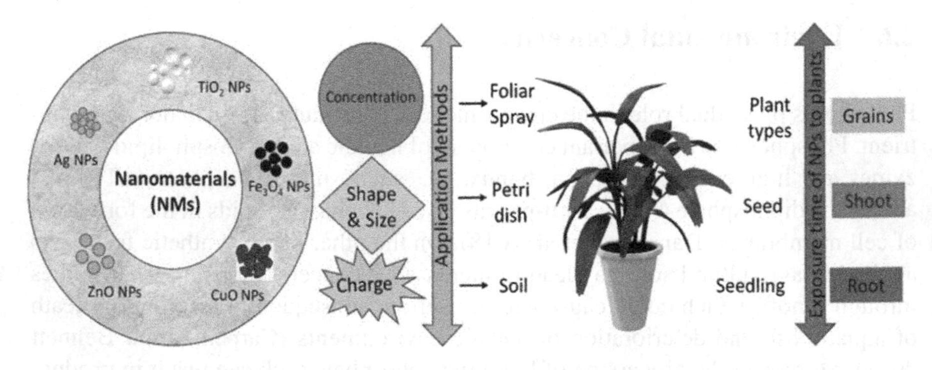

Fig. 2.5 Overview of ENM properties and their potential applications for plants

It is of utmost importance to explore innovative technologies like nanotechnology in the agricultural sector, for the cultivation of 'smart crops' that can grow and sustain in harsh climates, produce greater yields in shorter duration and with less agrochemicals inputs (Kumar et al. 2015; Sekhon 2014). ENMs could potentially provide the precise necessary quantities and maintain a sustained release of agrochemicals (i.e., herbicides, pesticides, and fertilizers) and targeted biomolecules (proteins, nucleotides) delivery, as catalysts with multifunctional abilities (Fraceto et al. 2016) (Fig. 2.6). Successful application of NPs was recently highlighted as an appropriate method for controlled and sustained release of nitrogen fertilizers (Kottegoda et al. 2017).

NPs are under increased scrutiny in certain areas of agricultural research. Various studies have investigated the behavior of NPs on specific plant species. Application of NPs to plants in different experimental conditions such as type of cultural media (soil and hydroponics), exposure time (long- and short-term), changes in nutrient levels of cultured medium, etc. have shown varied results. A detailed list of NP effects on plants, and their applications are presented in Table 2.1.

2.9 NP Amendments and Soil Nutrients

Improving phytoavailability of nutrients in soil is one of the primary goals of plant nutrition research. NP application can influence nutrient availability in soil by modifying the adsorption-desorption mechanism. Diffusional transportation and diffusion coefficients lie within low ranges (10–12 to 10–15 $cm^2\ s^{-1}$) in soils that ensure an adequate phosphorus supply to plants (Lynch et al. 2012). Phosphorus supply was reported to improve in *Brassica napus* by using P-loaded Al_2O_3 NP in a hydroponic system (Santner et al. 2012).

Furthermore, the precise application of NPs may improve use of soil pore water occurring in lower soil horizons, resulting in better dissolution of NP complexes thereby improving the diffusion of bound phosphorus to the rhizosphere. Various approaches could be used to improve the phytoavailability of phosphorus in soils such as changing the rhizosphere soil pH or improving the root exudation.

Some researchers have investigated NPs for providing slow release of nutrients from fertilizer materials (Corradini et al. 2010). A coating of synthetic NPs on fertilizer granules can provide controlled release of P. Release could extend over a longer period and provide for rapid growth and high crop yield during the entire growth period. Limited scientific investigations have been conducted to date, and have reported both negative and positive impacts of ENP use. Further research must be conducted to explore the maximum potential effects of these NPs.

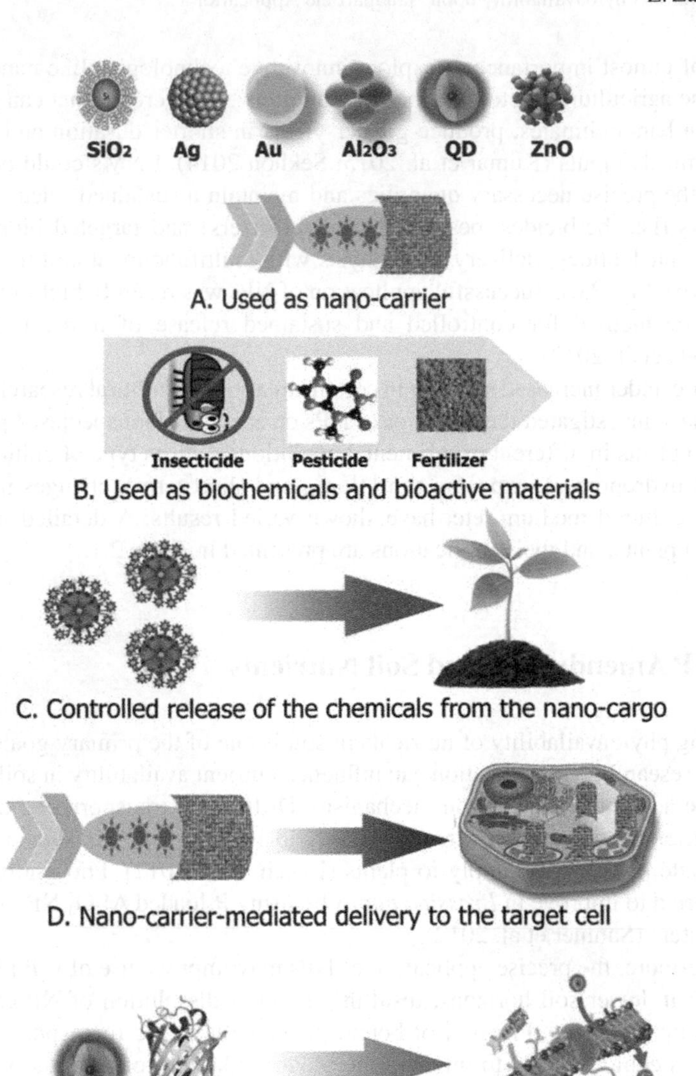

Fig. 2.6 Different applications of nanomaterials in plant science. (**a**) ENPs used as carriers for delivery of various compounds; (**b**) used as insecticides, pesticides, and nano-fertilizers in conjugation with ENPs; (**c**) controlled release of insecticides, pesticides, and fertilizers etc. from nano-carrier to the target; (**d**) transport of bioactive molecules using nano-carriers into plant cells; (**e**) nano-carriers used as indicators or biomarkers. (Source: Verma et al. 2018)

Table 2.1 Effects of different concentrations of NPs on several crop plants

NP Treatment	Concentration	Effects	Reference
TiO_2 Application medium: MS media	Levels: 10, 20, 30, 40 mg mL^{-1}	The 30 mg mL^{-1} rate significantly increased germination percentage, root-shoot length and biomass fresh weight of parsley seedlings. Chlorophyll content also improved.	Dehkourdi and Mosavi (2013)
TiO_2 Application medium: Petri dish Exposure period: 14 days	Levels: 0–80 mg L^{-1}	40 mg L^{-1} TiO_2 NPs improved mean germination time by 31.8% in fennel seeds relative to control. Germination percentage greatly improved at 60 mg L^{-1}.	Feizi et al. (2013)
TiO_2 Application medium: Soil Exposure period: 14 days	Levels: 0, 0.01%, 0.02%, and 0.03%	Under the water deficit conditions, the 0.02% concentration increased height, biomass and seed number of wheat along with the other traits such as gluten and starch content.	Jaberzadeh et al. (2013)
TiO_2 Application medium: Soil Exposure period: 150 days	Levels: 0–750 mg kg^{-1}	500 mg kg^{-1} to cucumber plants increased phosphorus concentration by 34% and K by 35% in cucumber fruit. Evidence of TiO_2 NP translocation from root to fruit was found, which suggests potential for introduction to food chain.	Servin et al. (2013)
TiO_2 Application medium: Petri dish and soil Exposure period: 15 days and 7 days	Levels: 0–5000 mg kg^{-1}	Results of germination and root elongation of tomato seedlings showed that short-term exposure of TiO_2 NPs did not induce phytotoxicity. In soil, the 5000 mg kg^{-1} rate after 1-week exposure induced phytotoxicity as increased superoxide dismutase activity in tomato seedlings.	Song et al. (2013)
TiO_2 Application medium: Petri dish and soil Exposure period: 21 days	Levels: 0, 2, 5, and 10 mg L^{-1}	No morphological effect on chickpea were observed. The 5 mg L^{-1} concentration minimized damage in sensitive and resistant chickpea genotypes induced by cold stress. TiO_2 NP activated some defensive mechanisms in chickpea seedlings, that help plants to cope with the cold stress.	Mohammadi et al. (2013)
TiO_2 Application medium: Nutrition pots Exposure period: 7 days	Levels: 0.05, 0.1 and 0.2 g L^{-1}	The 0.1 g L^{-1} rate increased photosynthesis rate in tomato leaves during mild heat stress.	Qi et al. (2013)

(continued)

Table 2.1 (continued)

NP Treatment	Concentration	Effects	Reference
TiO_2 Application medium: Spraying Exposure period: Up to vegetative and reproductive stage	Levels: 0.01 and 0.03%	Chlorophyll content (a and b), total chlorophyll (a + b), chlorophyll a/b, carotenoids and anthocyanins of maize significantly improved in response to NPs application. At the reproductive stage (emergence of male and female flowers), the maximum increase in pigment was observed relative to the control.	(Morteza et al. 2013)
TiO_2 Application medium: Petri dish Exposure period: 10 days	Levels: 0, 10, 20, 40 and 80 mg L^{-1}	For five medicinal plants *Alyssum homolocarpum*, *Salvia mirzayanii*, *Carum copticum*, *Sinapis alba*, *and Nigella sativa*, treatment at suitable concentrations improved characteristics of germination as well as vigor index.	Hatami et al. (2014)
TiO_2 and Fe_3O_4 Application medium: Soil Exposure period: 90 days	Levels: 0–250 mg kg^{-1}	Phosphorus accumulation varied in roots (TiO_2 > Fe_3O_4 > control) and shoots (Fe_3O_4 > TiO_2 > control). Increased growth of *Lactuca sativa* plants, and greater biomass and phosphorus content along with increased phytoavailable phosphorus in soil. Methionine and cystine levels were high with increase in rhizosphere soil solution P.	Zahra et al. (2015)
TiO_2 Application medium: Soil Exposure period: 14 days	Levels: 0–100 mg kg^{-1}	Shoot-root length of *Lactuca sativa* plants increased by 49% and 62%, respectively, with 100 mg kg^{-1} treatment. Phosphorus uptake per plant increased up to five-fold.	Hanif et al. (2015)
TiO_2 Application medium: Soil Exposure period: 90 days	Levels: 0–750 mg kg^{-1}	Phosphorus concentration in *Oryza sativa* roots, shoots, and grains increased by 2.6-, 2.4- and 1.3-fold, respectively, in response to 750 mg kg^{-1}. Grains harvested from plants possessed improved levels of glycerol content, palmitic acid, and amino acids.	Zahra et al. (2017)
TiO_2 Application medium: Soil Exposure period: 60 days	Levels: 0–100 mg kg^{-1}	Wheat plant length and phosphorus uptake increased compared to control with 20, 40 and 60 mg kg $^{-1}$ treatment. Chlorophyll content increased up to 32.3% at 60 mg kg^{-1}, whereas 11.1% decrease occurred at 100 mg kg^{-1}. Micronuclei cells increased by 53.6% and 62.5%, at 80 and 100 mg kg^{-1}.	Rafique et al. (2018a, b)

(continued)

Table 2.1 (continued)

NP Treatment	Concentration	Effects	Reference
TiO_2 Application medium: Soil Exposure period: 60 days	Levels: 0,100 and 500 mg kg^{-1}	Uptake and translocation of TiO_2 NPs by cultivated wheat plants. 100 and 500 mg kg^{-1} rates showed no signs of phytotoxicity in terms of plant growth, biomass or chlorophyll content.	Larue et al. (2018)
ZnO Application medium: Soil Exposure period: 60 days	Levels: 0,100 and 500 mg kg^{-1}	Increased phosphorus uptake up to 10.8% in mung bean plants in response to 10 mg L^{-1} treatment.	Raliya et al. (2016)
ZnO Application medium: Soil Exposure period: 30 days	Levels: 10, 100 and 1000 mg kg^{-1}	The concentration level of <100 mg kg^{-1} of ZnO induced positive effects on cucumber plants; 1000 mg kg^{-1} induced phytotoxic effects.	Moghaddasi et al. (2017)
Ag Application medium: Foliar Exposure period: 7 days	Levels: 40 mg per plant	In 4-week old cucumber plants, antioxidant defense systems were activated (increased phenolic compounds) and a decrease in photosynthesis (decrease in phytol) upon NPs exposure. NPs improved respiration (activation of TCA cycle intermediates), inhibited photorespiration, modified membrane properties decreased inorganic nitrogen fixation.	Zhang et al. (2018)
Apatite (nHA) Application medium: Peat moss and perlite (1:1) Exposure period: 20 weeks	Levels: 21.8 as P	Growth rate of soybean increased by 32.6% due to application of nHA (21.8 as P) relative to soybeans treated with ca$(H_2PO_4)_2$ fertilizer (a regular P fertilizer 21.8 as P). Shoot and root biomass improved by 18.2% and 41.2% respectively.	Liu and Lal (2014)
Urea-HA Application medium: Soil Exposure period: 4 weeks	Levels: Urea to hydroxyapatite (6:1)	Application of urea-HA NPs saved urea consumption up to 50% in rice plants compared to control group. The controlled release properties also improved rice crop yields.	Kottegoda et al. (2017)

2.9.1 Soil Properties and the Effect of NPs

The determination of NP behavior in soil is challenging, as soils are a complex medium. According to recent estimates (Larue et al. 2018; Sun et al. 2016), about 61 mg of TiO_2 per kg of soil is contaminated due to the sewage sludge. This is a result of sludge application to agricultural soils which is considered a source of fertilizer. This suggest that agricultural soils are an important sink of ENMs to the environment. These sinks allow for translocation of NPs to crop plants and possible entry to food chains. Presently, limited information is available on the interactions of these NPs with soil (Sun et al. 2016).

Other factors are involved in phosphorus availability, e.g., soil properties (texture, pH, EC) and nutrient status of the soil. NPs possess small size that could favour their diffusion into the plant roots and paved way for the improved absorption of P along with NPs activating different physiological processes. In the literature, it is reported that NPs could stimulate the plant root exudation process that lead towards the pH decrease in rhizosphere zone consequently solubilize the fixed soil P making it available to plants. However, change in rhizosphere pH depends on plant species (Rengel and Marschner 2005) as well as soil properties. Further study is required to reveal the possible mechanism(s) in detail.

In previous studies it was reported that the concentration of phytoavailable P increased by 56% after 14 days culturing of *Lactuca sativa* with TiO_2 NPs in sandy soil (Hanif et al. 2015). Larue et al. used TiO_2 NPs (100 and 500 mg kg^{-1}) on four different soils. NP mobility was greater in sandy soil relative to clayey soil. TiO_2 NP behavior in different soil types was found to be mainly related to the clay percentage, organic carbon content and cation exchange capacity of the soil (Larue et al. 2018).

2.9.2 *Effect of NPs on Phosphorus Availability*

NPs possess a large surface area for sorption activity. Ti^{3+} possesses cationic sorptive characteristics; the presence of TiO_2 NPs in soil provide more adsorption sites for PO_4^{3-} ions. Increased root exudation due to pH decrease in rhizosphere zone increased the desorption of PO_4^{3-} ions consequently improved P mobilization in soil (Zahra et al. 2015). In another study, wheat plants exposed to 60 mg kg^{-1} TiO_2 NPs over 60 days showed improved P uptake (Rafique et al. 2018a, b). TiO_2 NP treatment (500 mg kg^{-1}) to cucumber plants resulted in an increase of P by 34% and K by 35% in cucumber fruit (Servin et al. 2013).

2.10 Phytotoxic Effects of NPs on Plants

Increased use of ENMs in various fields has raised concerns about their release and influence on the environment. Potential toxicological effects of NPs to soil and plants have received attention in the past decade. Most people know little about nanotechnology; many consider that future benefits of nanotechnology will offset its risks, if any. Although in some studies NPs seem to be beneficial to plants, declines in plant growth have also been reported. To date, data on bioaccumulation, biomagnification, and biotransformation of NPs in food crops have not been well defined; only limited studies have been conducted. Toxic effects of ZnO and TiO_2 NPs were studied in regard to seed germination of rice. Significantly reduced seed germination was not observed in response to both types of NPs; however, ZnO NPs showed detrimental effects on rice roots at early seedling stage and stunted roots and reduced numbers of roots. TiO_2 NPs were reported to have no effect on root length (Boonyanitipong et al. 2011).

In another study, ZnO and TiO_2 NPs were reported to reduce wheat biomass, and were suggested to be harmful to this species. Moreover, small-sized TiO_2 NPs (approximately 20 nm) were able to penetrate the plant cell wall. ZnO NPs possess higher solubility than do TiO_2 NPs. Once dissolved in soil they are reported to be taken up by wheat at a higher rate (Du et al. 2011). In another study, nanosized Fe_3O_4 and TiO_2 were employed at different concentrations to assess the effects on seedling growth of tomato in hydroponic conditions. Morphological alterations caused by NPs as well as tissue internalization and possible upward translocation of Fe_3O_4 and TiO_2 NPs were documented. NP uptake and/or deposition in roots were observed using an Environmental Scanning Electron Microscope (ESEM) combined with energy dispersive X-ray spectroscopy (EDX) (Giordani et al. 2012).

Conflicting results were reported within the same species with reference to the effects of NPs. It must be kept in mind that NPs behave differently not only based on phase, size, and shape, but also with respect to plant species, experimental conditions, and applied concentrations (Castiglione et al. 2011). The food chain begins with plants, so at this stage, is critical to understand how plants respond to nanomaterials. The current literature indicates that knowledge of the environmental behavior of NPs is at the early stages. Thus far there are few conclusive reports on nanotoxicity to plants.

2.11 Mechanisms and Entry Routes of NPs in Plants

The plant cell wall does not allow the seamless entry of any external entities including NPs into the cell. The diameters of pores in the cell wall determine the size of NPs that may pass through. NPs and their aggregates within that size range could migrate across the cell membrane and be transferred to aerial parts of the plants. NPs might induce different morphological changes in root cell structures, resulting in widening of pores or stimulation of new cell wall pores (Wong et al. 2016) which ultimately enhance uptake of NPs, their aggregates or complexes. Further internalization takes place during endocytosis (Etxeberria et al. 2006) due to formation of a cavity-like structure across the NP by the plasma membrane. By using embedded transport carrier proteins (Nel et al. 2009) or ion channels, NPs might also be able to pass the membrane (Fig. 2.7).

Accumulation of NPs lead to foliar heating on the leaf surface, which causes stomatal obstruction and variation in gaseous exchange that affect cellular functions and physiological traits of plants (Nair et al. 2010). Accumulation of TiO_2 NPs measuring <35 nm was reported in wheat (*Triticum aestivum*) roots. NPs can be translocated to leaves if their diameter is <25 nm. Accumulation reached 109 mg Ti/kg dry weight in wheat roots, but the concentration in wheat leaves was below detection

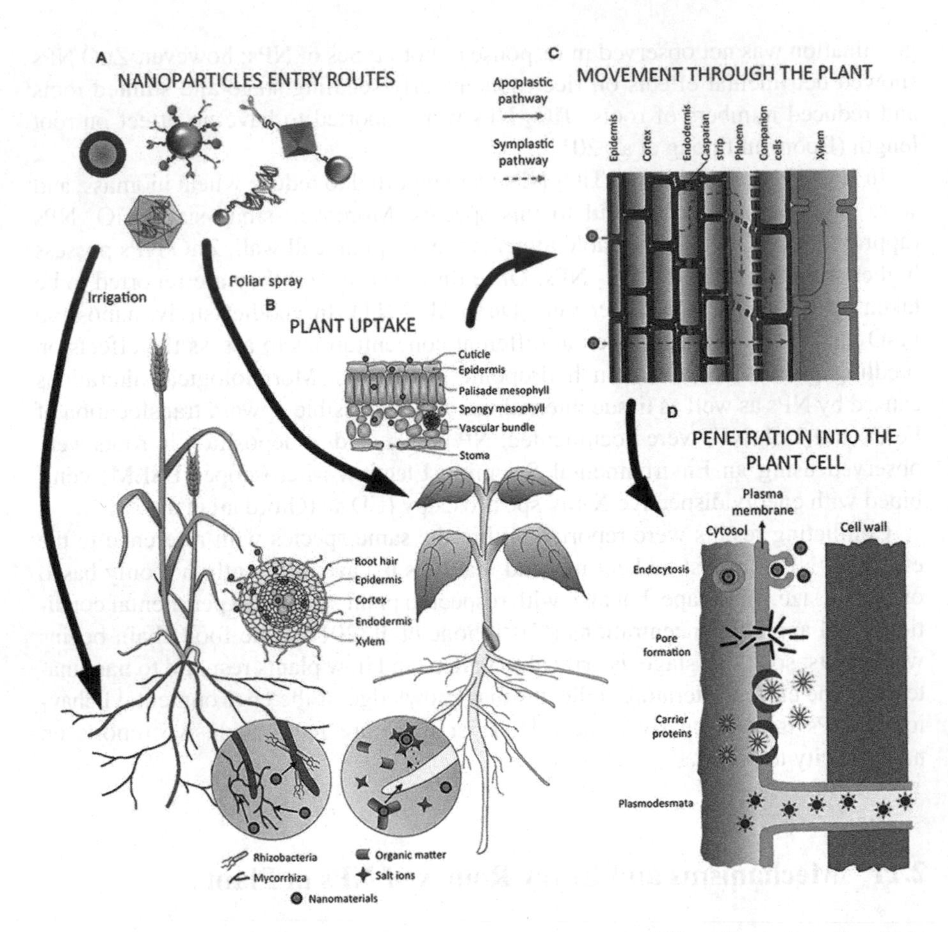

Fig. 2.7 NP entry routes and their translocation within plants. (**a**) NP attributes that affect their entry and translocation in the plant, as well as their application method; (**b**) NP behavior in soil and their interaction with microorganisms and other compounds which facilitate or inhibit absorption by plant roots or leaves; (**c**) NPs can follow the apoplastic and/or the symplastic pathways for their translocation into plant cells. They use radial movement for shifting from one pathway to another; (**d**) NP penetration into cells may involve several mechanisms, i.e., endocytosis, or via pore formation or across the plasmodesmata. (Pérez-de-Luque 2017)

limits (Larue et al. 2012). Another study demonstrated that nanostructures of TiO_2 (<5 nm size with 43% sucrose coating), translocated into the *Arabidopsis thaliana* cells and accumulated in distinct subcellular locations (Kurepa et al. 2010). In the cytoplasm, NPs attach to various organelles and obstruct metabolic processes at the site of attachment. Studies on translocation and influence of NPs within plants must be further investigated to underpin the mechanism of their behavior in plants.

2.12 Microscopic and Spectrometric Techniques Used to Determine NPs Effects

Plants offer a prospective route for NP transfer to the ecosystem and serve as an important factor in the food chain for assessing possible bioaccumulation. Various microscopic and spectrometric techniques are now available to investigate NP effects on different parts of plants and their possible mechanisms of transfer (Fig. 2.8).

Scanning electron microscopy (SEM) equipped with EDX can be used to analyze morphological changes induced by NPs in plant shoots and roots. Morphological alterations including tissue internalization and possible upward translocation of Fe_3O_4 and TiO_2 NPs in roots were examined using SEM equipped with EDX for chemical recognition. NPs agglomerates were observed on the root epidermis upon 500 mg L^{-1} treatment level (Giordani et al. 2012).

Fourier-transform infrared (FTIR) spectroscopy was employed to study alterations in functional groups of plants exposed to TiO_2 NPs. FTIR analysis of cucumber fruit exposed to TiO_2 NPs (250–750 mg kg^{-1}) showed significant alterations in amide, lipids, carbohydrates, and lignin (Servin et al. 2013). Another study using FTIR showed that NP-treated plants of *Lactuca sativa* exhibited stronger OH bonds (water) in shoots than the untreated group. Similarly, shoots of TiO_2 and Fe_3O_4 NP-treated plants had more phosphate, hydroxyl, and carbonyl groups (Zahra et al. 2015).

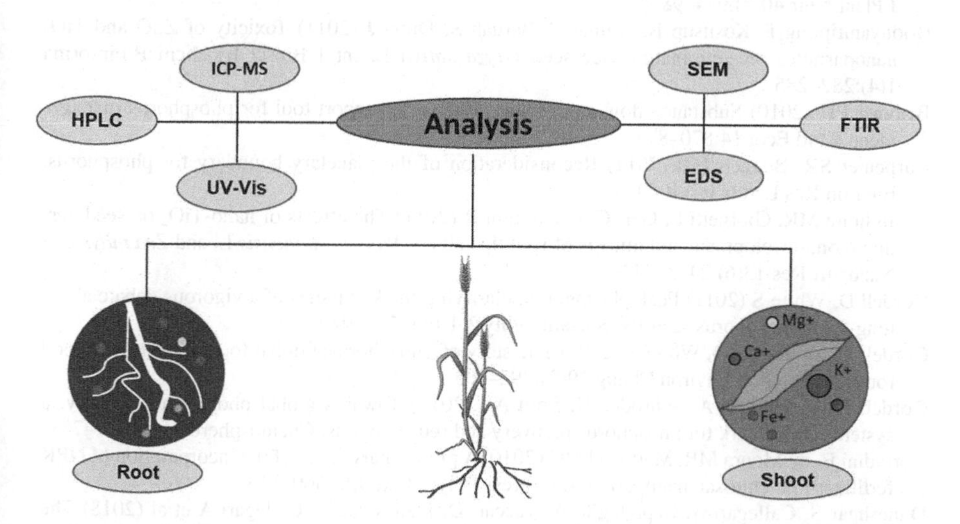

Fig. 2.8 Microscopic and spectrometric techniques for analysis of NPs in plants

2.13 Conclusions

Phosphorus is the second most essential nutrient required for the plant's growth and development and its deficiency in soil cause to limit the crop yield. Recently, NPs have been used as a factor to improve the nutritional content in plants such as improved P mobilization in soil and increased phytoavailability. These recent researches will help to estimate the potential of NPs to overcome the burden of nutrient deficit in soils for the provision of better crop yield. Furthermore, these studies helped to pave the way for future implication of these NPs in the form of nano-fertilizers at commercial level. To date, information gained from different studies has laid the foundation for basic understanding about the potential effects of NPs on plants. Based on different experimental and environmental conditions, both positive and negative impacts of NPs have been reported. However, controlled use of NPs is suggested in agro-environments to avoid environmental disturbances however, the potential benefits could only be obtained within the specified ranges.

References

Akhtar M, Yaqub M, Naeem A, Ashraf M, Hernandez VE (2016) Improving phosphorus uptake and wheat productivity by phosphoric acid application in alkaline calcareous soils. J Sci Food Agric 96(11):3701–3707

Askary M, Amirjani MR, Saberi T (2017) Comparison of the effects of nano-iron fertilizer with iron-chelate on growth parameters and some biochemical properties of *Catharanthus roseus*. J Plant Nutr 40(7):974–982

Boonyanitipong P, Kositsup B, Kumar P, Baruah S, Dutta J (2011) Toxicity of ZnO and TiO_2 nanoparticles on germinating rice seed *Oryza sativa* L. Int J Biosci Biochem Bioinforma 1(4):282–285

Brunner PH (2010) Substance flow analysis as a decision support tool for phosphorus management. J Ind Ecol 14:870–873

Carpenter SR, Bennett EM (2011) Reconsideration of the planetary boundary for phosphorus. Environ Res Lett 6(1):14009

Castiglione MR, Giorgetti L, Geri C, Cremonini R (2011) The effects of nano-TiO_2 on seed germination, development and mitosis of root tip cells of *Vicia narbonensis* L. and *Zea mays* L. J Nanopart Res 13(6):2443–2449

Cordell D, White S (2011) Peak phosphorus: clarifying the key issues of a vigorous debate about long-term phosphorus security. Sustainability 3(10):2027–2049

Cordell D, Drangert JO, White S (2009) The story of phosphorus: Global food security and food for thought. Glob Environ Chang 19(2):292–305

Cordell D, Rosemarin A, Schröder JJ, Smit AL (2011) Towards global phosphorus security: a systems framework for phosphorus recovery and reuse options. Chemosphere 84(6):747–758

Corradini E, de Moura MR, Mattoso LHC (2010) A preliminary study of the incorparation of NPK fertilizer into chitosan nanoparticles. Express Polym Lett 4(8):509–515

Daneshgar S, Callegari A, Capodaglio A, Vaccari D, Daneshgar S, Callegari A et al (2018) The potential phosphorus crisis: resource conservation and possible escape technologies: a review. Resources 7(2):37. https://doi.org/10.3390/resources7020037

Dehkourdi EH, Mosavi M (2013) Effect of anatase nanoparticles (TiO_2) on parsley seed germination (*Petroselinum crispum*) in vitro. Biol Trace Elem Res 155(2):283–286

Du W, Sun Y, Ji R, Zhu J, Wu J, Guo H (2011) TiO_2 and ZnO nanoparticles negatively affect wheat growth and soil enzyme activities in agricultural soil. J Environ Monit 13(4):822

Etxeberria E, Gonzalez P, Baroja-Fernandez E, Romero JP (2006) Fluid phase endocytic uptake of artificial nano-spheres and fluorescent quantum dots by sycamore cultured cells: evidence for the distribution of solutes to different intracellular compartments. Plant Signal Behav 1(4):196–200

FAO (2016) FAOSTAT, Statistics Division

Feizi H, Kamali M, Jafari L, Rezvani Moghaddam P (2013) Phytotoxicity and stimulatory impacts of nanosized and bulk titanium dioxide on fennel (*Foeniculum vulgare* mill). Chemosphere 91(4):506–511

Foresight UK (2011) The future of food and farming: challenges and choices for global sustainability. The government office for science, London

Fraceto LF, Grillo R, de Medeiros GA, Scognamiglio V, Rea G, Bartolucci C (2016) Nanotechnology in agriculture: which innovation potential does it have? Front Environ Sci 4:20

Giordani T, Fabrizi A, Guidi L, Natali L, Giunti G, Ravasi F et al (2012) Response of tomato plants exposed to treatment with nanoparticles. EQA - Int J Environ Qual 8(8):27–38

GoP (2017) Agriculture. In: *Pakistan Economic Survey*. Ministry of Finance, Government of Pakistan, Islamabad

Gyaneshwar P, James EK, Reddy PM, Ladha JK (2002) *Herbaspirillum* colanization increases growth and nitrogen accumulation in aluminium-tolerant rice varieties. New Phytol 154(1):131–145

Hanif HU, Arshad M, Ali MA, Ahmed N, Qazi IA (2015) Phyto-availability of phosphorus to *Lactuca sativa* in response to soil applied TiO_2 nanoparticles. Pak J Agric Sci 52(1):177–182

Hatami M, Ghorbanpour M, Salehiarjomand H (2014) Nano-anatase TiO_2 modulates the germination behavior and seedling vigority of some commercially important medicinal and aromatic plants. J Biol Environ Sci 8(22):53–59

Heffer P, Gruère A, Roberts T (2017) Assessment of fertilizer use by crop at the global level 2014–2014/15. International Fertilizer Association and International Plant Nutrition Institute, Paris

Hinsinger P (2001) Bioavailability of soil inorganic P in the rhizosphere as affected by root-induced chemical changes: a review. Plant and Soil 237:173–195

Holford ICR (1997) Soil phosphorus: its measurement, and its uptake by plants. Aus J Soil Res 35(2):227–239

IFA (2017) Short-term fertilizer outlook 2017–2018

IFA (2018) Fertilizer outlook 2018–2022. In: International Fertilizer Association (IFA) Annual Conference. International Fertilizer Association (IFA), Berlin

Jaberzadeh A, Moaveni P, Tohidi Moghadam HR, Zahedi H (2013) Influence of bulk and nanoparticles titanium foliar application on some agronomic traits, seed gluten and starch contents of wheat subjected to water deficit stress. Notulae Botanicae Horti Agrobotanici Cluj-Napoca 41(1):201–207

Kaiser DL, Standridge S, Friedersdorf L, Geraci CL, Kronz F, Meador MA, Stepp DM (2014) National nanotechnology initiative strategic plan. United States National Nanotechnology Initiative

Karaman MR, Ersahin S, Durak A (2006) Spatial variability of available phosphorus and site specific P fertilizer recommendations in a wheat field. Plant Nutrition 92:876–877

Kottegoda N, Sandaruwan C, Priyadarshana G, Siriwardhana A, Rathnayake UA, Berugoda Arachchige DM et al (2017) Urea-hydroxyapatite nanohybrids for slow release of nitrogen. ACS Nano 11(2):1214–1221

Kumar A, Pathak RK, Gupta SM, Gaur VS, Pandey D (2015) Systems biology for smart crops and agricultural innovation: filling the gaps between genotype and phenotype for complex traits linked with robust agricultural productivity and sustainability. OMICS 19(10):581–601

Kurepa J, Paunesku T, Vogt S, Arora H, Rabatic BM, Lu J et al (2010) Uptake and distribution of ultrasmall anatase TiO_2 *alizarin red* S nanoconjugates in *Arabidopsis thaliana*. Nano Lett 10(7):2296–2302

Larue C, Veronesi G, Flank A-MM, Surble S, Herlin-Boime N, Carrière M (2012) Comparative uptake and impact of TiO_2 nanoparticles in wheat and rapeseed. J Toxicol Environ Health A Curr Issues 75:722–734

Larue C, Baratange C, Vantelon D, Khodja H, Surblé S, Elger A, Carrière M (2018) Influence of soil type on TiO_2 nanoparticle fate in an agro-ecosystem. Sci Total Environ 630:609–617

Liu R, Lal R (2014) Synthetic apatite nanoparticles as a phosphorus fertilizer for soybean (*Glycine max*). Sci Rep 4:5686

Lynch J, Marschner P, Rengel Z (2012) Effect of internal and external factors on root growth and development. In: Marschner's mineral nutrition of higher plants. Academic, Amsterdam/Boston, pp 331–346

Marschner P, Rengel Z (2012) Nutrient Availability in Soils. In: Marschner's mineral nutrition of higher plants, 3rd edn. Elsevier/Academic, Amsterdam/Boston, pp 315–330

Maurer-Jones MA, Gunsolus IL, Murphy CJ, Haynes CL (2013) Toxicity of engineered nanoparticles in the environment. Anal Chem 85(6):3036–3049

Menzies N (2009) The science of phosphorus nutrition: forms in the soil, plant uptake, and plant response. Science 18:09

Moghaddasi S, Fotovat A, Khoshgoftarmanesh AH, Karimzadeh F, Khazaei HR, Khorassani R (2017) Bioavailability of coated and uncoated ZnO nanoparticles to cucumber in soil with or without organic matter. Ecotoxicol Environ Saf 144:543–551

Mohammadi R, Maali-Amiri R, Abbasi A (2013) Effect of TiO_2 nanoparticles on chickpea response to cold stress. Biol Trace Elem Res 152(3):403–410

Moraru CI, Panchapakesan CP, Huang Q, Takhistov P, Sean L, Kokini JL (2003) Nanotechnology: a new frontier in food science. Food Technology. Food Technol 57(12):24–29

Morteza E, Moaveni P, Farahani H, Kiyani M (2013) Study of photosynthetic pigments changes of maize (Zea mays L.) under nano TiO_2 spraying at various growth stages. Springerplus 2(1):247

Nair R, Varghese SH, Nair BG, Maekawa T, Yoshida Y, Kumar DS (2010) Nanoparticulate material delivery to plants. Plant Sci 179(3):154–163

Nel AE, Mädler L, Velegol D, Xia T, Hoek EMV, Somasundaran P et al (2009) Understanding biophysicochemical interactions at the nano–bio interface. Nat Mater 8(7):543–557

Pagliari PH, Kaiser DE, Rosen CJ (2018) Understanding phosphorus in Minnesota soils | UMN Extension. https://extension.umn.edu/phosphorus-and-potassium/understanding-phosphorus-minnesota-soils. Accessed 15 Sept 2018

Pérez-de-Luque A (2017) Interaction of nanomaterials with plants: what do we need for real applications in agriculture? Front Environ Sci 5:12

Qi M, Liu Y, Li T (2013) Nano-TiO_2 improve the photosynthesis of tomato leaves under mild heat stress. Biol Trace Elem Res 156(1–3):323–328

Rafique R, Zahra Z, Virk N, Shahid M, Pinelli E, Kallerhoff J et al (2018a) Data on rhizosphere pH, phosphorus uptake and wheat growth responses upon TiO_2 nanoparticles application. Data Brief 5(2):91

Rafique R, Zahra Z, Virk N, Shahid M, Pinelli E, Park TJ et al (2018b) Dose-dependent physiological responses of *Triticum aestivum L.* to soil applied TiO_2 nanoparticles: alterations in chlorophyll content, H_2O_2 production, and genotoxicity. Agric Ecosyst Environ 5(2):95–101

Raliya R, Tarafdar JC, Biswas P (2016) Enhancing the mobilization of native phosphorus in the mung bean rhizosphere using ZnO nanoparticles synthesized by soil fungi. J Agric Food Chem 64(16):3111–3118

Ramírez Avila JJ, Almansa Manrique EF, Ortega Achury SL (2011) Soil erosion and productivity losses in highly degraded soils of the Eastern Savannas of Colombia. In: ASABE – International Symposium on Erosion and Landscape Evolution 2011

Rengel Z, Marschner P (2005) Nutrient availability and management in the rhizosphere: exploiting genotypic differences. New Phytol 168(2):305–312

Santner J, Smolders E, Wenzel WW, Degryse F (2012) First observation of diffusion-limited plant root phosphorus uptake from nutrient solution. Plant Cell Environ 35(9):1558–1566

Sekhon BS (2014) Nanotechnology in agri-food production: an overview. Nanotechnol Sci Appl 7:31–53

Servin AD, Morales MI, Castillo-Michel H, Hernandez-Viezcas JA, Munoz B, Zhao L et al (2013) Synchrotron verification of TiO_2 accumulation in cucumber fruit: a possible pathway of TiO_2 nanoparticle transfer from soil into the food chain. Environ Sci Technol 47(20):11592–11598

Song U, Jun H, Waldman B, Roh J, Kim Y, Yi J, Lee EJ (2013) Functional analyses of nanoparticle toxicity: a comparative study of the effects of TiO_2 and Ag on tomatoes (*Lycopersicon esculentum*). Ecotoxicol Environ Saf 93:60–67

Stensberg MC, Wei Q, McLamore ES, Porterfield DM, Wei A, Sepúlveda MS (2011) Toxicological studies on silver nanoparticles: challenges and opportunities in assessment, monitoring and imaging. Nanomedicine 6(5):879–898

Sun TY, Bornhöft NA, Hungerbühler K, Nowack B (2016) Dynamic probabilistic modeling of environmental emissions of engineered nanomaterials. Environ Sci Technol 50(9):4701–4711

Vance CP, Uhde-Stone C, Allan Deborah L, Allan DL (2003) Phosphorus acquisition and use: critical adaptations by plants for securing a nonrenewable resource. New Phytol 157(3):423–447

Verma SK, Das AK, Patel MK, Shah A, Kumar V, Gantait S (2018) Engineered nanomaterials for plant growth and development: a perspective analysis. Sci Total Environ 630(July):1413–1435

Vindedahl AM, Strehlau JH, Arnold WA, Penn RL (2016) Organic matter and iron oxide nanoparticles: aggregation, interactions, and reactivity. Environ Sci Nano 3(3):494–505

Wong MH, Misra RP, Giraldo JP, Kwak S-Y, Son Y, Landry MP et al (2016) Lipid exchange envelope penetration (LEEP) of nanoparticles for plant engineering: a universal localization mechanism. Nano Lett 16(2):1161–1172

Yashveer S, Singh V, Kaswan V, Kaushik A, Tokas J (2014) Green biotechnology, nanotechnology and bio-fortification: perspectives on novel environment-friendly crop improvement strategies. Biotechnol Genet Eng Rev 30(2):113–126

Zahra Z, Arshad M, Rafique R, Mahmood A, Habib A, Qazi IA, Khan SA (2015) Metallic nanoparticle (TiO_2 and Fe_3O_4) application modifies rhizosphere phosphorus availability and uptake by *Lactuca sativa*. J Agric Food Chem 63(31):6876–6882

Zahra Z, Waseem N, Zahra R, Lee H, Badshah MA, Mehmood A et al (2017) Growth and metabolic responses of rice (*Oryza sativa* L.) cultivated in phosphorus-deficient soil amended with TiO_2 nanoparticles. J Agric Food Chem 65(28):5598–5606

Zhang H, Du W, Peralta-Videa JR, Gardea-Torresdey JL, White JC, Keller A et al (2018) Metabolomics reveals how cucumber (*Cucumis sativus*) reprograms metabolites to cope with silver ions and silver nanoparticle-induced oxidative stress. Environ Sci Technol 52(14):8016–8026

Zucker RM, Massaro EJ, Sanders KM, Degn LL, Boyes WK (2010) Detection of TiO_2 nanoparticles in cells by flow cytometry. Cytometry A 77A(7):677–685

Chapter 3
Synthesis of Metal/Metal Oxide Nanoparticles by Green Methods and Their Applications

Latifeh Pourakbar, Sina Siavash Moghaddam, and Jelena Popović-Djordjević

Abstract Nanotechnology is an exciting field of research; numerous versatile nanoparticles can be synthesized into a range of sizes, shapes, and chemical compositions, ultimately offering extensive applications for humans. Correct synthesis, manipulation, and use of metal NPs grant them with unique thermal, optical and electronic properties. In material science, 'green' synthesis has been considered a reliable, sustainable and environmentally-friendly protocol. Non-toxic and environmentally-friendly methods have been developed for synthesis of metal/metal oxide NPs. These techniques use live organisms such as bacteria, fungi, yeast, algae, and plants and their tissues and extracts. The biomolecules of natural extracts, such as enzymes, flavonoids, phenols, and terpenoids can be used as reducing agents of metal ions to metal NPs. Whilst the physical and chemical techniques used in traditional synthesis methods have raised environmental concerns due to use of hazardous chemicals and their possible emissions to the environment, green methods have made it possible to develop a simple, rapid, and environmentally-friendly means of synthesizing NPs. NPs produced by green methods are usually more stable and do not require application of chemical stabilizers; as a result, toxic residues do not enter the environment. Green-synthesized NPs have extensive applications for their antibacterial and antifungal properties and may be used as either plant growth stimulators or inhibitors, depending on their type, size, and shape, as well as the specific plant species.

L. Pourakbar (✉)
Department of Biology, Faculty of Science, Urmia University, Urmia, Iran
e-mail: l.pourakbar@urmia.ac.ir

S. Siavash Moghaddam
Department of Agronomy, Faculty of Agriculture, Urmia University, Urmia, Iran

J. Popović-Djordjević (✉)
University of Belgrade, Faculty of Agriculture, Chair of Chemistry and Biochemistry, Belgrade, Serbia
e-mail: jelenadj@agrif.bg.ac.rs

S. Hayat et al. (eds.), *Sustainable Agriculture Reviews 41*, Sustainable Agriculture Reviews 41, https://doi.org/10.1007/978-3-030-33996-8_3

Keywords Nanoparticle · Green synthesis · Natural extracts

3.1 Introduction

Nanotechnology, or simply nanotech, deals with objects on nanometer scales as the name implies. Nanotech is a leading-edge technology with special roles in disciplines such as chemistry, physics, and biology, and in applied science like electrical engineering, mechanical engineering, materials engineering, medical engineering, environment, and others (Nair et al. 2010).

As we move from microparticles to nanoparticles, changes occur in certain physical properties; two important ones are increased surface area to volume ratio and the introduction of particle size into the realm of quantum effects (Trindade et al. 2001). As a result of the gradual increase in surface area to volume ratio following size reduction, the behavior of surface atoms dominates the behavior of internal atoms. This phenomenon affects particle properties both in pure form and when interacting with other materials. As soon as particles become sufficiently small, they begin to display quantum mechanics behavior (Trindade et al. 2001). The property of quantum dots (QDs) is an example. These dots are sometimes called artificial atoms because their free electrons occupy discrete and virtual states of energy like electrons trapped within atoms (Christian et al. 2008).

Nanoparticles are formed of different materials such as metals and ceramics. The basic element at the nanoscale is the nanoparticle (NP). As the name implies, NP refers to particles which one of its dimensions is in nanometer size. The second basic form is the nanocapsule, which is a capsule of nanometer diameter in which materials are encapsulated (Pokropoivny and Skorokhod 2007). Another basic element is the carbon nanotube (CNT). This class of NPs was discovered by Nippon Electric Company, Ltd., Japan, in 1991. If a 2-D plate of graphite is bent to form a cylinder, CNTs are produced. CNTs are in various forms and sizes and may be single-walled or multi-walled (Kumar and Kumbhat 2016).

NPs can be broadly divided into metals and non-metals. Metal NPs are reactive, so they are prone to contamination with impurities during synthesis (Jeevandandam et al. 2018). Non-metal NPs tend to be environmentally friendly, so they have been widely used in healthcare, water purification, and green chemistry (and vice versa).

3.2 Methods of NP Synthesis

NPs are manufactured from a wide range of materials by three general groupings of methods:

(i) *Vapor phase chemical condensation*: This method is used to synthesize ceramic and metal oxide NPs. It involves vaporizing a metal followed by rapid condensation, which yields nanometric clusters as a precipitated powder.

(ii) *Chemical synthesis*: This involves ‘growing’ NPs in a liquid medium (a ‘liquid intermediate’) containing the required reagents. The sol-gel method is an example. In chemical synthesis methods, final particle size can be controlled by stopping the process when the optimal size is attained or by selecting chemicals constituting sustainable particles and halting growth at a desired size. This method is used to fabricate QDs (Christian et al. 2008). It should be mentioned that to control the size of nanoparticles is usually used phytochemicals materials which are reducing and capping agent.

(iii) *Solid-state processes (grinding or powdering)*: In this method, micrometer-size particles are converted to smaller particles by applying direct energy such as by powdering or grinding. The properties of the resulting NPs depend on the grinding material, grinding duration, and the atmospheric environment. This method can be used to produce NPs from materials that do not readily lend themselves to production of NPs by the other two methods (Geonmonond et al. 2018).

A key aspect of nanotech is the development of safe and environmentally protective methods for NP synthesis without the involvement of hazardous chemicals. In this regard, biological processes for NP synthesis are evolving and have become a major branch of nanotech.

Presently, metal NPs have found various applications in medical engineering and biological, chemical, optical, and electrical science. Among the NPs, metal NPs have attracted more attention due to their catalytic and antimicrobial properties. However, most NP synthesis methods, e.g., laser-aided synthesis, chemical reduction, photochemical reduction, the use of irradiance, and so on, are still at the development stage, so most encounter problems such as stability of synthesized NPs, control of crystal growth, and aggregation of nascent NPs due to slight variations in temperature, pH, and other environmental parameters (Bhainsa and Souza 2006) and, above all, the release of pollutants from utilization of hazardous chemicals. Recent attempts to develop nano green synthesis have given rise to biomimetic approaches that apply biological principles to NP formation (Tripathy et al. 2009). For these reasons, interest in biological methods of NP synthesis, known as the ‘green method’ are growing.

3.2.1 Nanoparticle Synthesis by Green Methods

As mentioned, NPs can be synthesized in numerous ways. Nonetheless, some chemical methods cannot avoid the use of toxic chemicals in the synthesis protocol. Therefore, the need for reliable and environmentally benign methods of NP synthesis has motivated researchers to consider using microorganisms (Klaus et al. 1999; Nair and Pradeep 2002), enzymes (Willner et al. 2006) and plant extracts (Jafari et al. 2015; Beyrami Miavaghi and Pourakbar 2016) as alternatives to chemical and physical methods. The application of green methods for bio-preparation of NPs is an exciting, yet not fully understood, option.

Most organisms, both single-cell and multi-cell, that are known for the production of inorganic NPs act intra- or extracellularly (Honary et al. 2012; Honary et al. 2013). In green synthesis, metal ions are reduced by combining with biomolecules occurring in cell extracts such as enzymes, proteins, amino acids, polysaccharides, and vitamins that are environmentally friendly; this is not the case for NP synthesis with chemicals (Vinopal et al. 2007).

3.2.2 Bacteria

Naturally occurring metals and non-metals are in constant interaction with biological components. The most abundant organisms in the biosphere are bacteria (Kushwaha et al. 2015). Bacterial species have been widely used for commercial applications of biotechnology, such as genetic engineering (Suresh et al. 2014). Bacteria are capable of reducing metal ions and are important candidates for production of NPs (Yuvakkumar et al. 2014). The biological components and reactions of both prokaryotes and eukaryotes are exploited for synthesis of NPs (Kushwaha et al. 2015).

Even slight changes to the local environment can be potentially detrimental to bacterial physiological processes but can be an advantage for NP production (Kushwaha et al. 2015). Bacteria have been explicitly targeted for NP synthesis because of their rapid proliferation, low culture costs, and easy control and manipulation of their culture media.

During synthesis of metal NPs, some bacterial species possess distinct mechanisms for suppressing the potential toxicity of metals. Bacteria can synthesize NPs inside and outside cells, and can reduce and precipitate metal ions for NP synthesis using reducing agents such as proteins and enzymes (Ayesha 2017).

Prokaryotic bacteria and actinomycetes have been used to produce metal/metal oxide NPs such as gold and silver nanoparticles (Sowani et al. 2016; Dong et al. 2017; Ali et al. 2018).

3.2.3 Fungi

Fungi perform better as biological agents in the synthesis of metal/metal oxide NPs due to the action of various intracellular enzymes (Sastry et al. 2003). Fungi can produce more NPs than bacteria (Mohanpuria et al. 2008). In addition, fungi are more versatile in NP synthesis than other organisms by virtue of the presence of enzymes, proteins, and reducing molecules on cell surfaces (Ahmad et al. 2003).

To synthesize NPs by fungal-nanotech processes, extracellular enzymes and secondary metabolites of fungi reduce metal ions to metallic NPs. As an example, in the extracellular synthesis of silver NPs, NADPH and NADH-dependent reductase enzymes are employed to reduce silver ions into silver NPs (Fayaz et al. 2010; Syed et al. 2013; Li et al. 2012). Various strains of fungi, including *Fusarium viz Fusarium oxysporum*, *Aspergillus fumigatus*, and *Aspergillus flavus* have been used to produce NPs (Mukherjee et al. 2002). Most recently, the powdery mildew fungus *Coriolus versicolor* has been exploited to synthesize stable silver NPs (Duran et al. 2005). In addition, yeasts such as *Candida glabrata*, *Torulopsis sp.*, *Schizosaccharomyces pombe* and MKY3 (a silver-resistant yeast strain) can synthesize NPs like CdS QDs (Reese and Winge 1988), nanocrystallites (Kowshik et al. 2002a), and silver NPs (Kowshik et al. 2002b).

3.2.4 Algae

Both live and dead algal biomass is used for biocombination of NPs; algae are consequently referred to as bionanofactor plants. Algae have a marked capacity for metal absorption and, therefore, the methods in which algae are used are cost-effective and practical (Bilal et al. 2018). Algae contain substantial quantities of reducing factors that reduce metal salts to their metallic NP counterparts without production of harmful byproducts. Aqueous extracts of algae contain secondary metabolites, e.g., polysaccharides, proteins, tannins, and steroids as bioactive molecules (Arya et al. 2018). The polysaccharides of algae impart both reducing and stabilizing roles in this respect (Venkatpurwar and Pokharkar 2011; Xia ct al. 2011).

3.2.5 Plants

The application of plant extracts to synthesize metal NPs has been advanced in recent years as a simple and effective alternative to chemical and physical methods. NP biosynthesis methods, in particular the plant-involved ones, are more efficient than other methods as they are inexpensive and the required materials are readily available (Mohanpuria et al. 2008). Therefore, application of plants as a sustainable and available source of production of eco-compatible NPs has interested researchers in recent years (Kelly et al. 2002). This method is environmentally protective and reduces the risk for humans, the atmosphere, and ecosystems (Donaldson et al. 2005; Forough and Farhadi 2010). Plant-based synthesis is also much more reliable and safer than using bacteria, fungi, and yeasts. Furthermore, NPs produced from plants are more stable than the chemically synthesized ones.

Various secondary metabolites, enzymes, proteins, or other reducing factors are involved in the production of metal NPs in plants. The locus of NP bioaccumulation depends on the presence of the necessary enzymes and proteins involved in their synthesis. The recovery of NPs from plant tissue is time-consuming and expensive and enzymes are therefore important for degrading the cellulosic tissue, so it is easier to use plant extracts at both small- and large-scale to generate different metal NPs (Gardea-Torresdey et al. 2005).

Many diverse plants have been evaluated in efforts to synthesize metal/metal oxide NPs. In early attempts, extracts of geranium were collected from leaves, stems, and roots for extracellular synthesis of NPs. Several studies reported the bioreduction of metal ions to metal NPs using this extract (Shankar et al. 2003; Jafari et al. 2015; Pourakbar et al. 2019,).

Leaf extracts contain useful compounds like ketones, aldehydes, flavones, amides, terpenoids, carboxylic acids, phenols, and ascorbic acids, which can reduce metal salts to metal NPs (Doble and Krtuhiventi 2007).. Plant extracts have biomolecules of carbohydrates and proteins that act as a metal-reducing agent in the formation of nanoparticles. The proteins in plant extracts are involved in the metal ions reduction by the amine agent (Li et al. 2007). Functional groups such as -C-O-C-, -C-O-, -C = C- and -C = O which are present in phytochemicals such as flavones, alkaloids, phenols and anthracenes, can participate in the production of nanoparticles (Huang et al. 2007).

To synthesize NPs using leaf extracts, the extract is reacted with metal precursor solutions under specified conditions (Mittal et al. 2013). The determinant parameters of plant leaf extracts (e.g., types of phytochemicals, concentration, metal salt concentration, pH, and temperature) are important to controlling the extent of nanoparticle formation and their yield and stability (Dwivedi and Gopal 2010). The phytochemicals present in leaf extracts of plants have an extremely high potential for reducing metal ions within a short period as compared with fungi and bacteria, which require longer incubation times (Jha et al. 2009). Therefore, plant leaf extracts are perceived as an excellent source of metals for the synthesis of metal oxide NPs.

It is now well-established that many phytochemicals such as alkaloids, terpenoids, phenolic acid, sugars, polyphenols, and proteins, can reduce metal salts to metal NPs. For example, Shankar et al. (2003) confirm that terpenoids occurring in geranium leaf extract are actively involved in converting silver ions to NPs. Eugenol is a major terpenoid compound in the extract of *Cinnamomum zeylanisum*. It has been used in the bioreduction of $HAuCl_4$ and $AgNO_3$ metal salts to their metal NP counterparts (Singh et al. 2010).

3.3 Synthesis of Metal Nanoparticles From Plants

3.3.1 *Plant Extract Preparation*

Place the plant (leaf, root, gum, or any other part) in a dark environment and allow to dry. After grinding the dry tissue, add 5 g–100 mL deionized water and place in an oven at 70 °C for 15 min to derive the extract. After purifying, centrifuge the extract at 5000 × g for 10 min.

3.3.2 *Metal/Metal Oxide Nanoparticle Preparation*

To obtain metal NPs, prepare 1 mM solution of the intended metal, e.g. $AgNO_3$ solution for Ag NPs (Forough and Farhadi 2010), $Zn(NO_3)_2$ solution for Zn NPs (Fakhari et al. 2019), TiO_2 solution for TiO_2 NPs (Dobrucka 2017), $Mg(NO_3)_2$ solution for MgO NPs (Sharma et al. 2017), $Cu(CH_3COO)_2$ or $Cu(NO_3)_2$ solution for Cu NPs (Arya et al. 2018), and tetrachloroauric acid ($HAuCl_4{\cdot}3H_2O$) solution for Au NPs (Aljabali et al. 2018). Adjust 2 mL of the plant extract with 1 mM prepared metal solution to a final volume of 100 mL, and store at 62 °C for 15 min. The first sign of NP formation is the color change from light yellow to brown at both 62 °C and room temperature (Forough and Farhadi 2010).

3.3.3 *Specifications of Synthesized Nanoparticles*

To recognize NP formation and its unique properties such as size and form, the following methods can be used.

3.3.3.1 Color Change

The first sign of NP formation is color change from light yellow to brown. For example, Fig. 3.1 displays color change due to formation of Ag NPs from extracts of *Althaea officinalis* L., *Mentha pulegium* L., and *Thymus vulgaris* L. (Jafari et al. 2015). Since NPs occur in different sizes, they differ in physical and chemical properties, which are in turn a source of differences in optical characteristics and color changes (Viau et al. 2003).

Fig. 3.1 Color variations during silver nanoparticle synthesis. The photos display silver nitrate (colorless solution) versus silver nanoparticle solution synthesized from (**a**) thyme, (**b**) pennyroyal, and (**c**) hibiscus. (Jafari et al. 2015)

3.3.3.2 Spectrophotometry

The UV-visible technique is an efficient means of analyzing and characterizing metal NPs (Henglein 1993). The brown color characteristic of NP solutions emanates from the stimulation of surface plasmon vibrations on NPs. The correct signature allows the use of spectrophotometry for assessing their formation. UV-vis analysis is performed using a standard tabletop spectrophotometer. The NP solution is centrifuged at 2800 g (5000 rpm) for 15 min and its absorption is recorded at 200–800 nm. The maximum absorption varies for metal NPs according to metal type. For example, maximum absorption is in the range of 420–450 nm for Ag (Fig. 3.2) (Jafari et al. 2015), 350 nm for Zn (Fakhari et al. 2019), and 525–540 nm for Au (Aljabali et al. 2018).

3.3.3.3 Transmission Electron Microscopy (TEM) and Scanning Electron Microscopy (SEM)

The size distribution, morphology, and shape of prepared NPs are recognized by TEM and SEM. In SEM analysis, after refinement, desolvation, and freeze-drying of the prepared NPs, they are subjected to SEM analysis. Figure 3.3 shows the TEM imagery of silver NPs synthesized from *Althaea officinalis* L., *Mentha pulegium* L., and *Thymus vulgaris* L. (Jafari et al. 2015).

3.4 Applications of Nanoparticles

NPs synthesized by biological methods offer a plethora of applications. For example, they may be used as a selective spectral coverage in optical receptors to intercept solar energy, as secondary materials to modify the performance of electrical batteries, as strong anti-microbial agents, or as catalysts of chemical reactions (Duran et al. 2005). Below are two major applications of NPs for their antibacterial and antifungal activity and their effects on plant growth and development.

3.4.1 Antibacterial and Antifungal Properties of Silver Nanoparticles

The on-going emergence of resistance to antibiotics among pathogenic and opportunistic microorganisms has compelled scientific communities to constantly seek more effective medicines and new targets (Demir et al. 2014). New antibiotics have been introduced by the pharmaceutical industry over the past two decades, but none have been modified to address the activity of multidrug-resistant bacteria. Given

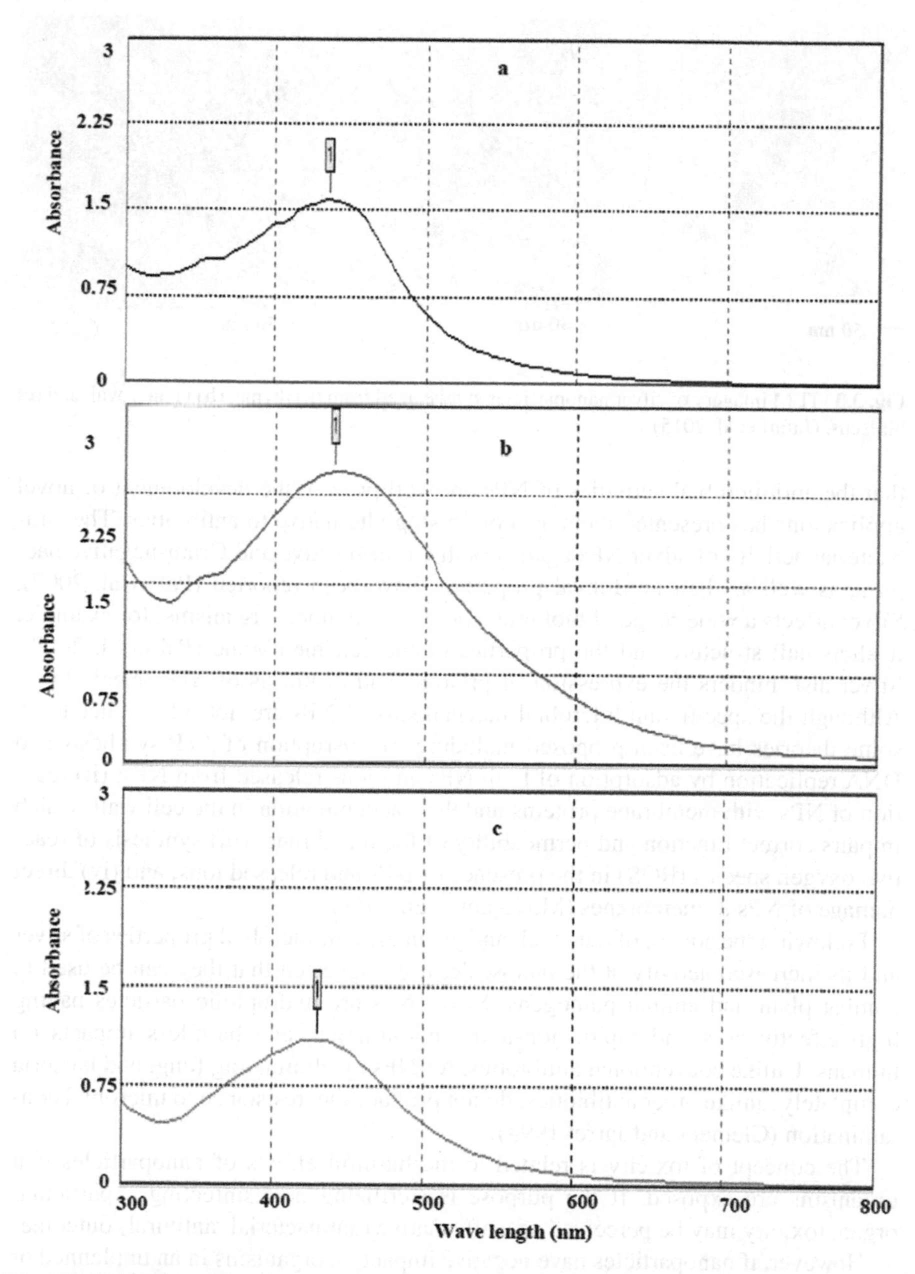

Fig. 3.2 UV-visible absorption spectrum of silver nanoparticles synthesized from (**a**) thyme, (**b**) pennyroyal, and (**c**) hibiscus by the green method. (Jafari et al. 2015)

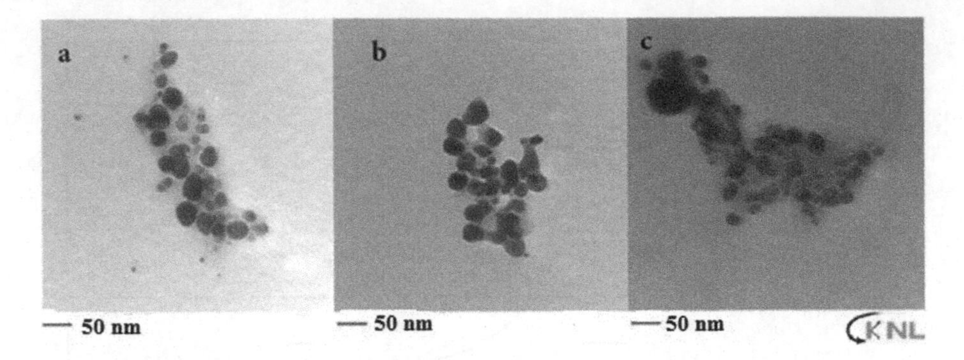

Fig. 3.3 TEM imagery of silver nanoparticles synthesized from (**a**) thyme, (**b**) pennyroyal, and (**c**) hibiscus. (Jafari et al. 2015)

that the antimicrobial activities of NPs are well proven, the development of novel applications has presented them as a promising alternative to antibiotics. The antibacterial activity of silver NPs against both Gram-positive and Gram-negative bacteria, as well as their antifungal properties, have been reported (Pal et al. 2007). Silver affects a wide range of biological processes in microorganisms; for example, it alters cell structure and the properties of the cell membrane (Pal et al. 2007). Silver also hinders the expression of proteins and synthesis of ATP (Neal 2008). Although the specific antimicrobial mechanisms of NPs are not fully understood, some theories have been proposed including: (i) disruption of ATP synthesis and DNA replication by adsorption of both NPs and ions released from NPs; (ii) reaction of NPs with membrane proteins and their accumulation in the cell wall, which impairs correct function and permeability of the membrane; (iii) synthesis of reactive oxygen species (ROS) in the presence of NPs and released ions; and (iv) direct damage of NPs to membranes (Mirzajani et al. 2011).

Following the advent of nanotech and given the antimicrobial properties of silver and its increased activity at the nanoscale, it is suggested that they can be used to combat plant and animal pathogens. Silver NPs are hydrophilic particles having high effectiveness and rapid, non-toxic, non-allergic, and harmless impacts on humans. Unlike conventional antibiotics, Ag NPs by eliminating fungi and bacteria completely, unlike other antibiotics, do not produce any resistance to microbial contamination (Clement and Jarret 1994).

The concept of toxicity is related to the harmful effects of nanoparticles that organisms are exposed. If the purpose is sterilizing or disinfecting a particular organ, toxicity may be perceived as a affirmative (antibacterial, antiviral) outcome.

However, if nanoparticles have negative impact on organisms in an unplanned or unwanted manner, such toxicity is a potential hazard (Park et al. 2007). Extensive research has focused on the effectiveness of silver NPs for control of microorganisms.

Many investigations have revealed the antifungal and antibacterial effects of silver NPs synthesized from plant extracts. Examples include antifungal effects of silver NPs synthesized from extracts of *Euphorbia hirta* L. (David et al. 2010), the antifungal (Figs. 3.4 and 3.5) and antibacterial (Figs. 3.6 and 3.7) effect of silver and copper NPs synthesized from extracts of *Althaea officinalis* L., leaf extracts of *Thymus vulgaris* L., and leaf extracts of *Mentha pulegium* L. (Jafari et al. 2015), and the antimicrobial effect of silver and copper NPs derived from the gum of *Cerasus avium* and *Prunus armeniaca* (Pourakbar et al. 2019).

It has been established that NP size influences the extent of their antibacterial and antifungal activities. In a study on the effect of NP size on bacteria, Agnihotri et al. (2014) found that antibacterial effects were enhanced when particle size decreased below 10 nm.

It has been demonstrated that commercial formulations of silver NP powder at a concentration of 300 μg/ml could reduce numbers of *E. coli* and *S. aureus* colonies from 2×10^4 CFU/ml to 0 and less than 20 colonies, respectively (Pape et al. 2002). The Pape et al. (2002) study showed weaker antibacterial properties of copper NPs than silver NPs. Similarly, Pourakbar et al. (2019) reported antibacterial properties of copper and silver NPs synthesized from tree gums. Other NPs also exhibit antibacterial effects; for instance, it has been shown that carbon nanotubes influence the abundance of soil bacterial groups in tomato culture.

Further research is needed to fully evaluate the antimicrobial effects of NPs including effective concentration, size, and type of NPs. Additionally, it is necessary to compare commercial compounds with their biological counterparts.

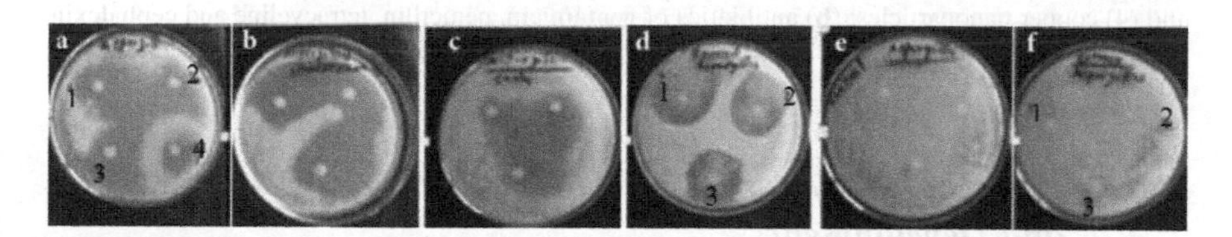

Fig. 3.4 The effect of different treatments on *Aspergillus flavus* fungus: (**a**) nanoparticles synthesized from hibiscus, pennyroyal, and thyme in Petri dishes labeled 1, 2, and 3, respectively, and (4) copper nanoparticles, (**b**) silver nitrate, (**c**) copper chloride, (**d**) benomyl, (**e**) control, (**f**) plant extracts of hibiscus, pennyroyal, and thyme in Petri dishes labeled 1, 2, and 3, respectively. (Jafari et al. 2015)

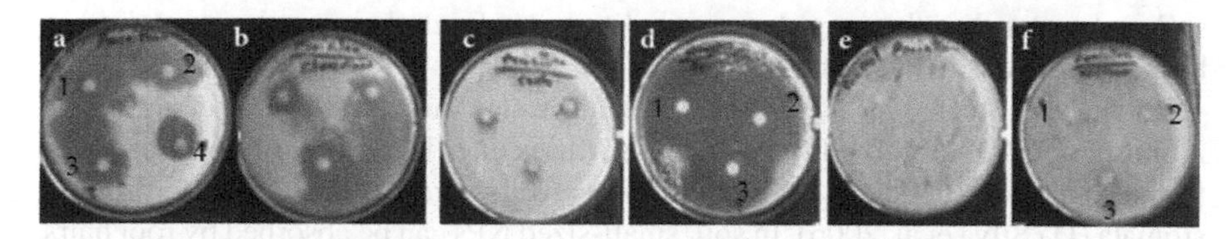

Fig. 3.5 The effect of different treatments on *Penicillium chrysogenum* fungus: (**a**) nanoparticles synthesized from hibiscus, pennyroyal, and thyme in Petri dishes labeled 1, 2, and 3, respectively, and (4) copper nanoparticles, (**b**) silver nitrate, (**c**) copper chloride, (**d**) benomyl, (**e**) control, (**f**) plant extracts of hibiscus, pennyroyal, and thyme in Petri dishes labeled 1, 2, and 3, respectively. (Jafari et al. 2015)

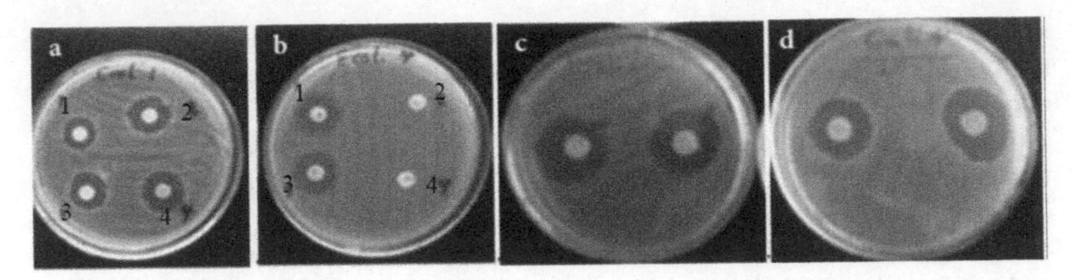

Fig. 3.6 The effect of different treatments on *E. coli* bacterum: (**a**) nanoparticles synthesized from hibiscus, pennyroyal, and thyme in Petri dishes labeled 1, 2, and 3, respectively and (4) copper nanoparticles; (**b**) antibiotics of gentamicin, penicillin, tetracycline and cephalexin in Petri dishes labeled 1, 2, 3 and 4 respectively; (**c**) copper chloride; (**d**) silver nitrate. (Jafari et al. 2015)

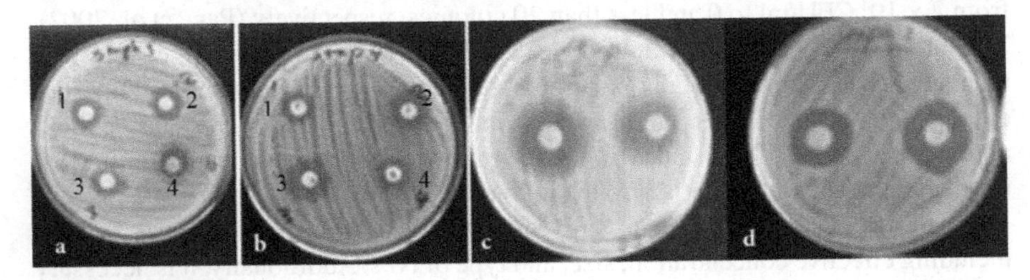

Fig. 3.7 The effect of different treatments on *Staphylococcus aureus* bacterum: (**a**) nanoparticles synthesized from hibiscus, pennyroyal, and thyme in Petri dishes labeled 1, 2, and 3, respectively and (4) copper nanoparticles; (**b**) antibiotics of gentamicin, penicillin, tetracycline and cephalexin in Petri dishes labeled 1, 2, and 3, respectively; (**c**) copper chloride; (**d**) silver nitrate. (Jafari et al. 2015)

3.4.2 Effects of Nanoparticles on Plant Growth and Development

As a result of the widespread use of NPs, significant quantities find their way into aquatic, terrestrial and atmospheric environments (Nowack and Bucheli 2007). To understand the risks and/or potential benefits of NPs, it is necessary to investigate their biological impacts including degradability and bioavailability in the environment (Juhel et al. 2011). Little information is available regarding the fate of NPs in soil and water, but they are documented to enter food chains and accumulate in higher organisms (Zhu et al. 2008).

Plants are affected by NPs present in land, water, and air (Navarro et al. 2008). Atmospheric NPs absorbed through leaf surfaces alter leaf morphological parameters such as the structure of the thricum; NPs are also present in the hypodermis and stomata (Da Silva et al. 2006). In soil, small-sized NPs can be absorbed by root hairs (Ovečka et al. 2005). It has been suggested that NPs are also absorbed into plant cells via endocytosis.

There is little information about the biological consequences of interactions and/or uptake of NPs by plants and many questions remain unanswered about the fate and behavior of NPs in plants (Ma et al. 2010).

Both positive and negative impacts of NPs on plants have been reported (Yuhui et al. 2009). It has been documented that NPs enhance seed germination and plant growth (Zheng et al. 2005; Lin et al. 2009). A comparative study on the use of nano-iron chelate and iron chelate on mung bean (*Vigna radiate* L.) showed that the nano-iron chelate increased the activity of ascorbate peroxidase and catalase enzymes, chlorophyll content, and consequently, photosynthesis rate to a greater extent than iron chelate (Karimi et al. 2014). Multi-walled carbon nanotubes can penetrate the hard coatings of tomato seeds, improving their growth. In fact, the key to increasing germination rate is penetration of nanomaterials into seeds, which augments water uptake and germination (Khodakovskaya et al. 2009). Acting as a carrier on the root surface, carbon nanomaterials enhanced nutrient uptake and increased root biomass of tobacco (Khot et al. 2012). Nano-TiO_2 can enhance photosynthesis and metabolism in spinach plants, thereby improving growth (Hong et al. 2005; Alia et al. 2012). Other positive effects of NPs on growth and chemical properties of plants have been reported (Zhang et al. 2004; Gao et al. 2008; Moradi Rikabad et al. 2019).

Early reports of the negative impacts of NPs on various plants (corn, cucumber, soybean, and carrot) has shown that even low concentrations may affect growth (Yang and Watts 2005). Nano-Al_2O_3 was found to inhibit root elongation in several species including corn, cucumber, and cabbage (Yang and Watts 2005; Lin and Xing 2007). Phytotoxicity of five NP types (multi-walled CNTs, nano-Al, nano-Al_2O_3, nano-Zn, and nano-ZnO) has been observed, specifically on seed germination, growth, and root elongation in six higher plant species. Multi-walled CNTs, nano-Al_2O_3, and nano-Al did not influence root elongation at a rate of 2000 mg/L, but 2000 mg/L nano-Zn and nano-ZnO reduced root growth significantly. Nano-ZnO at high dosages caused toxicity to seedlings of radish (*Raphanus sativus*), canola (*Brassica napus*), rye (*L. perenne*), lettuce (*Lactuca sativa*), barley (*Hordeum sativum*), and cucumber (*Cucumis sativus*) (Lin and Xing 2007). Franklin et al. (2007) investigated the toxicity of nano-ZnO and Zn^{2+} and found that the negative effects of nano-ZnO may partially emanate from the dissolution of Zn nanoparticles.

Metal NPs such as nano-TiO_2 are toxic to green algae *Desmodesmus* and hinder their growth (Hund-Rinke and Simon 2006). Lee et al. (2008) used plant agar to homogenize and make NPs available and found that Cu nanoparticles dispersed in agars were toxic to wheat (*Triticum aestivum*) and mung bean (*Phaseolus radiates*) at high doses.

These varied effects of NPs on plant growth raise the question as to what measurable biochemical/molecular factors can explain the observed differences in effects of NPs among different species (Atha et al. 2012; Moradi Rikabad et al. 2019). The mechanism of NP toxicity at higher concentrations is still unknown, but is usually explained by the release of heavy metal ions, synthesis of ROS intermediates, and oxidative stress (Oberdörster et al. 2005; Franklin et al. 2007; Limbach et al. 2007). NPs may increase lipid membrane peroxides during contact with cells owing to generation of ROS intermediates. This phenomenon may be related to the chemical composition, structure, and size of NPs, their contact surface, and other factors (Yuhui et al. 2009).

It is well established that a wide range of chemical stressors actually enhance plant growth (Kovács et al. 2009) and there is a question as to whether this growth stimulation results from the NPs or its specific ion. The diversity in NP responses is not surprising; different NPs possess varied chemical compositions often with unique crystalline structures. They furthermore have different particle sizes and contact areas. In addition, plant response to stresses is related to a number of components, e.g., whether NPs are sharp or spherical, the concentration used, plant species, and environmental parameters. Normally, stresses create a complicated mix of injuries and acclimatization response and the balance between these depends on conditions and exposure (Juhel et al. 2011). Acclimatization responses of plants may reduce damage (e.g., by increasing tolerance) including enhancement of antioxidant defense and signaling (Foyer and Noctor 2005) as well as stress-induced morphological responses (Potters et al. 2007).

A concern in the application of NPs to seed germination is potential phytotoxicity. Toxicity level may depend on NP type. Certain nano-silica and cadmium-selenide QDs have been tested for germination improvement. Nano-silica improved germination, but cadmium-selenide QDs had an inhibitory effect on germination (Monica and Cremonini 2009). The authors found that root growth depended on NP type and concentration (Lin and Xing 2007; Nair et al. 2011). On the other hand, comparison of the effect of nano-Cu and nano-$CuCl_2$ on the germination of basil (*Ocimum basilicum* L.) revealed that both impaired germination, but nano-Cu had a less toxic effect than nano-$CuCl_2$ on germination and growth (Yusefzaei et al. 2017).

Future research on the effects of nanomaterials on germination and growth response faces several challenges (Nair et al. 2010): (i) unpredictability of nanomaterial reactions in different plant species; (ii) phytotoxicity arising from the high toxicity of nanomaterials; and (iii) the limited uptake of nutrients and reduction in photosynthesis in plants by large-sized NPs.

3.5 Conclusion

Synthesis of metal/metal oxide NPs by the green method is more environmentally friendly than chemical synthesis methods – the green method minimizes the side effects encountered with chemical and physical methods, such as occurrence of residues of toxic substances and the formation of harmful and hazardous byproducts. Green NP synthesis involves the use of bacteria, fungi, yeasts, plants, and plant extracts both intra- and extracellularly. Among these, plant extracts are of greater importance because synthesis from extracts is rapid and poses less risk compared to the use of plants, bacteria, and fungi. In addition, the biomolecules occurring in plant extracts are used both as the agent of reducing metal ions to metal NPs and as the NP stabilizing agent. In recent years intensive research has focused on NPs synthesized by green methods because of their unique antibacterial and antifungal properties and possible applications as growth stimulators and occasionally as growth inhibitors. Future research is likely to focus on synthesis of NPs for treatment of severe diseases like cancer, or for transport of medications in the body. Other work will address NPs for use in agriculture, for example in the translocation of herbicides in plants or release of nano-fertilizers.

References

Agnihotri S, Mukherji S, Mukherji S (2014) Size controlled silver nanoparticles synthesized over the range 5–100 nm using the same protocol and their antibacterial efficacy. RSC Adv:3974–3983. https://doi.org/10.1039/C3RA44507K

Ahmad A, Mukherjee P, Senapati S, Mandal D, Khan MI, Kumar R, Sastry M (2003) Extracellular biosynthesis of silver nanoparticles using the fungus *Fusarium oxysporum*. Colloids Surf B 28:313–318. https://doi.org/10.1016/S0927-7765(02)00174-1

Ali I, Qiang TY, Ilahi N, Adnan M, Sajjad W (2018) Green synthesis of silver nanoparticles by using bacterial extract and its antimicrobial activity against pathogens. Int J Biosci 13(5):1–5

Alia DS, Castillo-Michel H, Hernandez-Viezcas JA, Diaz BC, Jose R, Peralta-Videa JR, Gardea-Torresdey JL (2012) Synchrotron micro-XRF and micro-XANES confirmation of the uptake and translocation of TiO_2 nanoparticles in cucumber (*Cucumis sativus*) plants. Environ Sci Technol 14:7637–7643

Aljabali AA, ID Akkam Y, Al Zoubi MS, Al-Batayneh KM, Al-Trad B, Abo Alrob O, Alkilany AM, Benamara M, Evans DJ (2018) Synthesis of gold nanoparticles using leaf extract of *Ziziphus zizyphus* and their antimicrobial activity. Nanomaterials 8:1–15. https://doi.org/10.3390/nano8030174

Arya A, Gupta K, Chundawat TS, Vaya D (2018) Biogenic synthesis of copper and silver nanoparticles using green alga *Botryococcus braunii* and its antimicrobial activity. Hindawi Bioinorg Chem Appl 2018:1–9

Atha DH, Wang H, Petersen EJ, Cleveland D, Holbrook RD, Jaruga P, Dizdaroglu M, Xing B, Nelson BC (2012) Copper oxide nanoparticle mediated DNA damage in terrestrial plant models. Environ Sci Technol 46:1819–1827

Ayesha A (2017) Bacterial synthesis and applications of nanoparticles. Nano Sci Nano Technol Indian J 11:119–126

Beyrami Miavaghi M, Pourakbar L (2016) Phytosynthesis of silvern by medicinal plant *Malva neglecta*. Qom Univ Med Sci J 10:38–44. (English abstract)

Bhainsa KC, Souza SFD (2006) Extracellular biosynthesis of silver nanoparticles using the fungus *Aspergillus fumigates*. Colloids Surf B: Biointerfaces 47:160–164

Bilal M, Rasheed T, Sosa-Hernández JE, Raza A, Nabeel F, Iqbal HMN (2018) Biosorption: an interplay between marine algae and potentially toxic elements—a review. Mar Drugs 16:1–16

Christian F, Von der Kammer F, Baalousha M, Hofmann T (2008) Nanoparticles: structure, properties, preparation and behavior in environmental media. Ecotoxicology 17:326–343

Clement JL, Jarret PS (1994) Antimicrobial silver. Metal Based Drugs 1:467–482. https://doi.org/10.1155/MBD.1994.467

Da Silva LC, Oliva MA, Azevedo AA, De Araújo JM (2006) Responses of resting plant species to pollution from an iron pelletization factory. Water Air Soil Pollut 175:241–256

David E, Elumalai EK, Prasad TN, Venkata K, Nagajyothi PC (2010) Green synthesis of silver nanoparticle using *Euphorbia hirta* L and their antifungal activities. Arch Appl Sci Res 2:76–81

Demir E, Kaya N, Kaya B (2014) Genotoxic effects of zinc oxide and titanium dioxide nanoparticles on root meristem cells of *Allium cepa* by comet assay. Turk J Biol 38:31–39. https://doi.org/10.3906/biy-1306-11

Doble M, Kruthiventi AK (2007) Green chemistry and engineering. Academic, Cambridge

Dobrucka R (2017) Synthesis of titanium dioxide nanoparticles using *Echinacea purpurea* Herba. Iran J Pharm Res 16:753–759
Donaldson K, Tran L, Jimenez L, Duffin A, Newby R, Mills D, MacNee W, Stone V (2005) Combustion-derived nanoparticles: a review of their toxicology following inhalation exposure. Part Fibre Toxicol 2:10
Dong ZY, Manik PNR, Xiao M, Hong-Fei W, Wael N, Hozzein Wei C, Wen-Jun L (2017) Antibacterial activity of silver nanoparticles against Staphylococcus warneri synthesized using endophytic bacteria by photo-irradiation. Front Microbiol 8:1–8
Duran N, Marcato P, Alves O, Desouza G, Esposito E (2005) Mechanistic aspects of biosynthesis of silver nanoparticles by several *Fusarium oxysporum* strains. NanoBiotechnology 13:3–8
Dwivedi AD, Gopal K (2010) Biosynthesis of silver and gold nanoparticles using *Chenopodium album* leaf extract. Colloids Surf A Physicochem Eng Asp 369:27–33. https://doi.org/10.1016/j.colsurfa.2010.07.020
Fakhari SH, Jamzad M, Kabiri Fard H (2019) Green synthesis of zinc oxide nanoparticles: a comparison. Green Chem Lett Rev 12:19–24
Fayaz AM, Balaji K, Girilal M, Yadav R, Kalaichelvan PT, Venketesan RS (2010) Biogenic synthesis of silver nanoparticles and their synergistic effect with antibiotics: a study against gram-positive and gram-negative bacteria. Nanomedicine 6:103–109
Forough M, Farhadi KH (2010) Biological and green synthesis of silver nanoparticles. Turk J Eng Environ Sci 34:281–287
Foyer CH, Noctor G (2005) Oxidant and antioxidant signalling in plants: a reevaluation of the concept of oxidative stress in a physiological context. Plant Cell Environ 28:1056–1071
Franklin N, Rogers N, Apte S, Batley G, Gadd G, Casey P (2007) Comparative toxicity of nanoparticulate ZnO, bulk ZnO, and $ZnCl_2$ to a freshwater microalga (*Pseudokirchneriella subcapitata*): the importance of particle solubility. Environ Sci Technol 41:8484–8490
Gao F, Liu C, Qu C, Zheng L, Yang F, Su M, Hong F (2008) Was improvement of spinach growth by nano-TiO_2 treatment related to the changes of Rubisco activase? Biometals 21:211–217
Gardea-Torresdey J, Peralta-Videa R, Rosaa G, Parson GJ (2005) Phytoremediation of heavy metals and study of the metal coordination by X-ray absorption spectroscopy. Coord Chem Rev 249:797–810
Geonmonond RS, Da Silva AGM, Camargo PH (2018) Controlled synthesis of noble metal nanomaterials: motivation, principles, and opportunities in nanocatalysis. Anais Da Academia Brasileira De Ciencias 90:719–744. https://doi.org/10.1590/0001-3765201820170561
Henglein A (1993) Physicochemical properties of small metal particles in solution. "Microelectrode" reactions, chemisorptions, composite metal particles, and the atom-to-metal transition. J Phys Chem B 97:5457–5471
Honary S, Barabadi H, Gharaei – Fathabad E, Naghibi F (2012) Green synthesis of copper oxide nanoparticles using *Penicillium aurantiogriseum*, *Penicillium citrinum* and *Penicillium waksmanii*. Dig J Nanomater Biostruct 7:999–1005
Honary S, Barabadi H, Gharaei Fathabad E, Naghibi F (2013) Green synthesis of silver nanoparticles induced by the *Fungus Penicillium citrinum*. Trop J Pharm Res 12:7–11
Hong FS, Zhou J, Liu C, Yang F, Wu C, Zheng L, Yang P (2005) Effect of nano-TiO_2 on photochemical reaction of chloroplasts of spinach. Biol Trace Elem Res 105:269–279
Huang Q, Li D, Sun Y, Lu Y, Su Y, Yang X, Wang H, Wang Y, Shao W, He N (2007) Biosynthesis of silver and gold nanoparticles by novel sundried *Cinnamomum camphora* leaf. Nanotechnology 18(10):1–11. https://doi.org/10.1088/0957-4484/18/10/105104
Hund-Rinke K, Simon M (2006) Ecotoxic effect of photocatalytic active nanoparticles (TiO_2) on algae and daphnids. Environ Sci Pollut Res Int 13:225–232
Jafari A, Pourakbar L, Farhadi KH, Mohamad Golizad L (2015) Biological synthesis of silver nanoparticles and evalution of antibacterial and antifungal properties of silver and copper nanoparticles. Turk J Biol 39:1–6. https://doi.org/10.3906/biy-1406-81

Jeevanandam J, Barhoum A, Chan YS, Dufresne A, Danquah MK (2018) Review on nanoparticles and nanostructured materials: history, sources, toxicity and regulations. Beilstein J Nanotechnol 9:1050–1074. https://doi.org/10.3762/bjnano.9.98

Jha AK, Prasad K, Kumar V, Prasad K (2009) Biosynthesis of silver nanoparticles using eclipta leaf. Biotechnol Prog 25:1476–1479. https://doi.org/10.1002/btpr.233

Juhel G, Batisse E, Hugues Q, Daly D, Van Pelt FNAM, O'Halloran J, Jansen MAK (2011) Alumina nanoparticles enhance growth of *Lemna minor*. Aquat Toxicol 105:328–336

Karimi Z, Pourakbar L, Feizi H (2014) Comparison effect of nano-iron chelate and iron chelate on growth parameters and antioxidant enzymes activity of mung bean (*Vigna radiate L.*). Adv Environ Biol 8:916–930

Kelly KL, Coronado E, Zhao LL, Schatz GC (2002) The optical properties of metal nanoparticles: the influence of size, shape, and dielectric environment. J Phys Chem B 107:668–677

Khodakovskaya M, Dervishi E, Mahmood M, Xu Y, Li Z, Watanabe F, Biris AS (2009) Carbon nanotubes are able to penetrate plant seed coat and dramatically affect seed germination and plant growth. ACS Nano 3:3221–3227. https://doi.org/10.1021/nn900887m

Khot LR, Sankaran S, Maja JM, Ehsani R, Schuster EW (2012) Applications of nanomaterials in agricultural production andcrop protection: a review. Crop Prot 35:64–70

Klaus T, Joerger R, Olsson E, Granqvist CG (1999) Silver-based crystalline nanoparticles, microbially fabricated. Proc Natl Acad Sci U S A 96:13611–13614

Kovács E, Nyitrai P, Czövek P, Òvári M, Keresztes A (2009) Investigation into the mechanism of stimulation by low-concentration stressors in barley seedlings. J Plant Physiol 166:72–79

Kowshik M, Vogel W, Urban J, Kulkarni SK, Paknikar KM (2002a) Microbial synthesis of semiconductor PbS nanocrystallites. Adv Mater 14:815–818

Kowshik M, Vogel W, Urban J, Kulkarni SK, Paknikar KM (2002b) Extracellular synthesis of silver nanoparticles by a silver-tolerant yeast strain MKY3. Nanotechnology 14:95–100

Kumar N, Kumbhat S (2016) Carbon-based nanomaterials. In: Essentials in Nanoscience and Nanotechnology. Wiley, Hoboken, pp 189–236. https://doi.org/10.1002/9781119096122.ch5

Kushwaha A, Kumar Singh V, Bhartariya J, Singh P, Yasmeen K (2015) Isolation and identification of *E. coli* bacteria for the synthesis of silver nanoparticles: characterization of the particles and study of antibacterial activity. Eur J Exp Biol 5:65–70

Lee WM, An YJ, Yoon H, Kweon HS (2008) Toxicity and bioavailability of copper nanoparticles to the terrestrial plants mung bean *(Phaseolus radiatus)* and wheat *(Triticum aestivum)*: plant agar test for water-insoluble nanoparticles. Environ Toxicol Chem 27:1915–1921

Li S, Shen Y, Xie A, Yu X, Qui L, Zhang L, Zhang Q (2007) Green synthesis of silver nanoparticles using *Capsicum annuum* L. extract. Green Chem 9:852–858

Li G, He D, Qian Y, Guan B, Gao S, Cui Y, Wang L (2012) Fungus-mediated green synthesis of silver nanoparticles using *Aspergillus terreus*. Int J Mol Sci 13:466–476. https://doi.org/10.3390/ijms13010466

Limbach L, Wick P, Manser P, Grass R, Bruinink A, Stark W (2007) Exposure of engineered nanoparticles to human lung epithelial cells: influence of chemical composition and catalytic activity on oxidative stress. Environ Sci Technol 41:4158–4163

Lin D, Xing B (2007) Phytotoxicity of nanoparticles: inhibition of seed germination and root growth. Environ Pollut 150:243–250

Lin S, Reppert J, Hu Q, Hudson JS, Reid ML, Ratnikova TA, Rao AM, Luo H, Ke PC (2009) Uptake, translocation, and transmission of carbon nanomaterials in rice plants. Small 5:1128–1132

Ma X, Geiser-Lee J, Deng Y, Kolmakov A (2010) Interactions between engineered nanoparticles (ENPs) and plants: phytotoxicity, uptake and accumulation. Sci Total Environ 408:3053–3061

Mirzajani F, Ghassempour A, Aliahmadi A, Esmaeili MA (2011) Antibacterial effect of silver nanoparticles on *Staphylococcus aureus*. Res Microbiol 162:542–550. https://doi.org/10.1016/j.resmic.2011.04.009

Mittal AK, Chisti Y, Banerjee UC (2013) Synthesis of metallic nanoparticles using plant extracts. Biotechnol Adv 31:346–356

Mohanpuria P, Rana NK, Yadav SK (2008) Biosynthesis of nanoparticles: technological concepts and future applications. J Nanopart Res 10:507–517

Monica RC, Cremonini R (2009) Nanoparticles and higher plants. Caryologia 62:161–165

Moradi Rikabad M, Pourakbar L, Siavash Moghaddam S, Popović-Djordjević J (2019) Agrobiological, chemical and antioxidant properties of saffron (*Crocus sativus* L.) exposed to TiO_2 nanoparticles and ultraviolet-B stress. Ind Crop Prod 137:137–143. https://doi.org/10.1016/j.indcrop.2019.05.017

Mukherjee P, Senapati S, Mandal D, Ahmad A, Khan MI, Kumar R, Sastry M (2002) Extracellular synthesis of gold nanoparticles by the fungus *Fusarium oxysporum*. Chembiochem 3:461–463. https://doi.org/10.1002/1439-7633(20020503)3:5<461:AID-CBIC461>3.0.CO;2-X

Nair B, Pradeep T (2002) Coalescense of nanoclusters and formation of submicron crystallites assisted by Lactobacillus strains. Cryst Growth Des 2:293–298

Nair R, Varghese SH, Nair BG, Maekawa T, Yoshida Y, Kumar DS (2010) Nanoparticulate material delivery to plants. Plant Sci 179:154–163

Nair R, Poulose AC, Nagaoka Y, Yoshida Y, Maekawa T, Sakthi Kumar D (2011) Uptake of FITC labeled silica nanoparticles and quantum dots by rice seedlings: effects on seed germination and their potential as biolabels for plants. J Fluoresc 21:2057–2068

Navarro E, Baun A, Behra R, Hartmann NB, Filser J, Miao AJ, Quigg A, Santschi PH, Sigg L (2008) Environmental behavior and ecotoxicity of engineered nanoparticles to algae, plants, and fungi. Ecotoxicology 17:372–386

Neal AL (2008) What can be inferred from bacterium nanoparticle interactions about the potential consequences of environmental exposure to nanoparticles? Ecotoxicology 17:362–371. https://doi.org/10.1007/s10646-008-0217-x

Nowack B, Bucheli TD (2007) Occurrence, behavior and effects of nanoparticles in the environment. Environ Pollut 150:5–22

Oberdörster G, Oberdörster E, Oberdörster J (2005) Nanotoxicology: an emerging discipline evolving from studies of ultrafine particles. Environ Health Perspect 113:823–839

Ovecka M, Lang I, Baluska F, Ismail A, Illes P, Lichtscheidl IK (2005) Endocytosis and vesicle trafficking during tip growth of root hairs. Protoplasma 226:39–54

Pal S, Tak YK, Song JM (2007) Does the antibacterial activity of silver nanoparticles depend on the shape of the nanoparticle? A study of the gram negative bacterium *Escherichia coli*. Appl Environ Microbiol 73:1712–1720. https://doi.org/10.1128/AEM.02218-06

Pape HL, Serena FS, Contini P, Devillers C, Maftah A, Leprat P (2002) Evaluation of the antimicrobial properties of an activated carbon fibre supporting silver using a dynamic method. Carbon 40:2947–2954

Park HJ, Kim SH, Kim HJ, Choi SH (2007) A new composition of nano sized silica silver for control of various plant diseases. Plant Pathol 22:295–302

Pokropivny VV, Skorokhod VV (2007) Classification of nanostructures by dimensionality and concept of surface forms engineering in nanomaterial science. Mater Sci Eng C 27:990–993. https://doi.org/10.1016/j.msec.2006.09.023

Potters G, Pasternak TP, Guisez Y, Jansen MAK (2007) Stress-induced morphogenic responses: growing out of trouble? Trends Plant Sci 12:98–105

Pourakbar L, Yosefzaei F, Farhadi K (2019) Biosynthesis of silver nanoparticles from tree gum extracts and evaluation of antibacterial properties of silver and copper nanoparticles. Sci J Ilam Univ Med Sci 26:1–9

Reese RN, Winge DR (1988) Sulfide stabilization of the cadmium–γglutamyl peptide complex of Schizosaccharomyces pombe. J Biol Chem 263:12832–12835

Sastry M, Ahmad A, Khan MI, Kumar R (2003) Biosynthesis of metal nanoparticles using fungi and actinomycete. Curr Sci 85:162–170

Shankar SS, Ahmad A, Pasricha R, Sastry M (2003) Bioreduction of chloroaurate ions by geranium leaves and its endophytic fungus yields gold nanoparticles of different shapes. J Mater Chem 13:1822–1826. https://doi.org/10.1039/b303808b

Sharma G, Soni R, Jasuja ND (2017) Phytoassisted synthesis of magnesium oxide nanoparticles with *Swertia chirayaita*. J Taibah Univ Sci 11:471–477

Singh AK, Talat M, Singh DP, Srivastava ON (2010) Biosynthesis of gold and silver nanoparticles by natural precursor clove and their functionalization with amine group. J Nanopart Res 12:1667–1675. https://doi.org/10.1007/s11051-009-9835-3

Sowani H, Mohite P, Munot H, Shouche Y, Bapat T, Kumar AR, Kulkarni M, Zinjarde S (2016) Green synthesis of gold and silver nanoparticles by an actinomycete Gordonia amicalis HS-11: mechanistic aspects and biological application. Process Biochem 51:374–383

Suresh J, Yuvakkumar R, Sundrarajan M, Hong SI (2014) Green synthesis of magnesium oxide nanoparticles. Adv Mater Res 952:141–144

Syed A, Saraswati S, Kundu GC, Ahmad A (2013) Biological synthesis of silver nanoparticles using the fungus Humicola sp. and evaluation of their cytoxicity using normal and cancer cell lines. Spectrochim Acta A Mol Biomol Spectrosc 114:144–147. https://doi.org/10.1016/j.saa.2013.05.030

Trindade T, O'Brien P, Pickett N (2001) Nanocrystalline semiconductors: synthesis, properties, and perspectives. Reviews. Chem Mater 13:3843–3858

Tripathy A, Raichur AM, Chandrasekaran N, Prathna AMTC (2009) Process variables in biomimetic synthesis of silver nanoparticles by aqueous extract oa *Azadirachta indica* (Neem) leaves. J Nanopart Res 12:237–246. https://doi.org/10.1007/s11051-009-9602-5

Venkatpurwar V, Pokharkar V (2011) Green synthesis of silver nanoparticles using marine polysaccharide: study of in-vitro antibacterial activity. Mater Lett 65:999–1002. https://doi.org/10.1016/j.matlet.2010.12.057

Viau G, Brayner R, Poul L, Chakroune N, Lacaze E, Fievet-Vincent F, Fievet F (2003) Ruthenium nanoparticles: size, shape, and self-assemblies. Chem Mater 15:486–494

Vinopal S, Runal T, Kotrba P (2007) Biosorption os Cd^{2+} and Zn^{2+} bycell surface engineered *Saccharomyces cerevisiae*. Int Biodeterior Biodegradation 60:96–102

Willner I, Baron R, Willner B (2006) Growing metal nanoparticles by enzymes. Adv Mater 18:1109–1120

Xia N, Cai Y, Jiang T, Yao J (2011) Green synthesis of silver nanoparticles by chemical reduction-with hyaluronan. Carbohydr Polym 86:956–961. https://doi.org/10.1016/j.carbpol.2011.05.053

Yang L, Watts D (2005) Particle surface characteristics may play an important role in phytotoxicity of alumina nanoparticles. Toxicol Lett 158:122–132

Yuhui M, Linglin K, Xiao H, Wei B, Yayun D, Zhiyong Z, Yuliang Z, Zhifang C (2009) Effects of rare earth oxide nanoparticles on root elongation of plants. Chemosphere 78:273–279

Yusefzaei F, Poorakbar L, Farhadi K, Molaei R (2017) The effect of copper nanoparticles and copper chloride solution on germination and solution some morphological and physiological factors *Ocimum basilicum* L. Iran J Plant Res 30:1–12. (*English abstract*)

Yuvakkumar R, Suresh J, Joseph Nathanael A, Sundrarajan M, Hong SI (2014) Novel green synthetic strategy to prepare ZnO nanocrystals using rambutan (*Nephelium lappaceum* L.) peel extract and its antibacterial applications. Mater Sci Eng C 41:17–27

Zhang Z, Li F, Xu L, Liu N, Xiao H, Chai Z (2004) Study of binding properties of lanthanum to wheat roots by INAA. J Radioanal Nucl Chem 259:47–49

Zheng L, Hong F, Lu S, Liu C (2005) Effect of nano-TiO(2) on strength of naturally aged seeds and growth of spinach. Biol Trace Elem Res 104:83–92

Zhu H, Han J, Xiao JQ, Jin Y (2008) Uptake, translocation, and accumulation of manufactured iron oxide nanoparticles by pumpkin plants. J Environ Monit 10:713–717

Chapter 4
Effects of Zinc Oxide Nanoparticles on Crop Plants: A Perspective Analysis

Mohammad Faizan, Shamsul Hayat, and John Pichtel

Abstract Nanotechnology is among the most innovative fields of twenty-first century. Nanoparticles (NPs) are organic or inorganic materials having sizes ranging from 1 to 100 nm; in recent years NPs have come into extensive use worldwide. The dramatic increase in use of NPs in numerous applications has greatly increased the likelihood of their release to the environment. Zinc oxide nanoparticles (ZnO-NPs) are considered a 'biosafe material' for organisms. Earlier studies have demonstrated the potential of ZnO-NPs for stimulation of seed germination and plant growth as well as disease suppression and plant protection by virtue of their antimicrobial activity. Both positive and negative effects of ZnO NPs on plant growth and metabolism at various developmental periods have been documented. Uptake, translocation and accumulation of ZnO-NPs by plants depend upon the distinct features of the NPs as well as on the physiology of the host plant. This review will contribute to current understanding the fate and behavior of ZnO-NPs in plants, their uptake, translocation and impacts on mitigating several negative plant growth conditions.

Keywords Antimicrobial activity · Biosafe · Seed germination · Translocation

4.1 Introduction

The "Nano-Era" that emerged in the 1990s is now a progressive field of research and technology. Nanotechnology is endowed with far-ranging applications in various sectors such as- cancer therapy, targeted drug delivery, biomedicines, waste-water treatment, cosmetic industries, electronics and biosensors (Nel et al. 2006; Peralta-

M. Faizan (✉)
Tree Seed Center, College of Forest Resources and Environment, Nanjing Forestry University, Nanjing, People's Republic of China

S. Hayat
Plant Physiology & Biochemistry Section, Department of Botany, Faculty of Life Sciences, Aligarh Muslim University, Aligarh, Uttar Pradesh, India

J. Pichtel
Environment, Geology and Natural Resources, Ball State University, Muncie, IN, USA

S. Hayat et al. (eds.), *Sustainable Agriculture Reviews 41*, Sustainable Agriculture Reviews 41, https://doi.org/10.1007/978-3-030-33996-8_4

Videa et al. 2011). The development of nanotechnology in conjunction with biotechnology has significantly expanded the application of nanomaterials in the above fields.

Materials with a particle size less than 100 nm in at least one dimension are generally classified as nanoparticles (NPs) (Khot et al. 2012). NPs divided into three main classes: natural, incidental, and engineered. Since the earth's formation, natural NPs have been present from sources such as volcanic ejecta and soil dispersal (Handy et al. 2008). The second class, incidental NPs, results from human industrial activities, for example- burning coal, welding and from diesel exhaust (Monica and Cremonini 2009). Of immediate interest is the third category which represents engineered nanoparticles (ENPs). These are designed and synthesized to possess unique physicochemical properties such as conductivity, reactivity, and optical sensitivity which may differ markedly compared to their bulk form (Lin and Xing 2007). Engineered nanoparticles are further divided into five subclasses: carbonaceous nanoparticles, metal oxide nanoparticles, semiconductors, metal nanoparticles, and nanopolymers (Handy et al. 2008; Monica and Cremonini 2009; Ma et al. 2010; Bhatt and Tripathi 2011).

Based on core material, NPs can be broadly divided into inorganic and organic forms. Inorganic NPs includes metals (Al, Bi, Co, Cu, Au, Fe, In, Mo, Ni, Ag, Sn, Ti, W, Zn), metal oxides (Al_2O_3, CeO_2, CuO, Cu_2O, In_2O_3, La_2O_3, MgO, NiO, TiO_2, SnO_2, ZnO, ZrO_2) and quantum dots, while fullerenes and carbon nanotubes comprise-organic NPs.

Metal-based NPs are widely used and presumably released to the environment; they must, therefore, be monitored for their potential toxic effects on activity, abundance and diversity among flora and fauna. To illustrate, some NPs are estimated to be absorbed 15–20 times more than their bulk particles by plants (Srivastav et al. 2016). It is estimated that 260,000–309,000 metric tons of NPs were produced globally in 2010 (Yadav et al. 2014). As per another estimate, worldwide consumption of NPs is likely to grow from 225,060 metric tons to nearly 585,000 metric tons between 2014–2019 (BCC Research 2014).

ZnO-NPs, with an estimated global annual production between 550 and 33,400 tons, are the third most commonly used metal-containing nanomaterial (Bondarenko et al. 2013; Connolly et al. 2016; Peng et al. 2017). Environmental levels of ZnO-NPs were reported to be in the range of 3.1–31 μg/kg soil and 76–760 μg/L water (Boxall et al. 2007; Ghosh et al. 2016). ZnO is a bio-safe material that possesses photo-oxidizing and photocatalytic capabilities for both chemicals and biota (Sirelkhatim et al. 2015; Vaseem et al. 2010).

The use of NPs in agriculture is a promising area which could potentially improve prevailing crop management over the long-term. For example, use of nano-encapsulated pesticides have been successfully applied for the release of chemicals in a controlled and specifically targeted manner which provides for a safer and easier pest control system (Beddington 2010; Nair et al. 2010). NPs are generally believed to increase profitability and sustainability, both of which are essential requirements for improved agricultural production (Som et al. 2010).

Zinc (Zn) is an essential micronutrient for normal growth in animals, humans, and plants. Numerous studies have focused on the effect of zinc on plant growth and metabolism (Auld 2001). It is vital to crop nutrition, as it is required in various enzymatic reactions, metabolic processes, and oxidation-reduction reactions. Zinc is essential for proper activity of enzymes such as dehydrogenases, aldolases, isomerases, transphosphorylases, and RNA and DNA polymerases, which are required for a range of critical physiologic functions (Lacerda et al. 2018). Zinc is also involved in synthesis of tryptophan, cell division, and maintenance of membrane structure and photosynthesis. It acts as a regulatory cofactor in protein synthesis (Lacerda et al. 2018; Marschner 2011). Thus, adequate Zn fertilization supports increase cereal, vegetable and forage production (Prasad et al. 2012).

Zinc insufficiency is characterized by reduced leaf size with interveinal necrosis and rippled leaf margins. Low values of leaf area, SPAD values, and total N and NO_3 concentration were observed under conditions of severe zinc deficiency (Castillo-Gonzalez et al. 2019). With worsening deficiency, activities of superoxide dismutase, catalase and glutathione peroxidase increased. Under severe Zn deficiency, decreases in trunk cross-sectional area, yield and percentage kernel are observed. Increased activity of superoxide dismutase, catalase and peroxidase enzymes is associated with detoxification of reactive oxygen species.

Zinc deficiency not only retards growth and yield of plants, but also imparts adverse effects to humans (Singh 2009; Shukla et al. 2014). More than 3 billion people worldwide suffer from Fe and Zn deficiencies. This condition is particularly widespread in areas where population is heavily dependent on a regular diet of cereal-based foods, in which Fe and Zn are stored almost exclusively in the husk and are therefore lost during milling and polishing (Cakmak 2002; Graham et al. 2001).

Zinc oxide (ZnO) is an amphoteric oxide which is almost insoluble in water and alcohol, but is soluble in most acids (Spero et al. 2000). ZnO crystallizes into two main forms, hexagonal wurtzite and cubic zincblende. The wurtzite structure is the most stable form at ambient conditions. Due to its unique properties including high thermal conductivity, refractive index, binding energy, UV protection and antibacterial capabilities, ZnO-NPs are widely applied in numerous products and materials including medicine, cosmetics, solar cells, rubber and concrete, and-foods (Uikey and Vishwakarma 2016). ZnO-NPs are the most common Zn-NPs used as a UV protector (e.g., in personal care products, coatings and paints), biosensors, electronics, and in rubber manufacture (Brayner et al. 2010; Kool et al. 2011). The wide range of industrial applications for ZnO-NPs can be used to predict future increases, in their production. The economical application of ZnO-NPs as Zn fertilizers can eventually become practical in large-scale agriculture globally.

4.2 Synthesis of ZnO-NPs

Several methods have been reported for synthesis of ZnO-NPs. The primary objective of each method is development of stable and uniform NPs.

4.2.1 Chemical Synthesis

NPs can be chemically synthesized by a number of techniques including spray pyrolysis, thermal decomposition, molecular beam epitaxy, chemical vapor deposition, and laser ablation. Chemical synthesis methods are among the most commonly used techniques, and can be performed using a range of precursors and variations of temperature, time and concentration of reactants. Modification of one or more parameters results in morphological differences in size and structures of the resulting NPs. Several popular chemical methods for synthesis of ZnO-NPs are described below.

4.2.1.1 Reaction of Metabolic Zinc with Alcohol

Several types of alcoholic media such as ethanol, methanol, or propanol are used for chemical synthesis of ZnO-NPs. At the laboratory scale, synthesis typically involves addition of 5 mg of zinc metal powder to 10 mL of ethanol. The reaction mixture is sonicated for 20 min and transferred to a stainless steel autoclave and sealed under an inert gas. The reaction mixture is heated slowly (2–200 °C per minute) and maintained at this temperature for 24–48 h. The resulting suspension is then centrifuged to retrieve the product, washed, and finally vacuum-dried. In alcoholic media the growth of oxide particles is slow and controllable (Koch et al. 2000).

4.2.1.2 Vapor Transport Synthesis

The vapor transport process is the most common method for synthesis of ZnO-NPs. In this process, zinc and oxygen (or oxygen mixtures) are reacted resulting in formation of ZnO-NPs. Numerous methods are available for generation of Zn and oxygen vapor. Decomposition of ZnO is an easier, direct, and simple method; however, it is limited to very high temperatures (e.g., 1400 °C).

4.2.1.3 Hydrothermal Technique

The hydrothermal technique is an efficient synthetic method because of the low process temperature required; furthermore, it is easier to control particle size.

Particle morphology and size are carefully controlled by adjusting reaction temperature, time, and concentrations of precursor'-s.

As stock solution of Zn $(CH_3COO)_2$.-$2H_2O$ (0.1 M) is first prepared. To this stock solution, 25 mL of NaOH (from 0.2 M to 0.5 M) prepared in methanol, is added under stirring to attain a pH value between 8 and 11. The solutions is transferred to a Teflon-lined sealed stainless steel autoclave and maintained at temperatures in the range of 100–200 °C for 6–12 h under autogenous pressure. The resulting white solid product is washed with methanol, filtered, and dried in a laboratory oven at 60 °C. This is followed by structural characterization of synthesized samples via X-ray diffraction (Lee et al. 2006).

The hydrothermal technique offers several advantages such as utilization of sample equipment, catalyst-free growth, low cost, uniform production, ecofriendliness, and being less hazardous compared with other synthesis methods. The technique is attractive for microelectronics and plastic electronics manufacture due to low reaction temperatures needed. This hydrothermal technique has been successfully employed for preparation of ZnO-NPs and other luminescent materials.

4.2.1.4 Precipitation Method

In this method ZnO-NPs are synthesized using zinc nitrate and urea as precursors. In a typical synthesis, 4.735 g zinc nitrate ($Zn[NO_3]_2$-$6H_2O$) is dissolved in 50 mL of distilled water and kept under constant stirring for 30 min for complete dissolution. In a separate container, 3.002 g urea is added to 50 mL distilled water under constant stirring for 30 min. The urea solution acts as precipitating agent. The urea solution is added drop-wise into the zinc nitrate solution with vigorous stirring at 70 °C for 2 h to allow formation of NPs. The solution eventually turns cloudy white. This precursor product is centrifuged at 8000 rpm for 10 min and washed with distilled water for removal of any impurities or absorbed ions. Calcination of the product is carried out at 500 °C in air for 3 h using a muffle furnace (Chen et al. 2008).

4.2.2 Green Synthesis

In recent years, green synthesis of metal NPs has become appealing to nanoscience and nanobiotechnology. There is a growing interest in biosynthesis of metal NPs using plants, which appear to be the optimal candidate for large-scale biosynthesis of NPs. Products of green synthesized NPs tend to be stable.

Synthesis using plants and their extracts are advantageous over other biological synthesis processes, for example using microorganism, which involves complex procedures for maintaining viable cultures (Sastry et al. 2003). Experimentals synthesis of metal NPs have used fungi like *Fusarium oxysporum* (Nelson et al. 2005), *Penicillium* sp. (Hemanth et al. 2010) and bacteria such as *Bacillus subtilis* (Natarajan et al. 2010; Elumalai et al. 2010). However, synthesis using plant extracts is the most widely accepted method of green, eco-friendly production of NPs. It has the advantage that plants are widely distributed, readily available, and safe to handle (Ankamwar et al. 2005). Moreover, the NPs produced are more varied in shape and size in comparison with those produced by other organisms (Korbekandi et al. 2009).

4.3 Uptake of ZnO-NPs in Plants

Several possible mechanisms exist for entry of ZnO-NPs into plants (Capaldi Arruda et al. 2015). NPs may dissolve in soil water and produce ions which are incorporated into the plant. Alternatively, upon exposure to the root, NPs will penetrates the cell wall and cell membrane of the epidermis accompanied by a complex series of processes for entry into the vascular bundle (xylem) and movement to the stele. Due to their size, many NPs may not be capable of passing through the cell wall of an intact plant cell. Solute exclusion techniques provide data on restrictive pore sizes of cell walls. The largest pore size of a plant cell wall is typically in the range of a few nanometers; for example, 3.5–3.8 nm is common in root hairs and 4.5–5.2 nm in palisade parenchyma cells (Carpita et al. 1979). NP sizes less than 5 nm in diameter are capable of transversing the cell wall of undamaged cells e (Fleischer et al. 1999). The xylem serves as the most important vehicle for distribution and translocation of NPs to leaves. The epidermis, cortex, endodermis, cambium and xylem are known to accumulate higher concentrations of NPs than other plant tissues.

The mechanism of NP uptake is generally considered active transport that includes cellular processes such as signaling, and regulation of the plasma membrane (Etxeberria et al. 2006). NPs can move through tissues via: apoplastic and symplastic routes. Apoplastic transport occurs outside the plasma membrane and through the extracellular spaces, cell walls of adjacent cells, and xylem vessels (Sattelmacher 2001); in contrast, symplastic transport involves movement of water and substances between the cytoplasm of adjacent cells through specialized structures called plasmodesmata (Roberts and Oparka 2003) and sieve plates. The apoplastic pathway is important for radial movement within plant tissues; it allows nanomaterials to reach the root central cylinder and vascular tissue, for further movement upward to aerial parts (Sun et al. 2014). Once inside the central cylinder, NPs are translocated upward though the xylem, following the transpiration stream (Sun et al. 2014).

Uptake, translocation and accumulation of NPs depend on plant species and the size, charge, chemical configuration, stability and concentration of the NPs. The mobility of NPs is also determined by van der Waals forces, Brownian motion (diffusion), gravity, and double-layer forces (Handy et al. 2008; Biswas and Wu 2005).

4.4 Distinguishing Properties of ZnO-NPs

ZnO-NPs are a new type of cost-effective and low-toxicity NPs which have attracted substantial interest in different fields as they possess a number of distinct and useful properties.

4.4.1 Physical Properties

ZnO-NPs possess valuable physical properties. As the dimensions of semiconductor materials shrinks continuously to nanometer or even smaller scales, some of their physical properties undergo changes known as "quantum size effects."

4.4.2 Antibacterial Properties

The antibacterial activity of ZnO-NPs lies in their ability to induce oxidative stress. Zn^{2+} ions, released by dissolution of ZnO, interact with thiol groups of respiratory enzymes, thus inhibiting their action. It has been demonstrated that ZnO-NPs affect the properties of the cell membrane and lead to ROS formation. When bacterial cells come into contact with ZnO-NPs, they absorb Zn^{2+}, which inhibits the action of respiratory enzymes, generates ROS, and produces free radicals, causing oxidative stress. ROS irreversibly damage bacterial membranes, DNA, and mitochondria, resulting in cell death (Dwivedi et al. 2014).

Ghasemi and Jalal (2016) investigated the effect of ZnO-NPs on the efficiency of the conventional antibiotics ciprofloxacin and cefta-zidime as well as their mechanisms of action against resistant *Acinetobacter baumannii*, an opportunistic pathogen that causes a range of diseases including pneumonia and meningitis. The antibacterial activity of both anti-biotics increased in the presence of a sub-inhibitory concentration of ZnO-NPs. Combining ZnO-NPs with antibiotics increased uptake of antibiotics and changed bacterial cells from rod to cocci. ZnO-NPs used against *Vibrio cholerae* (a causative agent of severe diarrhea) were investigated by Sarwar et al. (2016). ZnO-NPs deformed cellular architecture, increased fluidity and caused depolarization of cell membranes and protein leakage. ROS production and DNA damage were also observed. These results suggest the synergistic action of ZnO-NPs and anti-biotics as an alternative treatment for certain bacterial diseases.

4.5 The Role of ZnO-NPs in Agriculture

Limited studies have been carried out to date to determine the effects of ZnO-NPs on plant growth and productivity (Lin and Xing 2007; Stampoulis et al. 2009). It is well recognized that ZnO-NPs affect crop development and yield and accumulate in plant tissue, including edible portions. The behavior of ZnO-NPs in plants is not completely clear; however, the optoelectrical, physical and antimicrobial activities of ZnO-NPs offer several positive effects to plants (Table 4.1) (Liu and Lal 2015;

Table 4.1 Positive effects of ZnO-NPs on different plant species

S. no	Concentration	Plant	Effects	References
1	800 mg kg^{-1}	*Cucumis sativus*	Enhanced growth and increased dry weight	Zhao et al. (2014)
2	1000 mg kg^{-1}	*Arachis hypogaea*	Enhanced germination, plant growth and chlorophyll content	Prasad et al. (2012)
3	500, 1000, 2000 mg kg^{-1}	*Vigna radiata*	Increased dry weight	Pradhan et al. (2013)
4	1 mg L^{-1}	*Cicer arietinum*	Increased plant growth	Mahajan et al. (2011)
5	2 mg L^{-1}	*Brassica napus*	Increased root length and enhanced seed germination	Lin and Xing (2007)
6	1.5 mg kg^{-1}	*Cicer arietinum*	Increased dry plant weight	Burman et al. (2013)
7	20 mg	*Vigna radiata*	Increase shoot, root length and dry weight	Mahajan et al. (2011)
8	2 mg L^{-1}	*Lolium perenne*	Enhanced growth of seedlings and root elongation	Lin and Xing (2007)
9	20 ppm	*Vigna radiata*	Increased biomass	Dhoke et al. (2013)
10	2 ppm	*Zea mays corn*	Enhanced shoot dry weight and leaf area	Taheri et al. (2015)
11	10 ppm	*Cyamopsis tetragonoloba*	Increased plant growth and protein content	Raliya and Tarafdar (2013)
12	60 mg L^{-1}	*Lupinus termis*	Decreased MDA and Na content; improved salt tolerance	Latef et al. (2017)
13	1000 mg L^{-1}	*Triticum aestivum*	Increased pigment and protein content	Ramesh et al. (2014a, b)
14	2 mg L^{-1}	*Raphanus sativus*	Enhanced seed germination and root growth	Lin and Xing (2007)
15	500 mg L^{-1}	*Glycine max*	Elongation of roots	Lopez-Moreno et al. (2010)
16	20 µg mL^{-1}	*Allium cepa*	Promotes seed germination	Lawre and Raskar (2014)
17	10 mg L^{-1}	*Pennisetum glaucum*	Increased length, chlorophyll content and grain yield	Tarafdar et al. (2014)
18	10 mg L^{-1}	*Brassica nigra*	Enhanced root hair and radical scavenging activity	Singh et al. (2013)

Hussain et al. 2016). ZnO-NPs up to a certain concentration have the capability to enhance growth (Faizan et al. 2018), where they provide Zn^{2+} as a micronutrient (Liu and Lal 2015). Several reports suggest that ZnO-NPs improve growth and development in soybean (Priester et al. 2012), cucumber (Zhao et al. 2013), peanut (Prasad et al. 2012) and green pea (Mukherjee et al. 2014) (Table 4.1). Prasad et al. (2012) report that peanut seeds, when treated with 1000 mg/kg of ZnO NPs (average size ~25 nm), exhibited enhanced germination rate and improved seedling vigor, along with early flowering and higher leaf chlorophyll content. Similar inductive effects of ZnO-NPs resulted in increase stem and root growth. Pod yield per plant was 34% higher with ZnO-NPs compared to chelated bulk $ZnSO_4$-exposed plants.

ZnO-NPs enhanced growth and biomass production of alfalfa, tomato, and cucumber plants (de la Rosa et al. 2013; Panwar et al. 2012). Application of ZnO-NPs increased photosynthetic pigment levels in pearl millet (Tarafdar et al. 2014). Ramesh et al. (2014a, b) reported that low concentrations of ZnO-NPs imparted a beneficial effect on seed germination in wheat.

Significant increases in root growth and dry weight in onion was observed after ZnO-NP application (Raskar and Laware 2014). In contrast, Zhao et al. (2013) observed that ZnO-NPs had no impact on growth of cucumber plants, gas exchange, or chlorophyll content. Lower concentrations of ZnO-NPs were not harmful to cell division and early seedling growth in onion (Raskar and Laware 2014).

4.6 Zinc Oxide Nanoparticles and Plants Under Abiotic Stress

ZnO-NPs play important roles in plants for minimizing the harmful effects of ROS to cell organelles. Apart from its well-documented damaging effects, ROS are also known to trigger various defense systems by activating a cell signaling cascade and inducing or suppressing expression of many genes (Hancock et al. 2001). Nonetheless, plants are equipped with enzymatic and nonenzymatic systems of antioxidants generation, which continuously scavenge harmful ROS. ZnO-NPs play an important role in the protection of plants against various abiotic stresses by stimulating the activities of antioxidant enzymes and accumulating osmolytes, free amino acids and nutrients (Fig. 4.1; Table 4.2) (Taran et al. 2017; Venkatachalam et al. 2017; Wang et al. 2018). Application of ZnO-NPs at low concentrations was found effective in alleviating various abiotic stresses and enhanced plant growth and development (Table 4.2) (Mahajan et al. 2011; Soliman et al. 2015).

In contrast, however, a significant number of studies show toxic effects of ZnO-NPs (Miralles et al. 2012; Husen and Siddiqi 2014), and application of ZnO-NPs to stressed plants could galvanize the process of ROS generation leading to oxidative damage (Lin et al. 2010; Wang et al. 2014; Chichiricco and Poma 2015).

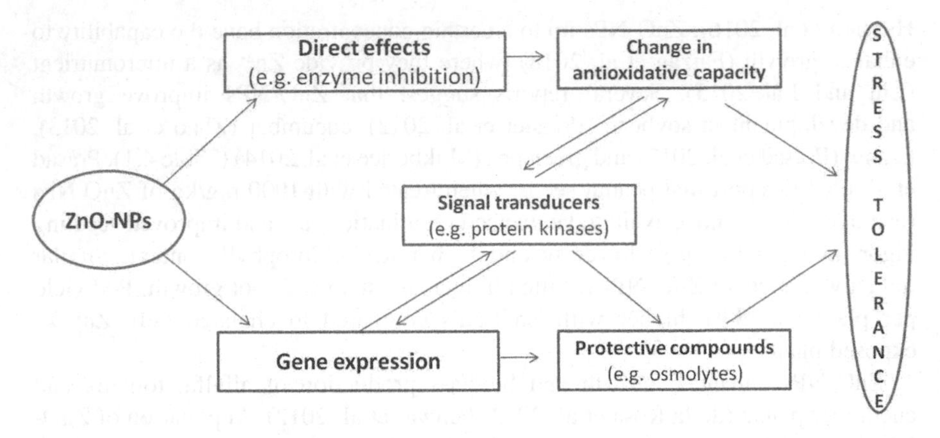

Fig. 4.1 Proposed model of ZnO-NPs for minimizing the adverse effects of abiotic stresses

4.6.1 Drought Stress

Drought stress is both a natural event, and also a consequence of anthropogenic climate change that limits crop production and distribution. Among various strategies adopted to counter drought-induced damage to plants, the use of NPs has proved promising.

Under conditions of water scarcity stomatal closure is the primary response that plants adopt to preserve water. When stomata are open, CO_2 is taken up at the cost of water transpired; when stomata are closed, water is conserved but uptake of CO_2 is compromised. Perception of drought stress triggers the activation of signal transduction cascades. This leads to stomatal closure and, in consequence, conservation of water through reduced transpiration coupled with reduced CO_2 uptake. According to Taran et al. (2017), foliar application of ZnO-NPs to wheat reduced the adverse effects of drought stress and improved yield components of wheat.

4.6.2 Metal Stress

Cadmium is considered quite toxic to plants (Garg and Kaur 2013). Cadmium induces phytotoxicity by disturbing many Zn-dependent physiological processes via displacement of Zn from the active sites of enzymes (Asmub et al. 2000). Venkatachalam et al. (2017) found that ZnO-NPs alleviated toxicity induced by Cd and Pb in *Leucaena leucocephala* seedlings. This finding is similar to results

Table 4.2 Alleviating effects of ZnO-NPs on abiotic stresses in plants

S. no.	Plant	Stress	Effects	References
1	*Glycine max*	Drought	Increased germination percentage and germination rate; decrease in seed residual fresh and dry weight	Sedghi et al. (2013)
2	*Triticum aestivum*	Drought	Decreased the negative effects of drought; increased activities of antioxidative enzymes; reduced accumulation of thiobarbituric acid reactive substances (TBARS); stabilized the content of photosynthetic pigments; increased relative water content in leaves	Taran et al. (2017)
3	*Moringa peregrina*	Salinity	Reduced salinity stress and increased content of chlorophyll, carotenoids, proline, carbohydrates, crude protein, and enzymatic and non-enzymatic antioxidants	Soliman et al. (2015)
4	*Helianthus*	Salt	Reduced the adverse effects of salt stress and increased shoot dry weight and SOD activity	Torabian et al. (2016)
5	Banana	Salt	Reduced protein content after salt treatment; increased protein content	Chaudhary et al. (2018)
6	*Solanum lycopersicum* MILL.	Salt	Mitigated the effects of NaCl; more effective lower concentrations (15 mg L^{-1}) than at higher concentration (30 mg L^{-1}).	Alharby et al. (2016)
7	*Triticum aestivum*	Cd	Plant height, spike length, and dry weights of shoots, roots, spikes, and grains increased	Rizwan et al. (2019)
8	*Triticum aestivum*	Cd	Increased chlorophyll content and gas exchange attributes of wheat under cd stress	Hussain et al. (2018)
9	*Leucaena leucocephala*	Cd and Pb	Enhanced level of antioxidant defense enzymes and related metabolites	Venkatachalam et al. (2017)
10	Sorghum bicolor	Cd	Reduced the negative effects of cd and enhanced plant growth and development	Wang et al. (2018)
11	*Triticum aestivum*	Heat	Stimulated wheat plants to tolerate heat stress	Hassan et al. (2018)

reported by Garg and Kaur (2013), who determined that the presence of Zn decreased Cd content in both roots and leaves of *Cajanus cajan*, thereby enhancing plant survival and growth. Zinc supplementation can protect plants against Cd-induced oxidative stress via modulating the redox status of the plant (Aravind and Prasad 2005).

4.6.3 Salinity Stress

The application of ZnO-NPs has led to mitigation of some adverse effects of soil salinity. According to Soliman et al. (2015), foliar application of 60 mg/L ZnO-NPs proved optimal for alleviating the effects of salt stress on *Moringa peregrina* plants (Table 4.2). Torabian et al. (2016) observed that alleviation of salt stress was greater in sunflower plants supplied with ZnO-NPs compared to plants treated with dissolved ZnO. The salt-alleviating effects of NPs was further confirmed when Almutairi (2016) reported that nano-Si differentially regulated the expression of salt stress genes. There was a positive response of SOD activity to foliar application of ZnO, particularly nanoparticles, under salt stress. Because Zn is present within the molecular structure of SOD, foliar application of ZnO promotes the formation and activity of this enzyme. Zinc deficiency probably increases ROS levels and, thus, requires higher SOD activity. Although salinity increased SOD activity, foliar application of ZnO contributed to its enhanced production. This may explain the role of Zn in salinity alleviation. In experiment by Sanaeiostovar et al. (2012), applied zinc increased the SOD activity of wheat cultivars. Foliar spray of ZnO-NPs reduced the negative effects of salinity on sunflower growth (Torabian et al. 2016).

4.7 Conclusion

The findings reported herein reveal that ZnO-NPs serve as a natural regulator for plants under both stressed and stress-free conditions. ZnO-NPs have the potential to enhance plant growth and development. ZnO-NPs play a pivotal role in modulating key plant physiological parameters under stressfull conditions such as lipid peroxidation, and production of proline and various antioxidant enzymes. ZnO-NPs can be efficiently synthesized by chemical and green synthesis methods.

The small size of ZnO-NPs facilitates easy penetration into plant cells and regulates water channels that assist seed germination and growth of plants. Moreover, different modes of ZnO-NP application successfully counter the adverse affects of ROS under abiotic stress through increased activity of CAT, POX and SOD. The role of ZnO-NPs in plants needs further investigation at both sub-cellular and molecular levels.

References

Alharby HF, Metwali EMR, Fuller MP, Aldhebiani AY (2016) Impact of zinc oxide nanoparticle application on callus induction, plant regeneration, element content and antioxidant enzyme activity in tomato (*solanum lycopersicum* mill.) under salt stress. Arch Biol Sci 68:723–735

Almutairi ZM (2016) Effect of nano-silicon application on the expression of salt tolerance genes in germinating tomato (*Solanum lycopersicum* L.) seedlings under salt stress. Plant Omics J 9:106–114

Ankamwar B, Damle C, Ahmad A, Sastry M (2005) Biosynthesis of gold and silver nanoparticles using Emblics Officinalis fruit extract and their phase transfer and transmetallation in an organic solution. J Nanosci Nanotechnol 5:1665–1671

Aravind P, Prasad MNV (2005) Cadmium–zinc interactions in a hydroponic system using Ceratophyllum demersum L.: adaptive ecophysiology, biochemistryand molecular toxicology. Braz J Plant Physiol 17:3–20

Asmub M, Mullenders LHF, Hartwig A (2000) Interference by toxic metal compounds with isolated zinc finger DNA repair proteins. Toxico Lett 15(112–113):227–231

Auld DS (2001) Zinc coordination sphere in biochemical zinc sites. Biometals 14:271–313

BCC Research (2014) Global markets for nanocomposites, nanoparticles, nanoclays, and nanotubes. https://www.bccresearch.com/market-research/nanotechnology/nanocomposites-market-nan021f.html?vsmaid=203/. Accessed 19 Oct 2017

Beddington J (2010) Food security: contributions from science to a new and greener revolution. Philos Trans R Soc B Biol Sci 365(1537):61–71

Bhatt I, Tripathi BN (2011) Interaction of engineered nanoparticles with various components of the environment and possible strategies for their risk assessment. Chemosphere 82:308–317

Biswas P, Wu CY (2005) Critical review: nanoparticles and the environment. J Air Waste Manage Assoc 55:708–746

Bondarenko O, Juganson K, Ivask A, Kasemets K, Mortimer M, Kahru A (2013) Toxicity of Ag, CuO and ZnO nanoparticles to selected environmentally relevant test organisms and mammalian cells in vitro: a critical review. Arch Toxicol 87:1181–1200

Boxall A, Chaudhry Q, Sinclair C, Jones A, Aitken R, Jefferson B, Watts C (2007) Current and future predicted environmental exposure to engineered nanoparticles. Technical Report, Central Science Laboratory, York

Brayner R, Dahoumane SA, Yepremian C, Djediat C, Meyer M, Coute A, Fievet F (2010) Zinc oxide nanoparticles: synthesis, characterization and ecotoxicological studies. Langmuir 26:6522–6528

Burman U, Saini M, Kumar P (2013) Effect of zinc oxide nanoparticles on growth and antioxidant system of chickpea seedlings. Toxicol Environ Chem 95:605–612

Cakmak I (2002) Plant nutrition research: priorities to meet human needs for food in sustainable ways. Plant Soil 247:3–24

Capaldi Arruda SC, Silva A, D L, Galazzi RM, Azevedo RA, Arruda MAZ (2015) Nanoparticles applied to plant science: a review. Talanta 131:693–705

Carpita N, Sabularse D, Montezinos D, Delmer DP (1979) Determination of the pore size of cell walls of living plant cells. Science 2059(4411):1144–1147

Castillo RR, Lozano D, Gonzalez B, Manzano M, Izquierdo-Barba I, Vallet-Regi M (2019) Advances in mesoporous silica nanoparticles for targeted stimuli-responsive drug delivery: an update. Expert Opin Drug Deliv 16(4):415–439

Chaudhary D, Basanti B, Singh DJ, Subhash K, Anil P (2018) An insight into in vitro micropropagation studies for banana- review. Int J Agri Sci 10(5):5346–5349

Chen C, Liu P, Lu C (2008) Investigation of photocatalytic degradation using nano-sized ZnO catalysts. Chem Eng J 144:509–513

Chichiricco G, Poma A (2015) Penetration and toxicity of nanomaterial in higher plants. Nanomaterials 5:851–873

Connolly M, Fernandez M, Conde E, Torrent F, Navas JM, Fernandez-Cruz ML (2016) Tissue distribution of zinc and subtle oxidative stress effects after dietary administration of ZnO nanoparticles to rainbow trout. Sci Total Environ 1(551–552):334–343

de la Rosa G, Lopez-Moreno ML, de Haro D, Botez CE, Peralta-Videa JR, Gardea- Torresdey JL (2013) Effects of ZnO nanoparticles in alfalfa, tomato, and cucumber at the germination stage: root development and X-ray absorption spectroscopy studies. Pure Appl Chem 85:2161–2174

Dhoke SK, Mahajan P, Kamble R, Khanna A (2013) Effect of nanoparticles suspension on the growth of mung (Vigna radiata) seedlings by foliar spray method. Nanotechnol Dev 3:e1

Dwivedi S, Wahab R, Khan F, Mishra YK, Musarrat J, Al-Khedhairy A (2014) Reactive oxygen species mediated bacterial biofilm inhibition via zinc oxide nanoparticles and their statistical determination. PLoS One 9:e111289

Elumalai EK, Prasad TNVKV, Hemachandran J, Vivivan Therasa S, Thirumalai T, David E (2010) Extracellular synthesis of silver nanoparticles using leaves of Euphorbia hirta and their antibacterial activities. J Pharm Sci Res 2:549–554

Etxeberria E, Gonzalez P, Baroja-Fernandez E, Romero JP (2006) Fluid phase endocytic uptake of artificial nano-spheres and fluorescent quantum dots by sycamore cultured cells: evidence for the distribution of solutes to different intracellular compartments. Plant Signal Behav 1:196–200

Faizan M, Faraz A, Yusuf M, Khan ST, Hayat S (2018) Zinc oxide nanoparticles-mediated changes in photosynthetic efficiency and antioxidant system of tomato plants. Photosynthetica 56:678–686

Fleischer A, O'Neill MA, Ehwald R (1999) The pore size of non-graminaceous plant cell wall is rapidly decreased by borate ester cross-linking of the pectic polysaccharide rhamnogalacturon II. Plant Physiol 121:829–838

Garg N, Kaur H (2013) Impact of cadmium-zinc interactions on metal uptake, translocation and yield in pigeonpea genotypes colonized by arbuscular mycorrhizal fungi. J Plant Nutr 36:67–90

Ghasemi F, Jalal R (2016) Antimicrobial action of zinc oxide nanoparticles in combination with ciprofloxacin and ceftazidime against multidrug-resistant Acinetobacter baumannii. J Glob Antimicrob Resist 6:118–122

Ghosh M, Jana A, Sinha S, Jothiramajayam M, Nag A, Chakraborty A, Mukherjee A, Mukherjee A (2016) Effects of ZnO nanoparticles in plants: cytotoxicity, genotoxicity, deregulation of antioxidant defenses, and cell-cycle arrest. Mutat Res Genet Toxicol Environ Mutagen 807:25–32

Graham RD, Welch RM, Bouis HE (2001) Addressing micronutrients malnutrition through enhancing the nutritional quality of staple foods principles, perspectives and knowledge gaps. Adv Agron 70:77–142

Hancock JT, Desikan R, Neill SJ (2001) Role of reactive oxygen species in cell signalling pathways. Biochem Soc Trans 29:345–350

Handy RD, Owen R, Valsami-Jones E (2008) The ecotoxicology of nanoparticles and nanomaterials: current status, knowledge gaps, challenges, and future needs. Ecotoxicology 17:315–325

Hassan N, Salah T, Hendawey MH, Borai IH, Mahdi AA (2018) Magnetite and zinc oxide nanoparticles alleviated heat stress in wheat plants. Curr Nanomaterials 3:32–43

Hemanth NKS, Kumar G, Karthik L, Bhaskara RKV (2010) Extracellular biosynthesis of silver nanoparticles using the filamentous fungus Penicillium sp. Arch Appl Sci Res 2:161–167

Husen A, Siddiqi KS (2014) Phytosynthesis of nanoparticles: concept, controversy and application. Nanoscale Res Lett 9:229–252

Hussain I, Singh NB, Singh A, Singh H, Singh SC (2016) Green synthesis of nanoparticles and its potential application. Biotechnol Lett 38:545–560

Hussain A, Ali S, Rizwan M, Rehman MZ, Javed MR, Imran M, Chatha SA, Nazir R (2018) Zinc oxide nanoparticles alter the wheat physiological response and reduce the cadmium uptake by plants. Environ Pollut 242:1518–1526

Khot LR, Sankaran S, Maja JM, Ehsani R, Schuster EW (2012) Applications of nanomaterials in agricultural production and crop protection: a review. Crop Protec 35:64–70

Kool PL, Ortiz MD, van Gestel CAM (2011) Chronic toxicity of ZnO nanoparticles, non-nano ZnO and ZnCL2 to Folsomia candida (Collembola) in relation to bioavailability in soil. Environ Pollut 159:2713–2719

Koch U, Fojtik A, Weller H, Henglein A (2000) Photochemistry of semiconductor colloids. Preparation of extremely small ZnO particles, fluorescence phenomena and size quantization effects. Chem Phys Lett 122:507–510

Korbekandi H, Iravani S, Abbasi S (2009) Production of nanoparticles using organisms production of nanoparticles using organisms. Crit Rev Biotechnol 29:279–306

Lacerda JS, Martinez HE, Pedrosa AW, Clemente JM, Santos RH, Oliveira GL, Jifon JL (2018) Importance of zinc for arabica coffee and its effects on the chemical composition of raw grain and beverage quality. Crop Sci 58:1360–1370

Latef AAHA, Alhmad MFA, Abdelfattah KE (2017) The possible roles of priming with ZnO nanoparticles in mitigation of salinity stress in lupine (Lupinus termis) plants. J Plant Growth Regul 36:60–70

Lawre S, Raskar S (2014) Influence of zinc oxide nanoparticles on growth, flowering and seed productivity in onion. Int J Curr Microbiol App Sci 3:874–881

Lee CY, Tseng TY, Li SY, Lin P (2006) Effect of phosphorus dopant on photoluminescence and field-emission characteristics of Mg0.1Zn0.9O nanowires. J Appl Phys 99:024303

Lin D, Xing B (2007) Phytotoxicity of nanoparticles: inhibition of seed germination and root growth. Environ Pollut 150:243–250

Lin C, Su YB, Takahiro M, Fugetsu B (2010) Multi-walled carbon nanotubes induce oxidative stress and vacuolar structure changes to Arabidopsis T87 suspension cells. Nano Biomed 2:170–181

Liu R, Lal R (2015) Potentials of engineered nanoparticles as fertilizers for increasing agronomic productions. Sci Total Environ 514:131–139

Lopez-Moreno ML, De La Rosa G, Hernandez-Viezcas JA, Castillo-Michel H, Botez CE, Peralta-Videa JR, Gardea-Torresdey JL (2010) Evidence of the differential biotransformation and genotoxicity of ZnO and CeO2 NPs on soybean (Glycine max) plants. Environ Sci Technol 44:7315–7320

Ma X, Geiser-Lee J, Deng Y, Kolmakov A (2010) Interactions between engineered nanoparticles (ENPs) and plants: phytotoxicity, uptake and accumulation. Sci Total Environ 408:3053–3061

Mahajan P, Dhoke SK, Khanna AS, Tarafdar JC (2011) Effect of nano-ZnO on growth of mung bean (Vigna radiata) and chickpea (Cicer arietinum) seedlings using plant agar method. Appl Biol Res 13:54–61

Marschner H (2011) Marschner's mineral nutrition of higher plants. Academic Press, San Diego, p 651

Miralles P, Johnson E, Church TL, Harris AT (2012) Multiwalled carbon nanotubes in alfalfa and wheat: toxicology and uptake. J R Soc Interface 9:3514–3527

Monica RC, Cremonini R (2009) Nanoparticles and higher plants. Caryol 62:161–165

Mukherjee A, Peralta-Videa JR, Bandyopadhyay S, Rico CM, Zhao L, Gardea- Torresdey JL (2014) Physiological effects of nanoparticulate ZnO in green peas (Pisum sativum L.) cultivated in soil. Metallomics 6:132–138

Nair R, Varghese SH, Nair BG, Maekawa T, Yoshida Y, Kumar DS (2010) Nanoparticulate material delivery to plants. Plant Sci 179(3):154–163

Natarajan K, Subbalaxmi SV, Ramchandra M (2010) Microbial production of silver nanoparticles. Dig J Nanomater Biostruct 5:135–140

Nel A, Xia T, Madler L, Li N (2006) Toxic potential of materials at the nano level. Science 311:622–627

Nelson D, Priscyla DM, Oswaldo LA, Gabriel IHDS, Elisa E (2005) Mechanical aspects of biosynthesis of silver nanoparticles by several Fusarium oxysporum strains. J Nanobiotechnol 3:8

Panwar J, Jain N, Bhargaya A, Akhtar M, Yun Y (2012) Positive effect of zinc oxide nanoparticles on tomato plants: a step towards developing nano-fertilizers. International Conference on Environmental Research and Technology (ICERT), Malaysia https://doi.org/10.13140/2.1.2697.8889
Peng YH, Tsai YC, Hsiung CE, Lin YH, Shih Y (2017) Influence of water chemistry on the environmental behaviors of commercial ZnO nanoparticles in various water and wastewater samples. J Hazard Mater 322:348–356
Peralta-Videa JR, Zhao L, Lopez-Morena ML, de la Rosa G, Hong J, Gardea-Torresdey JL (2011) Nanomaterials and the environment: a review for the biennium 2008–2010. J Hazard Mater 186:1–15
Pradhan S, Patra P, Das S, Chandra S, Mitra S, Dey KK, Akbar S, Palit P, Goswami A (2013) Photochemical modulation of biosafe manganese nanoparticles on vigna radiata: a detailed molecular, biochemical and biophysical study. Environ Sci Technol 47:13122–13131
Prasad TNVKV, Sudhakar P, Sreenivasulu Y, Latha P, Munaswamy V, Raja Reddy K, Sreeprasad TS, Sajanlal PR, Pradeep T (2012) Effect of nanoscale zinc oxide particles on the germination, growth and yield of peanut. J Plant Nutr 35:905–927
Priester JH, Ge Y, Mielke RE, Horst AM, Moritz SC, Espinosa K, Gelb J, Walker SL, Nisbet RM, An YJ (2012) Soybean susceptibility to manufactured nanomaterials with evidence for food quality and soil fertility interruption. Proc Natl Acad Sci U S A 109:E2451–E2456
Raliya R, Tarafdar JC (2013) ZnO nanoparticle biosynthesis and its effect on phosphorous-mobilizing enzyme secretion and gum contents in cluster bean (Cyamopsis tetragonoloba L.). Agric Res 2:48–57
Ramesh M, Palanisamy K, Babu K, Sharma NK (2014a) Effects of bulk & nano-titanium dioxide and zinc oxide on physio-morphological changes in Triticum aestivum Linn. J Glob Biosci 3:415–422
Ramesh P, Rajendran A, Meenakshisundaram M (2014b) Green synthesis of zinc oxide nanoparticles using flower extract Cassia auriculatas. J Nanosci Nanotechnol 1:41–45
Raskar S, Laware S (2014) Effect of zinc oxide nanoparticles on cytology and seed germination in onion. Int J Curr Microbiol App Sci 3:467–473
Rizwan M, Ali S, Ali B, Adrees M, Arshad M, Hussain A, Zia ur Rehman M, Waris AA (2019) Zinc iron oxide nanoparticles improved the plant growth and reduced the oxidative stress and cadmium concentration in wheat. Chemosphere 214:269–277
Roberts AG, Oparka KJ (2003) Plasmodesmata and the control of symplastic transport. Plant Cell Environ 26:103–124
Sanaeiostovar A, Khoshgoftarmanesh AH, Shariatmadari H, Afyuni M, Schulin R (2012) Combined effect of zinc and cadmium levels on root antioxidative responses in three different zinc-efficient wheat genotypes. J Agron Crop Sci 198:276–285
Sarwar S, Chakraborti S, Bera S, Sheikh IA, Hoque KM, Chakrabarti P (2016) The antimicrobial activity of ZnO nanoparticles against Vibrio cholerae: variation in response depends on biotype. Nanomedicine 12:1499–1509
Sastry M, Ahmad A, Islam NI, Kumar R (2003) Biosynthesis of metal nanoparticles using fungi and actinomycetes. Curr Sci 85:162–170
Sattelmacher B (2001) The apoplast and its significance for plant mineral nutrition. New Phytol 149:167–192
Sedghi M, Hadi M, Toluie SG (2013) Effect of nano zinc oxide on the germination parameters of soybean seeds under drought stress. Ann West Univ Timişoara Ser Biol XVI(2):73–78
Shukla AK, Tiwari PK, Prakash C (2014) Micronutrients deficiencies vis-avis food and nutritional security of India. Indian J Fertil 10:94–112
Sirelkhatim A, Shahrom M, Azman S, Noor HMK, Chuo AL, Siti KMB, Habsah H, Dasmawati M (2015) Review on zinc oxide nanoparticles: antibacterial activity and toxicity mechanism. Micro Nano Lett 7(3):219–242
Singh MV (2009) Micronutrient nutritional problems in soils of India and improvement for human and animal's health. Indian J Fertil 5(4):11–26

Singh NB, Amist N, Yadav K, Singh D, Pandey JK, Singh SC (2013) Zinc oxide nanoparticles as fertilizer for the germination, growth and metabolism of vegetable crops. J Nanoeng Nanomanuf 3:1–12

Soliman AS, El-feky SA, Darwish E (2015) Alleviation of salt stress on Moringa peregrina using foliar application of nanofertilizers. J Hortic For 7:36–47

Som C, Berges M, Chaudhry Q, Dusinska M, Fernandes TF, Olsen SI, Nowack B (2010) Toxicology 269:160–169

Spero JM, Devito B, Theodore L (2000) Regulatory chemical handbook. CRC press, Boca Raton

Srivastav AK, Kumar M, Ansari NG, Jain AK, Shankar J, Arjaria N, Jagdale P, Singh D (2016) A comprehensive toxicity study of zinc oxide nanoparticles versus their bulk in wistar rats: toxicity study of zinc oxide nanoparticles. Hum Exp Toxicol 35(12):1286–1304

Stampoulis D, Sinha SK, White JC (2009) Assay-dependent phytotoxicity of nanoparticles to plants. Environ Sci Technol 43:9473–9479

Sun D, Hussain HI, Yi Z, Siegele R, Cresswell T, Kong L, Cahill DM (2014) Uptake and cellular distribution, in four plant species, of fluorescently labeled mesoporous silica nanoparticles. Plant Cell Rep 33:1389–1402

Taheri M, Qarache HA, Qarache AA, Yoosefi M (2015) The effects of zinc-oxide nanoparticles on growth parameters of corn (SC704). STEM Fellowship J 1:17–20

Tarafdar JC, Raliya R, Mahawar H, Rathore I (2014) Development of zinc nanofertilizer to enhance crop production in pearl millet (Pennisetum americanum). Agric Res 3:257–262

Taran N, Storozhenko V, Svietlova N, Batsmanova L, Shvartau V, Kovalenko M (2017) Effect of zinc and copper nanoparticles on drought resistance of wheat seedlings. Nanoscale Res Lett 12:60

Torabian S, Zahedi M, Khoshgoftarmanesh A (2016) Effect of foliar spray of zinc oxide on some antioxidant enzymes activity of sunflower under salt stress. J Agric Sci Technol 18:1013–1025

Uikey P, Vishwakarma K (2016) Review of zinc oxide (ZnO) nanoparticles applications and properties. Int J Emerg Tech Com Sci Elec 21(2):239

Vaseem M, Umar A, Hahn YB (2010) ZnO Nanoparticles: growth, properties and applications. In: Umar A, Hahn Y-B (eds) Metal oxide nanostructures and their applications, 4th edn. American Scientific Publishers, Los Angeles, pp 1–36

Venkatachalam P, Jayaraj M, Manikandan R, Geetha N, Rene ER, Sharma NC, Sahi SV (2017) Zinc oxide nanoparticles (ZnONPs) alleviate heavy metalinduced toxicity in Leucaena leucocephala seedlings: a physiochemical analysis. Plant Physiol Biochem 110:59–69

Wang C, Liu H, Chen J, Tian Y, Shi J, Li D, Guo C, Ma Q (2014) Carboxylated multi-walled carbon nanotubes aggravated biochemical and subcellular damages in leaves of broad bean (Vicia faba L.) seedlings under combined stress of lead and cadmium. J Hazard Mater 274:404–412

Wang F, Jin X, Adams CA, Shi Z, Sun Y (2018) Decreased ZnO nanoparticles phytotoxicity to maize by arbuscular mycorrhizal fungus and organic phosphorus. Environ Sci Pollut Res 25:23736–23747

Yadav T, Mungray AA, Mungray AK (2014) Fabricated nanoparticles. Rev Environ Contam Toxicol 230:83–110

Zhao L, Sun Y, Hernandez-Viezcas JA, Servin AD, Hong J, Niu G, Peralta-Videa JR, Duarte-Gardea M, Gardea-Torresdey JL (2013) Influence of CeO2and ZnO nanoparticles on cucumber physiological markers and bioaccumulation of Ce and Zn: a life cycle study. J Agric Food Chem 61:11945–11951

Zhao L, Peralta-Videa JR, Rico CM, Hernandez-Viezcas JA, Sun Y, Niu G, Duarte-Gardea M, Gardea-Torresdey JL (2014) CeO_2 and ZnO nanoparticles change the nutritional qualities of cucumber (*Cucumis sativus*). J Agric Food Chem 62:2752–2759

Chapter 5
Response of Titanium Nanoparticles to Plant Growth: Agricultural Perspectives

Ahmad Faraz, Mohammad Faizan, Qazi Fariduddin, and Shamsul Hayat

Abstract Utilization of nanoparticles (NPs) has increased tremendously in recent years by virtue of their unique properties, which can be applied for numerous purposes. Titanium (Ti)/titanium dioxide (TiO_2) NPs are among the most widely used NPs for applications including the agriculture sector. Titanium is considered a beneficial element for plant growth and its nano form can be used to improve growth and yield of plants. Research has shown that TiO_2 NPs generate both positive as well as a negative impact to plant growth. This review discusses current knowledge of TiO_2 NPs including their interactions, transport, and translocation within plants, and future perspectives regarding their use.

Keywords Agriculture · Nanoparticles · Plants · Titanium dioxide

5.1 Introduction

Nanotechnology is one of the most important fields in modern science, which deals with the manipulation of matter at the nanometer scale. It is an interdisciplinary science which plays multifunctional roles in varied fields such as medicine, textiles, energy, automobiles, and agriculture (Chandran et al. 2006). Particles having a size between 1–100 nm are termed nanoparticles (NPs); however, in the context of potential for utilization, sizes greater than 100 nm can also be included within this class of materials.

A. Faraz (✉) · Q. Fariduddin · S. Hayat
Plant Physiology Section, Department of Botany, Faculty of Life Sciences, Aligarh Muslim University, Aligarh, Uttar Pradesh, India

M. Faizan
Tree Seed Center, College of Forest Resources and Environment, Nanjing Forestry University, Nanjing, People's Republic of China

S. Hayat et al. (eds.), *Sustainable Agriculture Reviews 41*, Sustainable Agriculture Reviews 41, https://doi.org/10.1007/978-3-030-33996-8_5

Recent studies suggest that many metallic nanomaterials like those of silver (Ag), titanium dioxide (TiO_2), zinc (Zn), copper (Cu), iron (Fe), etc. have proved beneficial as well as harmful to plants; however, in-depth research is required to attain further details. Plants, being primary producers in food webs, create key channels for bioaccumulation of engineered nanomaterials (ENMs). Inappropriate use or overuse of ENMs may result in adverse effects as these materials enter to air, water bodies and soil, ultimately creating primary environmental reservoirs (Batley et al. 2013).

In recent years, titanium dioxide nanoparticles (TiO_2 NPs) have become one of the most common commercially used nanomaterials (Piccinno et al. 2012). Worldwide Domestic production of TiO_2 pigment in 2014 was approximately 1.31 million tons (U.S. Geological Survey 2015). An estimated 10,000 tons of TiO_2 NP is used per year in the cosmetic, solar cell, paint, cement and coatings industries (Piccinno et al. 2012). Numerous products are in daily use which are functionalized with TiO_2 NPs. Household products such as toothpaste, sun protecting creams, shaving creams, shampoos, conditioners contain TiO_2 NPs (Weir et al. 2012). TiO_2 NPs are also commonly used as a food additive to enhance the colour, brightness, and flavour of a variety of food products (Peters et al. 2014).

Due to their extensive use in industry, TiO_2 NPs are released to air, water and soil, with the consequent urgency to determine potential impacts to organisms and the biosphere. A number of studies have revealed that NPs generate toxic effects on microorganisms, invertebrates, and higher plants (Menard et al. 2011). Interactions of TiO_2 NPs and plants have been explored on a number of species – impacts could be deleterious or advantageous depending upon numerous factors (Mukherjee et al. 2016). TiO_2 NPs caused toxicity to the marine microalga *Nitzschia closterium* due to production of increased levels of reactive oxygen species (ROS) (Xia et al. 2015). Similarly, toxic effects of TiO_2 NPs have been reported in maize (Castiglione et al. 2011). However, TiO_2 NPs have generated beneficial impacts in various plant species like *Vigna radiata*, *Spinacia oleracea* and *Vicia faba* (Raliya et al. 2015; Yang et al. 2006; Latef et al. 2018).

With attention to the sometimes-contradictory roles of TiO_2 NPs, current resources have been compiled regarding TiO_2 NPs and their interaction with plants. This chapter will discuss the uptake, translocation, and transformation of TiO_2 NPs in plants, and the various techniques used for detection of TiO_2 NPs in plant cells.

5.2 Biosynthesis of TiO_2 Nanoparticles

TiO_2 NPs are synthesized by a variety of means including chemical vapor deposition, microemulsion, chemical precipitation, hydrothermal crystallization, and sol-gel methods (Muhd Julkapli et al. 2014; Valencia et al. 2013). So-called 'green synthesis' of TiO_2 NPs is also a promising practice. With this technology TiO_2 NPs of varying sizes and structures can be produced using plants and microorganisms (Fig. 5.1). NPs produced via biological methods are useful for industries while being eco-friendly (Waghmode et al. 2019).

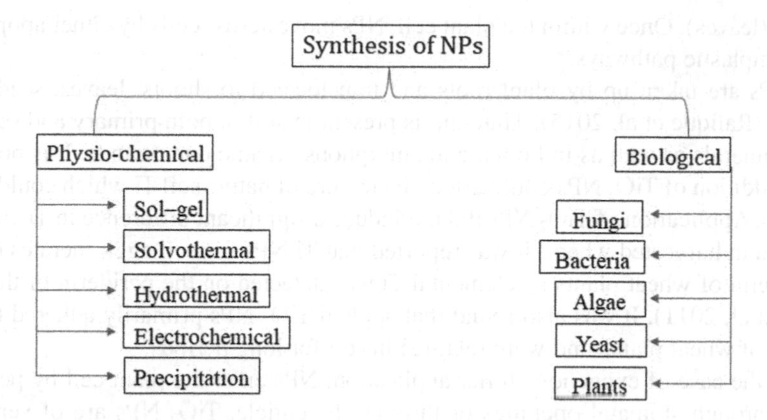

Fig. 5.1 Synthesis of nanoparticles by chemical and biological methods

Production of TiO_2 NPs using plant extracts has been reported (Subhapriya and Gomathipriya 2018). TiO_2 NPs have also been synthesized via the reaction between ethanolic leaf extracts of *Nyctanthes arbor-tristis* and titanium tetraisoproxide (TTIP) (Sundrarajan and Gowri 2011). TiO_2 NPs measuring 100–200 nm has been synthesized from titanium hydroxide by latex of *Jatropha curcas* L. (Hudlikar et al. 2012). Rajakumar et al. (2012) produced TiO_2 NPs by using the salt of titanium hydroxide along with an aqueous extract of *Eclipta prostrata*. Other workers utilized rice straw in powder form as a soft templating agent for the synthesis of TiO_2 NPs using TTIP and acetic acid (Ramimoghadam et al. 2014). TiO_2 NPs were produced from the aqueous extract of guava (Kumar et al. 2014). The rutile form of TiO_2 NPs was synthesized by using the salt of titanium hydroxide as a precursor along with a water extract of *Annona squamosa* (Roopan et al. 2012). NPs obtained through these sources had a spherical shape with size 23 ± 2 nm. Moreover, leaf extracts *of C. roseus* was used as a reducing agent for synthesis of TiO_2 NPs; however, the procedure produced irregularly-shape NPs having a size between 25–110 nm (Velayutham et al. 2012). Other sources of leaf extract, for example from *Morinda citrifolia,* was used for the production of TiO_2 NPs and were characterized by methods like XRD, FTIR, UV-vis and Raman spectroscopy (Sundrarajan et al. 2017). Polycrystalline TiO_2 NPs can be synthesized using *Justicia gendarussa* leaf extract as an oxidizing agent (Senthilkumar and Rajendran 2018). The crystalline nature of bio-synthesized TiO_2 NPs was confirmed by X-ray diffraction (XRD), and functional groups were confirmed by FT-IR spectroscopy. Cost-effective and eco-friendly TiO_2 NPs were produced from the leaf extract of *Sesbania grandiflora* (Srinivasan et al. 2019). Particles measured in the range of 43–56 nm which was further corroborated by SEM and TEM studies

5.3 Uptake, Transport, and Translocation of Ti/TiO_2 NPs

Uptake and translocation of NPs in plants is complex and variable, and not clearly understood. Uptake and transport vary with growth stage of plant, species and cultivar (Schwabe et al. 2015). NPs enter the plant cell either through roots or aerial parts (leaves). Once within the plant cell, NPs move across cells by either apoplastic or symplastic pathways.

NPs are taken up by plant roots and translocated to shoots, leaves, seeds and fruits (Rafique et al. 2015). Titanium is present in soil in both primary and secondary minerals as well as in bound and amorphous organic compounds. It is possible that addition of TiO_2 NPs could affect the texture of native soil Ti which could enter plants. Application of TiO_2 NPs did not induce a significant difference in Ti concentration in harvested wheat. It was reported that Ti NPs formed agglomerates on the periderm of wheat plants as elemental Ti was detected on the periderm of the root (Du et al. 2011). It was also found that applied TiO_2 NPs primarily adhered to cell walls of wheat plants and were retained in soil for long periods.

In the case of exogenous foliar application, NPs enter the plant cell by penetration through stomatal openings or through the cuticle. TiO_2 NPs are of very low solubility; however, their small size helps them enter cells. It was reported that smaller-sized NPs (2.8 ± 1.4 nm) readily penetrated cells and were translocated to the vacuole and nucleus (Chichiriccò and Poma 2015). Other studies have revealed that TiO_2 NPs enter the chloroplast where they bind with photosystem II and trigger the basic reactions of photosynthesis (Miralles et al. 2012). Larue et al. (2012) revealed that TiO_2 NPs penetrate leaf cells of wheat and rapeseed-mustard. Large NPs can also enter plants by foliar application. ZnO and TiO_2 NPs measuring 25 ± 3.5 nm entered leaves of tomato plants (Raliya et al. 2015).

Once inside the plant, NPs are redistributed in the stem and reach root cells via the phloem (Larue et al. 2014; Deepa et al. 2014). TiO_2 NPs (100 mg/L) applied to leaves of watermelon were translocated to leaves, stem and roots (Wang et al. 2015). Burke et al. (2014) reported that soybean and maize plants supplied with TiO_2 NPs concurrently with nitrogen had improved translocation.

Numerous techniques are available for identification of NPs in plant cells. Servin et al. (2012) used micro-XRF to reveal the translocation of TiO_2 NPs in cucumber. It was reported that TiO_2 NPs were transported as Ti from the root to trichomes of leaves in cucumber, which apparently act as a sink for Ti. Translocation of Ti in wheat plants was deduced by SEM-EDS studies. Roots of wheat exposed to TiO_2 NPs at different concentrations revealed the presence of Ti in shoots which was confirmed by SEM-EDS studies (Rafique et al. 2015).

5.4 Titanium Nanoparticles in Plants

Use of TiO_2 and related Ti NPs in various fields raises the concern about potential impacts generated in the plant. As evident from previous studies, TiO_2 NPs could induce both beneficial and toxic impacts to plants. The type and extent of impact generated by TiO_2 NPs to plants depends on their size and their concentration.

5.4.1 Beneficial Role of Ti/TiO_2 NPs in Plants

Numerous studies have clarified the beneficial impact of TiO_2 NPs to plants, which may directly or indirectly improve their growth (Fig. 5.2). Spinach seeds treated with low concentrations of TiO_2 NPs resulted in plants with greater shoot fresh/dry weight. In addition, chlorophyll content increased by approximately 17% and photosynthesis rate by 29% as compared with untreated plants (Zheng et al. 2005). Song et al. (2013) examined the effects of TiO_2 NPs on % germination of seeds and root elongation of seedlings in three plant species and determined no toxic effects. Canola seeds treated with different concentrations of TiO_2 NPs experienced improved germination and growth of radicles and plumules (Mahmoodzadeh et al. 2013). Rafique et al. (2018) investigated the effects of different concentrations (0, 20, 40, 60, 80, or 100 mg/kg) of TiO_2 NPs on wheat. Lower concentrations were beneficial while higher concentrations were inhibitory. The 60 mg/kg TiO_2 NP dosage was found most effective in increasing shoot and root length, and biomass. Frazier et al. (2014) reported that TiO_2 NPs generated a positive impact in tobacco: plants exposed to TiO_2 NPs (1000–25,000 mg/L) for 3 weeks significantly increased leaf count, root length and biomass which improved further with increasing Ti concentration. Moreover, treatment of TiO_2 NPs to maize improved chlorophyll content (a and b), total chlorophyll (a + b), carotenoids and anthocyanins in comparison with control and also improved crop yield (Morteza et al. 2013). Foliar application of biosynthesized NPs to 14-day-old mung bean improved shoot and root length, and content of biochemicals including chlorophyll and total soluble leaf protein as compared to control plants.

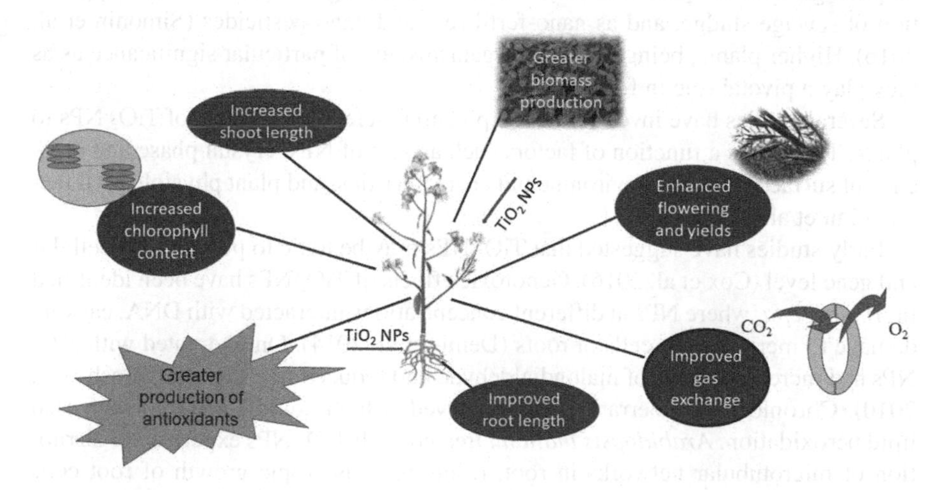

Fig. 5.2 Positive effects of TiO_2 NPs on plants in relation to growth, biomass production, enhanced chlorophyll content, and other parameters

Marchiol et al. (2016) investigated the uptake and translocation of TiO_2 NPs in barley. Plants raised from seeds treated with TiO_2 NPs experienced a prolonged growth cycle as compared to untreated plants. At physiological maturity, plants treated with TiO_2 NPs at 1000 mg kg^{-1} showed improved photosynthetic rate, stomatal conductance, and transpiration. Net photosynthetic rate and chlorophyll content increased significantly in rice plants treated with TiO_2 NPs (Ji et al. 2017). Genome-wide transcriptome analysis in leaves of *Arabidopsis thaliana* treated with TiO_2 NPs revealed that the NPs induced genes related to photosynthetic metabolism (Tumburu et al. 2017). Improved photosynthetic rate and other physiological processes due to TiO_2 NPs could be a function of increased light absorption by chloroplasts through upregulation of genes related to light-harvesting complex II (Ze et al. 2011). Foliar application of TiO_2 NPs boosted total phenolic content and biosynthesis of leaf flavanols in *Vitis vinifera* (Korosi et al. 2019).

TiO_2 NPs have been associated with reducing metal toxicity in plants. Deleterious effects of Cd in soybean were overcome by application of TiO_2 NPs (Singh and Lee 2016), and treated plants had improved photosynthetic rate and growth parameters. This suggests that TiO_2 NPs could be used in phytoremediation as it increases uptake of Cd and decreases their toxicity to soybean.

5.4.2 *Negative Impacts of TiO_2 NPs to Plants*

Based on empirical data and computer models NPs have been released into ecosystems in substantial quantities, which has triggered concern about possible impacts on plant growth. TiO_2 NPs are released into soil through irrigation or land application of sewage sludge, and as nano-fertilizers and nano-pesticides (Simonin et al. 2016). Higher plants, being producer organisms, are of particular significance as as they play a pivotal role in food webs.

Several studies have investigated the potential deleterious effects of TiO_2 NPs to plants. Toxicity is a function of factors such as size of NPs, crystal phase and presence of surface coatings, environmental characteristics, and plant physiological factors (Tan et al. 2018).

Early studies have suggested that TiO_2 NPs may be toxic to plants at the cellular and gene level (Cox et al. 2016). Genotoxic effects of TiO_2 NPs have been identified in *Allium cepa*, where NPs at different concentrations interacted with DNA, causing damage to meristematic cells of roots (Demir et al. 2014). Onion treated with TiO_2 NPs had increased levels of malondialdehyde and reduced root growth (Ghosh et al. 2010). Chromosomal aberration was observed, which could be due to increased lipid peroxidation. *Arabidopsis thaliana* treated with TiO_2 NPs experienced disruption of microtubular networks in root, resulting in isotropic growth of root cells (Wang et al. 2011). Pakrashi et al. (2014) observed negative impacts of TiO_2 NPs in a dose-dependent manner in *Allium cepa*. The particles decreased the mitotic index and increased the number of chromosomal aberrations in root tips of *A. cepa* exposed to 12.5–100 μg/mL TiO_2 NPs. Similarly, Fellmann and Eichert (2017) reported that treatment with TiO_2 NPs led to reduction in germination rate, and root and shoot growth in corn in a dose-dependent manner. Korenkova et al. (2017) found adverse effects of TiO_2 NPs with increasing concentration on root growth of *Hordeum vulgare*.

5.5 Conclusions and Future Perspectives

A vast range of nanomaterials are being produced and applied, offering great potential in diverse fields including manufacturing industries, energy, cosmetics, and agriculture. Due to overuse of nanoparticles, releases to soil, water, and air raise concerns due to their potential negative effects to plants, animals and humans.

Of the various NPs, TiO_2 NPs are becoming popular in agriculture for improving plant productivity. Uptake, translocation, and accumulation of TiO_2 NPs occurs in the plant via underground parts (root system) or by shoots (stem, leaves, cuticles). Once inside the plant TiO_2 NPs move through tissue either by symplastic or apoplastic pathway. TiO_2 NPs may affect plant growth, physiological processes (e.g., photosynthesis, respiration) as well as biochemical processes (antioxidant system). More work is required to determine the mechanisms regarding how nanoparticles interact with plants at both cellular and molecular levels. NPs used for any purpose should be eco-friendly and non-hazardous to biota; furthermore, an environmental assessment should be carried out before use of a specific nanomaterial.

References

Abdel Latef AAH, Srivastava AK, El-sadek MSA, Kordrostami M, Tran LSP (2018) Titanium dioxide nanoparticles improve growth and enhance tolerance of broad bean plants under saline soil conditions. Land Degrad Dev 29(4):1065–1073

Batley G, Kirby JK, McClaughlin MJ (2013) Fate and risks of nanomaterials in aquatic and terrestrial environments. Acc Chem Res 46:854–862

Burke DJ, Zhu S, Pablico-Lansigan MP, Hewins CR, Samia ACS (2014) Titanium oxide nanoparticle effects on composition of soil microbial communities and plant performance. Biol Fertil Soils 50(7):1169–1173

Castiglione MR, Giorgetti L, Geri C, Cremonini R (2011) The effects of nano-TiO_2 on seed germination, development and mitosis of root tip cells of Vicia narbonensis L. and Zea mays L. J Nanopart Res 13(6):2443–2449

Chandran SP, Chaudhary M, Pasricha R, Ahmad A, Sastry M (2006) Synthesis of gold nanotriangles and silver nanoparticles using Aloe vera plant extract. Biotechnol Prog 22:577–583

Chichiriccò G, Poma A (2015) Penetration and toxicity of Nanomaterials in higher plants. Nano 5(2):851–873

Cox A, Venkatachalam P, Sahi S, Sharma N (2016) Silver and titanium dioxide nanoparticle toxicity in plants: a review of current research. Plant Physiol Biochem 107:147–163

Deepa K, Singha S, Panda T (2014) Doxorubicin nanoconjugates. J Nanosci Nanotechnol 14:892–904

Demir E, Kaya N, Kaya B (2014) Genotoxic effects of zinc oxide and titanium dioxide nanoparticles on root meristem cells of Allium cepa by comet assay. Turk J Biol 38(1):31–39

Du W, Sun Y, Ji R, Zhu J, Wu J, Guo H (2011) TiO2 and ZnO nanoparticles negatively affect wheat growth and soil enzyme activities in agricultural soil. J Environ Monit 13(4):822

Fellmann S, Eichert T (2017) Acute effects of engineered nanoparticles on the growth and gas exchange of Zea mays L. what are the underlying causes? Water Air Soil Pollut 228(5):176

Frazier TP, Burklew CE, Zhang B (2014) Titanium dioxide nanoparticles affect the growth and microRNA expression of tobacco (*Nicotiana tabacum*). Funct Integr Genomics 14(1):75–83

Ghosh M, Bandyopadhyay M, Mukherjee A (2010) Genotoxicity of titanium dioxide (TiO2) nanoparticles at two trophic levels: plant and human lymphocytes. Chemosphere 81(10):1253–1262

Hudlikar M, Joglekar S, Dhaygude M, Kodam K (2012) Green synthesis of TiO_2 nanoparticles by using aqueous extract of *Jatropha curcas* L. latex. Mater Lett 75:196–199

Ji Y, Zhou Y, Ma C, Feng Y, Hao Y, Rui Y, Xing B (2017) Jointed toxicity of TiO2 NPs and Cd to rice seedlings: NPs alleviated Cd toxicity and Cd promoted NPs uptake. Plant Physiol Biochem 110:82–93

Korenkova L, Sebesta M, Urik M, Kolen cik M, Kratosova G, Bujdos M, Vavra I, Dobrocka E (2017) Physiological response of culture media-grown barley (*Hordeum vulgare* L.) to titanium oxide nanoparticles. Acta Agric Scand 67:285–291

Korosi L, Bouderias S, Csepregi K, Bognár B, Teszlák P, Scarpellini A, Jakab G (2019) Nanostructured TiO_2-induced photocatalytic stress enhances the antioxidant capacity and phenolic content in the leaves of *Vitis vinifera* on a genotype-dependent manner. J Photochem Photobiol B Biol 190:137–145

Kumar PV, Pammi SV, Kollu P, Satyanarayan KV, Shameem U (2014) Green synthesis and characterization of silver nanoparticles using *Boerhaavia diffusa* plant extract and their antibacterial activity. Ind Crop Prod 52:562–566

Larue C, Laurette J, Herlin-Boime N, Khodja H, Fayard B, Flank A-M, Brisset F, Carriere M (2012) Accumulation, translocation and impact of TiO2 nanoparticles in wheat (Triticum aestivum spp.): influence of diameter and crystal phase. Sci Total Environ 431:197–208

Larue C, Castillo-Michel H, Sobanska S, Cécillon L, Bureau S, Barthès V, Ouerdane L, Carrière M, Sarret G (2014) Foliar exposure of the crop *Lactuca sativa* to silver nanoparticles: evidence for internalization and changes in Ag speciation. J Hazard Mater 264:98–106

Mahmoodzadeh H, Nabavi M, Kashefi H (2013) Effect of nanoscale titanium dioxide particles on the germination and growth of canola (*Brassica napus*). J Ornam Hortic Plants 3:25–32

Marchiol L, Mattiello A, Poscic F, Fellet G, Zavalloni C, Carlino E, Musetti R (2016) Changes in physiological and agronomical parameters of barley (*Hordeum vulgare*) exposed to cerium and titanium dioxide nanoparticles. Int J Environ Res Public Health 13:332

Menard A, Drobni D, Jemec A (2011) Ecotoxicity of nanosized TiO_2. Review of in vivo data. Environ Pollut 159:677–684

Miralles P, Church TL, Harris AT (2012) Toxicity, uptake, and translocation of engineered Nanomaterials in vascular plants. Environ Sci Technol 46(17):9224–9239

Morteza E, Moaveni P, Farahani HA, Kiyani M (2013) Study of photosynthetic pigments changes of maize (*Zea mays* L.) under nano TiO_2 spraying at various growth stages. Springer Plus 2(1):247

Muhd Julkapli N, Bagheri S, Bee Abd Hamid S (2014) Recent advances in heterogeneous photocatalytic decolorization of synthetic dyes. Sci World J 2014:1–25

Mukherjee A, Majumdar S, Servin AD, Pagano L, Dhankher OP, White JC (2016) Carbon nanomaterials in agriculture: a critical review. Front Plant Sci 7(16):172

Pakrashi S, Jain N, Dalai S, Jayakumar J, Chandrasekaran PT, Raichur AM, Chandrasekaran N, Mukherjee A (2014) In vivo genotoxicity assessment of titanium dioxide nanoparticles by *Allium cepa* root tip assay at high exposure concentrations. PLoS One 9(2):87789

Peters RJ, van Bemmel G, Herrera-Rivera Z, Helsper HP, Marvin HJ, Weigel S et al (2014) Characterization of titanium dioxide nanoparticles in food products: analytical methods to define nanoparticles. J Agric Food Chem 62:6285–6293

Piccinno F, Gottschalk F, Seeger S, Nowack B (2012) Industrial production quantities and uses of ten engineered nanomaterials in Europe and the world. J Nanopart Res 14:1109

Rafique R, Arshad M, Khokhar MF, Qazi IA, Hamza A, Virk N (2015) Growth response of wheat to titania nanoparticles application. NUST J Eng Sci 7(1):42–46

Rafique R, Zahra Z, Virk N, Shahid M, Pinelli E, Park TJ, Kallerhoff J, Arshad M (2018) Dose-dependent physiological responses of *Triticum aestivum* L. to soil applied TiO2 nanoparticles: alterations in chlorophyll content, H2O2 production, and genotoxicity. Agric Ecosyst Environ 255:95–101

Rajakumar G, Rahuman AA, Priyamvada B, Khanna VG, Kumar DK, Sujin PJ (2012) Eclipta prostrate leaf aqueous extract mediated synthesis of titanium dioxide nanoparticles. Mater Lett 68:115–117

Raliya R, Biswas P, Tarafdar JC (2015) TiO_2 nanoparticle biosynthesis and its physiological effect on mung bean (*Vigna radiata* L.). Biotechnol Rep 5:22–26

Ramimoghadam D, Bagheri S, Bee S, Hamid A (2014) Biotemplated synthesis of anatase titanium dioxide nanoparticles via lignocellulosic waste material. Biomed Res Int 2014:7

Roopan SM, Bharathi A, Prabhakarn A, Rahuman AA, Velayutham K, Rajakumar G, Padmaja RD, Lekshmi M, Madhumitha G (2012) Efficient phyto-synthesis and structural characterization of rutile TiO_2 nanoparticles using *Annona squamosa* peel extract. Spectrochim Acta Part A 98:86–90

Schwabe F, Tanner S, Schulin R, Rotzetter A, Stark W, von Quadt A, Nowack B (2015) Dissolved cerium contributes to uptake of Ce in the presence of differently sized CeO-nanoparticles by three crop plants. Metallomics 7(3):466–477

Senthilkumar S, Rajendran A (2018) Biosynthesis of TiO_2 nanoparticles using Justicia gendarussa leaves for photocatalytic and toxicity studies. Res Chem Intermed 44(10):5923–5940

Servin AD, Castillo-Michel H, Hernandez-Viezcas JA, Diaz BC, Peralta-Videa JR, Gardea-Torresdey JL (2012) Synchrotron micro-XRF and micro-XANES confirmation of the uptake and translocation of TiO_2 nanoparticles in cucumber (*Cucumis sativus*) plants. Environ Sci Technol 46(14):7637–7643

Simonin M, Richaume A, Guyonnet JP, Dubost A, Martins JM, Pommier T (2016) Titanium dioxide nanoparticles strongly impact soil microbial function by affecting archaeal nitrifiers. Sci Rep 6:33643

Singh J, Lee BK (2016) Influence of nano-TiO_2 particles on the bioaccumulation of Cd in soybean plants (*Glycine max*): a possible mechanism for the removal of Cd from the contaminated soil. J Environ Manag 170:88–96

Srinivasan M, Venkatesan M, Arumugam V, Natesan G, Saravanan N, Murugesan S, Pugazhendhi A (2019) Green synthesis and characterization of titanium dioxide nanoparticles (TiO_2 NPs) using Sesbania grandiflora and evaluation of toxicity in zebrafish embryos. Process Biochem 80:197–202

Song U, Jun H, Waldman B, Roh J, Kim Y, Yi J, Lee EJ (2013) Functional analyses of nanoparticle toxicity: a comparative study of the effects of TiO2 and Ag on tomatoes (*Lycopersicon esculentum*). Ecotoxicol Environ Saf 93:60–67

Subhapriya S, Gomathipriya P (2018) Green synthesis of titanium dioxide (TiO2) nanoparticles by *Trigonella foenum-graecum* extract and its antimicrobial properties. Microb Pathog 116:215–220

Sundrarajan M, Gowri S (2011) Green synthesis of titanium dioxide nanoparticles by *Nyctanthes arbor-tristis* leaves extract. Chalcogenide Lett 8:447–451

Sundrarajan M, Bama K, Bhavani M, Jegatheeswaran S, Ambika S, Sangili A, Sumathi R (2017) Obtaining titanium dioxide nanoparticles with spherical shape and antimicrobial properties using M. citrifolia leaves extract by hydrothermal method. J Photochem Photobiol B Biol 171:117–124

Tan W, Peralta-Videa JR, Gardea-Torresdey JL (2018) Interaction of titanium dioxide nanoparticles with soil components and plants: current knowledge and future research needs–a critical review. Environ Sci Nano 5(2):257–278

Tumburu L, Andersen CP, Rygiewicz PT, Reichman JR (2017) Molecular and physiological responses to titanium dioxide and cerium oxide nanoparticles in Arabidopsis. Environ Toxicol Chem 36(1):71–82

U.S. Geological Survey (2015) Mineral commodity summaries 2015. U.S. Government Printing Office, Washington, DC. https://doi.org/10.3133/70140094

Valencia S, Vargas X, Rios L, Restrepo G, Marín JM (2013) Sol–gel and low-temperature solvothermal synthesis of photoactive nano-titanium dioxide. J Photochem Photobiol A Chem 251:175–181

Velayutham K, Rahuman AA, Rajakumar G, Santhoshkumar T, Marimuthu S, Jayaseelan C, Bagavan A, Kirthi AV, Kamaraj C, Zahir AA, Elango G (2012) Evaluation of *Catharanthus roseus* leaf extract-mediated biosynthesis of titanium dioxide nanoparticles against *Hippobosca maculata* and *Bovicola ovis*. Parasitol Res 111:2329–2337

Waghmode MS, Gunjal AB, Mulla JA, Patil NN, Nawani NN (2019) Studies on the titanium dioxide nanoparticles: biosynthesis, applications and remediation. SN Appl Sci 1(4):310

Wang S, Kurepa J, Smalle JA (2011) Ultra-small TiO_2 nanoparticles disrupt microtubular networks in *Arabidopsis thaliana*. Plant Cell Environ 34(5):811–820

Wang TY, Jiang HT, Wan L, Zhao QF, Jiang TY, Wang B, Wang SL (2015) Potential application of functional porous TiO_2 nanoparticles in light-controlled drug release and targeted drug delivery. Acta Biomater 13:354–363

Weir A, Westerhoff P, Fabricius L, Hristovski K, von Goetz N (2012) Titanium dioxide nanoparticles in food and personal care products. Environ Sci Technol 46(4):2242–2250

Xia B, Chen B, Sun X, Qu K, Ma F, Du M (2015) Interaction of TiO 2 nanoparticles with the marine microalga *Nitzschia closterium*: growth inhibition, oxidative stress and internalization. Sci Total Environ 508:525–533

Yang F, Hong F, You W, Liu C, Gao F, Wu C, Yang P (2006) Influence of nano-anatase TiO_2 on the nitrogen metabolism of growing spinach. Biol Trace Elem Res 110(2):179–190

Ze Y, Liu C, Wang L, Hong M, Hong F (2011) The regulation of TiO_2 nanoparticles on the expression of light-harvesting complex II and photosynthesis of chloroplasts of *Arabidopsis thaliana*. Biol Trace Elem Res 143:1131

Zheng L, Hong F, Lu S, Liu C, Hong J, Niu G, Peralta-Videa JR, Duarte-Gardea M, Gardea-Torresdey JL (2005) Effect of nano-TiO_2 on strength of naturally aged seeds and growth of spinach. Biol Trace Elem Res 104(1):83–91

Chapter 6
Impact of Silver Nanoparticles on Plant Physiology: A Critical Review

Fareen Sami, Husna Siddiqui, and Shamsul Hayat

Abstract Nanotechnology is a rapidly growing field of science and technology that focuses on the production and utilization of materials measuring <100 nm in at least one dimension. The unique physicochemical properties of nanoparticles are a result of their high surface area and high reactivity, which renders them beneficial in biotechnology industries and in agriculture. In recent years, researchers have focused on the beneficial effects of silver nanoparticles (Ag-NPs) on plant growth and development. Ag-NPs, when applied at low concentrations, enhance shoot and root growth of many species. Also, Ag-NPs enhance the activities of antioxidant enzymes which limit production of reactive oxygen species in plant cells. Lower doses of Ag-NPs are also beneficial in enhancing chlorophyll production as well as enhancing chlorophyll florescence parameters. This review highlights the current understanding as well as the future possibilities of Ag-NP research in plant systems.

Keywords Abiotic stress · Antioxidative defense system · Compatible solutes · Flooding stress · Growth · Heat stress · Nanotechnololgy · Photosynthesis · ROS homeostasis · Salt stress · Seed germination

6.1 Introduction

Nanotechnology is a branch of science that synthesizes, utilizes and examines nanomaterials, i.e., particles which possess unique properties by virtue of their small size (less than 100 nm). Nanotechnology embraces the manipulation of the size, structure and dimension of these particles (Savithramma et al. 2011).

F. Sami (✉) · H. Siddiqui · S. Hayat
Plant Physiology & Biochemistry Section, Department of Botany, Faculty of Life Sciences, Aligarh Muslim University, Aligarh, Uttar Pradesh, India

S. Hayat et al. (eds.), *Sustainable Agriculture Reviews 41*, Sustainable Agriculture Reviews 41, https://doi.org/10.1007/978-3-030-33996-8_6

Nanomaterials are used in almost every field of science due to their distinct chemical and physical characteristics. Silver nanoparticles (Ag-NPs) are the most extensively studied nanoparticles as a result of their diverse applications including as an antimicrobial agent, for optics and electronics, as coatings for stainless steel, and for water purification (Duran et al. 2007). Global production of Ag-NPs has reached 500 tons per year (Mueller and Nowack 2008). The effectiveness of Ag-NPs is related to their small size that results in maximum exposure of total surface area in solution (Panyala et al. 2008).

The positive and negative impacts of nanoparticles (NPs) to plants are a function of size, stability, shape, concentration and presence of coatings (Rastogi et al. 2017). In agriculture, Ag-NPs are extensively used to improve crop production (Almutairi 2016). Low doses of Ag-NPs are reported to enhance seed germination in various plant species (Yin et al. 2012; Sharma et al. 2012; Savithramma et al. 2012; Almutairi and Alharbi 2015). Improved chlorophyll content as well as photosynthetic efficiency was observed in various species at moderate doses of Ag-NPs (Salama 2012; Sharma et al. 2012; Rani et al. 2016; Mohamed et al. 2017). It was reported that application of Ag-NPs enhances carbohydrate content in plants (Salama 2012; Mirzajani et al. 2013; Nair and Chung 2014; Rani et al. 2016; Mohamed et al. 2017).

Lower concentrations of Ag-NPs accelerate growth of several species (Salama 2012; Sharma et al. 2012; Mirzajani et al. 2013; Vannini et al. 2013; Mehta et al. 2016; Jasim et al. 2017; Rani et al. 2016; Tomacheski et al. 2017; Mohamed et al. 2017). However, a marked inhibition in different growth biomarkers has been reported at higher concentrations (Dimkpa et al. 2013; Qian et al. 2013; Hojjat and Hojjat 2015; Abd-Alla et al. 2016; Al-Huqail et al. 2018). Reduction of photosynthesis was reported at high Ag-NP concentrations (Qian et al. 2013; Nair and Chung 2014; Sosan et al. 2016; Tripathi et al. 2017; Vinkovic et al. 2018). A marked decline in reactive oxygen species (ROS) was reported by the application of different concentrations of Ag-NPs (Sharma et al. 2012; Olchowih et al. 2017; Mohamed et al. 2017).

Exposure to Ag-NPs has been found to enhance the antioxidative defense system and proline content of several crops (Sharma et al. 2012; Yasur and Rani 2013; Mehrian et al. 2015; Rani et al. 2016; Yang et al. 2017; Mohamed et al. 2017; Tripathi et al. 2017). Enhanced abiotic stress tolerance has also been reported by application of Ag-NPs. For example, recent studies have revealed the positive role of Ag-NPs under salt stress (Ekhtiyari et al. 2011; Almutairi 2016; Ghavam 2018; Mohamed et al. 2017; Hojjat and Kamyab 2017; Hojjat 2019). The ameliorative function of Ag-NPs in wheat under heat stress was demonstrated by Iqbal et al. (2019). Some studies have reported the beneficial role of Ag-NPs in saffron and soybean under flooding stress (Rezvani et al. 2012; Mustafa et al. 2015).

The intent of the chapter is to present both the beneficial and harmful effects of Ag-NPs to plant physiological and developmental processes. This chapter summarizes data depicting concentration-dependent effects of Ag-NPs on plant growth, photosynthesis, ROS homeostasis, accumulation of compatible solutes and the antioxidative defense system under normal as well as abiotic stress conditions.

6.2 Effect of Silver Oxide Nanoparticles in Plant Physiological Processes

6.2.1 Seed Germination

Germination percentage in oat, lettuce and radish was higher in Ag-NPs treated plants as compared to control (Tomacheski et al. 2017). Coated Ag-NPs also enhanced germination rate in several plant species *(Lolium multiflorum, Carex spp., Eupatorium fistulosum and Phytolacca americana)* (Yin et al. 2012). In *Brassica juncea* seedlings, application of 25 and 50 mg/L Ag-NPs improved percent germination (Sharma et al. 2012) (Table. 6.1). Application of Ag-NPs enhance germination rate in corn *(Zea mays),* watermelon *(Citrullus lanatus)* and zucchini *(Cucurbita pepo)* (Almutairi and Alharbi 2015) (Table. 6.1). In *Boswellia ovailifoliolata,* 10, 20 and 30 μg/ml Ag-NPs increased percent germination (Savithramma et al. 2012) (Table 6.1). Exogenously sourced Ag-NPs (20–50 mg/l) increased seed germination in *Pennisetum glaucum* (Parveen and Rao 2015) (Table 6.1). In fenugreek, 0–40 μg ml^{-1} Ag-NPs promoted percent seed germination and germination rate (Hojjat and Hojjat 2015) (Table 6.2). This increase may be due to enhanced germination potential in fenugreek.

Negative effects of Ag-NPs on germination have also been identified. In *Lolium perenne,* application of 10–20 mg/L Ag-NPs led to reduction in germination percentage (El-Temsah and Joner 2012), and in *Hordeum vulgare,* 10–20 mg/L Ag-NPs reduced germination percentage (El-Temsah and Joner 2012). In *Vicia faba,* application of 500–900 μg/kg resulted in a decline in seed germination (Abd-Alla et al. 2016) (Table 6.2). In *Arabidopsis thaliana,* 75–300 μg/l Ag-NPs decreased germination rate (Geisler-Lee et al. 2014). In *Oryza sativa,* varied concentrations of Ag-NPs (0.1, 1, 10, 100 and 1000 mg/L) caused a decline in seed germination and seedling growth (Thuesombat et al. 2014). Exogenous application of 40 mg/L Ag-NPs increase germination in *Eupatorium fistulosum* while inhibiting germination of *Scirpus cyperinus, Juncus effusus* and *Phytolacca americanum* (Yin et al. 2012).

6.2.2 Photosynthesis

Exogenously sourced Ag-NPs modulates photosynthesis (Fig. 6.1). Ag-NPs function as catalysts in redox reactions and hence affect photosynthetic and respiratory processes (Tripathi et al. 2017) (Table 6.2). In water hyacinth, exogenously sourced Ag-NPs increased chlorophyll content (Rani et al. 2016) (Table 6.2). In *Brassica juncea,* 25, 50 and 100 mg/L Ag-NPs enhanced chlorophyll content as well as PSII efficiency. However, 200 and 400 mg/L reduced the same (Sharma et al. 2012) (Table 6.1). Pandey et al. (2014) found that application of 1000 mg/L Ag-NPs

Table 6.1 Effects of exogenous application of Ag-NPs in different plant species

Plant species	Dose-dependent effect of Ag-NPs		References
	Positive effects	Negative effects	
Brassica juncea	25, 50 and 100 mg/L Ag-NPs enhance chlorophyll content and PSII efficiency; Application of 25 and 50 mg/L Ag-NPs enhanced shoot fresh mass, length of shoots and roots as well as vigor index; 25–50 mg/L Ag-NPs decrease H_2O_2 and MDA content; 50,100, 200 and 400 mg/L Ag-NPs increase antioxidant enzymes as well as proline content	200 and 400 mg/L Ag-NPs reduce chlorophyll content and PSII efficiency; >100 mg/L Ag-NPs decrease percent seed germination, vigor index, fresh weight and root and shoot length; 100–400 mg/L Ag-NPs increase H_2O_2 and MDA content	Sharma et al. (2012)
Zea mays	0.05–2.5 mg/ml Ag-NPs positively enhance germination rate, germination percentage and mean germination time; 0.5–2.5 mg/ml Ag-NPs enhance seedling fresh weight; 0.5 and 1 mg/ml Ag-NPs increase seedling dry weight	0.05–2.5 mg/ml Ag-NPs negatively affect root length; 0.05 and 0.1 mg/ml Ag-NPs decrease seedling fresh weight and dry weight; 1.5–2.5 mg/ml Ag-NPs decrease seedling dry weight	Almutairi and Alharbi (2015)
Citrullus lanatus	0.05–2.5 mg/ml Ag-NPs positively enhance germination rate, germination percentage and mean germination time; 0.5–2 mg/ml Ag-NPs increase seedling fresh weight	0.05 and 0.1 mg/ml Ag-NPs decrease root length and seedling fresh weight; 0.05–2.5 mg/ml Ag-NPs decrease seedling dry weight	Almutairi and Alharbi (2015)
Cucurbita pepo	0.05–2.5 mg/ml Ag-NPs enhance germination rate, germination percentage and mean germination time; 0.05–2.5 mg/ml Ag-NPs increase root length; 0.05–1.5 and 2.5 mg/ml Ag-NPs positively increase seedling fresh weight; 0.05–2.5 mg/ml Ag-NPs increase seedling dry weight	2 mg/ml Ag-NPs decrease seedling fresh weight	Almutairi and Alharbi (2015)
Boswellia ovailifoliolata	10–30 μg/ml Ag-NPs enhance percent of seed germination and seedling growth	10–30 μg/ml Ag-NPs decrease germination period	Savithramma et al. (2012)
Pennisetum glaucum	20 and 50 mg/L Ag-NPs increase percentage of seed germination	20 and 50 mg/L Ag-NPs decrease shoot, root and seedling length	Parveen and Rao (2015)

Table 6.2 Effects of exogenous application of Ag-NPs in different plant species

Plant species	Dose-dependent effect of Ag-NPs		References
	Positive effects	Negative effects	
Trigonella foenum-graecum	10–40 μg/mL Ag-NPs enhance percent and speed of germination, root length, root fresh and dry weight		Hojjat and Hojjat (2015)
Vicia faba		500–900 μg/kg Ag-NPs decline seed germination; 100–900 μg/kg Ag-NPs decrease leaf area, shoot/root ratio, length and dry weight of root and shoot	Abd-Alla et al. (2016)
Pisum sativum	Exogenous application of 1000 and 3000 μM Ag-NPs inhibited the activities of GR and DHAR	1000 and 3000 μM Ag-NPs decline photosynthetic content and chlorophyll fluorescence	Tripathi et al. (2017)
Eichhornia crassipes	10 mg/L Ag-NPs increase carbohydrate content; 100 mg/L Ag-NPs increase protein content; 10 mg/L Ag-NPs increase phenol content; 100 mg/L Ag-NPs increase chlorophyll content; 1–10 mg/L Ag-NPs increased carbohydrate content; 1–100 mg/L Ag-NPs increase SOD activity	1 and 100 mg/L Ag-NPs decrease carbohydrate content; 1 and 10 mg/L Ag-NPs decrease protein content; 1 and 10 mg/L Ag-NPs decrease chlorophyll content; a reduction in CAT and POX activity was observed	Rani et al. (2016)
Brassica juncea	1000 mg/L Ag-NPs increase chlorophyll content	100 and 500 mg/L Ag-NPs decrease chlorophyll content	Pandey et al. (2014)
Phaseolus vulgaris *Zea mays*	20,40 and 60 mg/L Ag-NPs increase chlorophyll content; 20–60 mg/L Ag-NPs increase leaf area, length of root and shoot; 20–60 mg/L Ag-NPs enhance carbohydrate content in *Zea mays and Phaseolus vulgaris*	80 and 100 mg/L Ag-NPs reduce chlorophyll content; 80–100 mg/L Ag-NPs decrease leaf area, length of root and shoot; 80–100 mg/L Ag-NPs inhibits carbohydrate content in *Zea mays and Phaseolus vulgaris*	Salama (2012)
Psychomitrella patens	An increase in chlorophyll content was observed		Liang et al. (2018)
Arabidopsis thaliana		300, 500, 1500 and 3000 mg/L Ag-NPs decreased PSII efficiency	Sosan et al. (2016)
Oryza sativa	1 mg/L Ag-NPs enhance sugar content; 1 mg/L Ag-NPs enhance proline accumulation	0.2, 0.5 and 1 mg/L Ag-NPs reduce total chlorophyll and carotenoid content	Nair and Chung (2014)

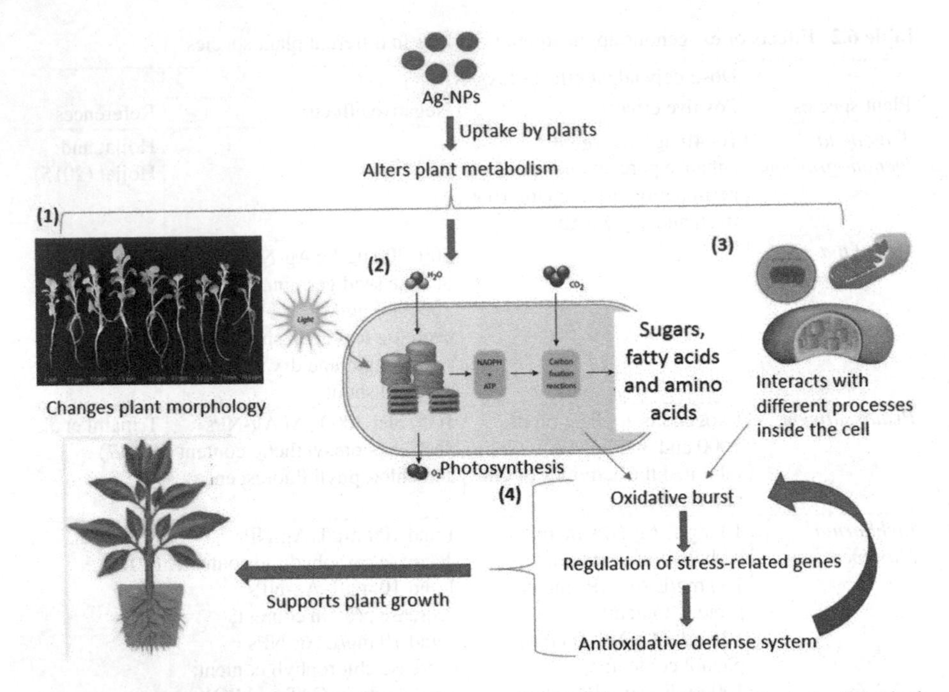

Fig. 6.1 Diagrammatic representation showing the effects of silver Ag-NPs on plant physiological processes. (1) Different concentrations of Ag-NPs alter plant morphology; (2) Ag-NPs negatively affect photosynthesis which leads to oxidative burst within the cell; (3) Ag-NPs alter processes in the mitochondria, chloroplast and peroxisomes that leads to oxidative burst; and (4) Ag-NPs regulate stress-related genes that enhance the antioxidative defense system

increased chlorophyll content in the same species (Table 6.2). In corn and pea, application of 20–60 mg/L Ag-NPs resulted in an increase in chlorophyll content whereas a marked inhibition in chlorophyll was observed at 80–100 mg/L (Salama 2012) (Table 6.2).

In many cases Ag-NPs caused detrimental effects to the photosynthetic system. Ag-NPs caused a decline in chlorophyll and chlorophyll fluorescence parameters and subsequently altered the photosynthetic system. In *Physcomitrella patens*, Ag-NPs inhibited thylakoid and chlorophyll content (Liang et al. 2018) (Table 6.2). Exogenous application of 1000 and 3000 μM Ag-NPs caused a decline in photosynthetic pigments and chlorophyll fluorescence in pea (Tripathi et al. 2017) (Table 6.2). Decreased photosynthetic efficiency was observed at 100–5000 mg/L Ag-NPs in *Arabidopsis thaliana* (Sosan et al. 2016) (Table 6.2). In *Oryza sativa*, doses of Ag-NPs (0–1 mg/L) enhanced total chlorophyll and carotenoid contents (Nair and Chung 2014) (Table 6.2). Varied doses of Ag-NPs (0.2, 0.5 and 3 mg/L) disrupted thylakoid membrane structure and decreased chlorophyll content (Qian et al. 2013) (Table 6.3).

Table 6.3 Effects of exogenous application of Ag-NPs in different plant species

Plant species	Dose-dependent effect of Ag-NPs		References
	Positive effects	Negative effects	
Arabidopsis thaliana		Ag-NPs (0.2,0.5 and 3 mg/L) disrupted thylakoid membrane structure and decreased chlorophyll content	Qian et al. (2013)
Oryza sativa	30 μg/ml Ag-NPs accelerated root growth; an enhanced production in reducing sugar content	60 μg/ml Ag-NPs restricted root growth; 30–60 μg/ml Ag-NPs reduce total carbohydrate content	Mirzajani et al. (2013)
Tritlcum aestlvum Vlgna unguiculata Brassica	50 mg/L Ag-NPs increase length and dry weight of shoot in wheat; in cowpea, 50 mg/L Ag-NPs increase shoot and root length, fresh and dry weight of roots and shoots and root nodulation In Brassica, 50 and 75 mg/L Ag-NPs increase shoot length, fresh and dry weight of root and shoot	75 mg/L Ag-NPs decrease shoot and root length, fresh and dry weight of root and shoot in cowpea; In wheat, 50 and 75 mg/L Ag-NPs decrease root length, fresh and dry weight of root	Mehta et al. (2016)
Trigonella foenum-graecum	20–60 mg/kg Ag-NPs increased leaf area as well as length of roots and shoots.	higher concentrations (>60 mg/kg Ag-NPs) were inhibitory	Jasim et al. (2017)
Eruca sativa	1–20 mg/L Ag-NPs enhance root length	0.1 and 100 mg/L decrease root length	Vannini et al. (2013)
Lupinus termis	100 mg/L Ag-NPs increase germination percentage; 100 mg/L Ag-NPs increase root, shoot and seedling fresh weight; 100 mg/L Ag-NPs increase root, shoot and seedling dry weight; 100 mg/L Ag-NPs increase root, shoot and seedling length	150–900 mg/L Ag-NPs decrease germination percentage; 300–500 mg/L Ag-NPs decrease root, shoot and seedling fresh weight; 300–500 mg/L Ag-NPs decrease root, shoot and seedling dry weight; 300–500 mg/L Ag-NPs decrease root, shoot and seedling length	Al-Huqail et al. (2018)
Triticum aestivum	0–5 mg/kg Ag-NPs led to accumulation of oxidized glutathione disulfide (GSSG) in a dose-dependent manner	2.5 mg/kg reduce root and shoot length	Dimkpa et al. (2013)
Triticum aestivum		20, 200 and 2000 mg/kg Ag-NPs resulted in lower biomass, shorter plant height and lower plant weight	Yang et al. (2018)
Allium cepa		0–80 mg/L increase ROS generation	Panda et al. (2011)

6.2.3 Plant Growth

Exogenous application of Ag-NPs imparts varied effects on plant growth (Fig. 6.1). In *Oryza sativa,* application of 30 μg/ml Ag-NPs accelerated root growth whereas 60 μg/ml restricted root growth (Mirzajani et al. 2013) (Table 6.3). Higher percentage root growth and dry mass production was observed in Ag-NP-treated plants (Tomacheski et al. 2017). In wheat, application of 50 mg/L Ag-NPs increased length and dry weight of shoots (Mehta et al. 2016) (Table 6.3). In cowpea, 50 mg/L Ag-NPs increased most growth biomarkers, such as shoot and root length, fresh and dry weight of roots and shoots and root nodulation (Mehta et al. 2016) (Table 6.3). In *Brassica,* 75 mg/L Ag-NPs provided optimum results for almost all growth biomarkers (Mehta et al. 2016) (Table 6.3). In water hyacinth, an increase in protein content was observed with 100 mg/L Ag-NPs (Rani et al. 2016) (Table 6.2), an improved growth was observed. Application of 30 mg/L Ag-NPs promoted growth in rice, maize and bean whereas growth was inhibited at higher concentrations (Salama 2012; Mirzajani et al. 2013) (Tables 6.2 and 6.3). Exogenous application of Ag-NPs (0.2 mg/seedling) increased root and shoot length, wet weight and leaf number in fenugreek (Jasim et al. 2017) (Table 6.3). In maize and bean, 20–60 mg/kg Ag-NPs increased leaf area as well as length of roots and shoots. However, higher concentrations (>60 mg/kg Ag-NPs) were inhibitory (Salama 2012) (Table 6.2). In *Brassica juncea* seedlings, application of 25 and 50 mg/L enhanced shoot fresh mass, length of shoots and roots as well as vigor index (Sharma et al. 2012) (Table 6.1). In *Eruca sativa,* supplementation of 1–20 mg/L Ag-NPs enhance root length (Vannini et al. 2013) (Table 6.3). Exogenously sourced Ag-NPs were correlated with increased root length, fresh and dry weight of corn, watermelon and zucchini in a concentration-dependent manner (Almutairi and Alharbi 2015) (Table 6.1). Exogenous application of 20–60 mg/L Ag-NPs increased shoot growth, root length and leaf area in *Zea mays* and *Phaseolus vulgaris.* However, a marked inhibition was observed at 80–100 mg/L Ag-NPs (Salama 2012) (Table 6.2). In *Pennisetum glaucum,* 20–50 mg/L Ag-NPs increased seedling growth (Parveen and Rao 2015) (Table 6.1). In *Arabidopsis thaliana,* an improved growth as reported in Ag-NP-treated plants (Geisler-Lee et al. 2014). Also in *Arabidopsis thaliana*, application of 0.01–100 mg/L Ag-NPs was found to increase root length and biomass production (Wang et al. 2013). In *Zea mays*, enhanced root length and percent germination rate was observed under different concentrations of Ag-NPs (Pokhrel and Dubey 2013). In fenugreek, 0–40 μg ml^{-1} Ag-NPs enhanced root length, root fresh and dry weight (Hojjat and Hojjat 2015) (Table 6.2). In *Lupinus termis,* application 100 mg/L Ag-NPs increased root and shoot length, root and shoot fresh weight, and fresh and dry weight of seedlings (Al-Huqail et al. 2018) (Table 6.3). In *Triticum aestivum*, application of Ag-NPs enhance root branching (Dimkpa et al. 2013) (Table 6.3). However, at 300 and 500 mg/L Ag-NPs all parameters were reduced (Al-Huqail et al. 2018) (Table 6.3).

Negative effects on growth of numerous species from Ag-NP application have been documented. In *Triticum aestivum*, 2.5 mg/kg Ag-NPs led to reduction in root and shoot length in a dose-dependent manner (Dimkpa et al. 2013) (Table 6.3). Also in *T. aestivum*, 20, 200 and 2000 mg/kg Ag-NPs resulted in lower biomass, shorter plant height and lower plant weight (Yang et al. 2018) (Table 6.3). In soybean, application of 1, 10 and 30 mg/L Ag-NPs significantly reduced plant biomass (Li et al. 2017) (Table 6.4). Application of 1000 and 3000 μM led to a decline in growth in pea (Tripathi et al. 2017) (Table 6.2). In *Vicia faba*, supplementation with 100–900 μg/kg Ag-NPs decreased shoot/root ratio, leaf area and length and dry weight of shoot and root (Abd-Alla et al. 2016) (Table 6.2). Exogenous application of 0.2 μg/l Ag-NPs inhibited root hair development in *Arabidopsis thaliana* (Garcia-Sanchez et al. 2015). Different concentrations of Ag-NPs (0.2, 0.5 and 3 mg/L) inhibited root growth in this species (Qian et al. 2013) (Table 6.3). Also in *A. thaliana,* 75–300 μg/l Ag-NPs led to prolongation in vegetative growth and curtailment of reproductive growth (Geisler-Lee et al. 2014). Application of 67–535 μg/l Ag-NPs inhibited seedling root elongation in *A. thaliana* (Geisler-Lee et al. 2012). In *Capsicum annuum*, application of 1 mg/L decreased plant height and biomass (Vinkovic et al. 2018). In cowpea, 0–20 mg/L Ag-NPs reduced growth by 52% (Wang et al. 2015).

6.2.4 Reactive Oxygen Species Generation

In tobacco, 500 μM Ag-NPs decreased ROS and MDA content (Cvjetko et al. 2018). In contrast, exogenously sourced 25 and 50 mg/L Ag-NPs increased MDA and H_2O_2 content in *Brassica juncea* seedlings (Sharma et al. 2012) (Table 6.1). In *Allium cepa,* 0–80 mg/L Ag-NPs led to ROS generation that damages DNA structure and ultimately led to cell death (Panda et al. 2011) (Table 6.3). In soybean, application of 1, 10 and 30 mg/L Ag-NPs significantly increased MDA and H_2O_2 content in leaves (Li et al. 2017) (Table 6.4). In rice leaves, 0.1, 0.5 and 1 mg/L Ag-NPs enhanced MDA and H_2O_2 content (Li et al. 2017) (Table 6.4). In *Solanum tuberosum*, different concentrations of Ag-NPs (2, 10 and 20 mg/L) increased total ROS (Homaee and Ehsanpour 2016) (Table 6.4). Exogenous application of 100–5000 mg/L Ag-NPs induced ROS accumulation in *Arabidopsis thaliana* (Sosan et al. 2016) (Table 6.2).

6.2.5 Accumulation of Compatible Solutes

In water hyacinth, application of 1–10 mg/L Ag-NPs increased carbohydrate content (Rani et al. 2016) (Table 6.2). Application of 20–60 mg/L Ag-NPs enhanced carbohydrate content in *Zea mays* and *Phaseolus vulgaris* in a concentration-dependent manner. However, a significant inhibition was observed at higher concentrations (80–100 mg/L) (Salama 2012) (Table 6.2). In *Oryza sativa*, different concentrations of Ag-NPs (1 mg/L) enhanced sugar content (Nair and Chung 2014)

Table 6.4 Effects of exogenous application of Ag-NPs in different plant species

	Dose-dependent effect of Ag-NPs		
Plant species	Positive effects	Negative effects	References
Glycine max		1,10 and 30 mg/L increase MDA and H_2O_2 content	Li et al. (2017)
Oryza sativa		0.1, 0.5 and 1 mg/L Ag-NPs enhanced MDA and H_2O_2 content	Li et al. (2017)
Solanum tuberosum	2,10 and 20 mg/L Ag-NPs increase the activities of SOD, CAT, APX and GR	Ag-NPs (2,10 and 20 mg/L) increased total ROS	Bagherzadeh homaee and Ehsanpour (2016)
Triticum aestivum	Ag-NPs increase fresh and dry weight, total chlorophyll content, soluble sugar content and antioxidant enzymes under salt stress; Ag-NPs lowered MDA and H_2O_2 content; an enhancement in total soluble sugars and proline content was reported		Mohamed et al. (2017)
Wolffia globosa	Enhanced activities of CAT and POX were observed at 10 mg/L Ag-NPs		Zou et al. (2017)
Lycopersicon esculentum	Ag-NPs enhance germination under salinity stress		Almutairi (2016)
Lathyrus sativus	5–15 mg/L Ag-NPs enhance germination speed index		Hojjat (2019)
Thymus vulgaris and T. daenensis	0–10 mm/L Ag-NPs increase germination percentage, shoot and root length, and seed vigor in under salinity stress		Ghavam (2018)
Foeniculum vulgare	20 mg/kg of nano-silver particles showed highest germination percentage, germination speed and vigor of stem		Ekhtiyari et al. (2011)
Lycopersicon esculentum	20 mg/L Ag-NPs increase percentage plant survival at different levels of salinity; Varied concentrations of Ag-NPs improved average fruit diameter and weight, number of branches and plant height		Younes and Nassef (2015)
Cuminum cyminum	10–30 mm/L Ag-NPs enhance germination percentage, shoot length, root length and seed vigor		Ekhtiyari and Moraghebi (2011)

(Table 6.2). Exogenous application of 30–60 μg/ml Ag-NPs reduce total carbohydrate content in *Oryza sativa* (Mirzajani et al. 2013) (Table 6.3). However, an enhanced production in reducing sugar content was observed (Mirzajani et al. 2013) (Table 6.3).

6.2.6 Enzymatic and Non-enzymatic Antioxidants

Exogenous application of Ag-NPs results in an improved antioxidative defense system (Fig. 6.1). Ag-NPs have the potential to modify oxidation and antioxidant homeostasis in plants (Yang et al. 2017) (Table 6.3). In *Triticum aestivum,* Ag-NPs limited oxidative stress via adjustment of the antioxidative defense system in a concentration-dependent manner (Mohamed et al. 2017) (Table 6.4). Also in *T. aestivum,* 0–5 mg/kg Ag-NPs led to accumulation of oxidized glutathione disulfide (GSSG) in a dose-dependent manner (Dimkpa et al. 2013) (Table 6.3). In water hyacinth, supplementation with 1–100 mg/L Ag-NPs increased SOD activity whereas a reduction in CAT and POX activity was observed at similar concentrations (Rani et al. 2016) (Table 6.2). Enhanced activities of antioxidant enzymes were observed at high concentrations of Ag-NPs (50, 100, 200 and 400 mg/L) in *Brassica juncea*, and increased proline content was noted (Sharma et al. 2012) (Table 6.1). In *Wolffia globosa,* enhanced activities of CAT and POX were observed at 10 mg/L Ag-NPs (Zou et al. 2017) (Table 6.4). Exogenous application of 1000 and 3000 μM Ag-NPs inhibited the activities of glutathione reductase (GR) and dehydroascorbate reductase (DHAR) in pea (Tripathi et al. 2017) (Table 6.2). Different concentrations of Ag-NPs (2, 10 and 20 mg/L) increased activities of SOD, CAT, ascorbate peroxidase (APX) and GR (Homaee and Ehsanpour 2016) (Table 6.4). In *Oryza sativa,* Ag-NPs enhanced proline accumulation (Nair and Chung 2014; Table 6.2).

6.3 Effect of Silver Nanoparticles Under Plant Abiotic Stress

6.3.1 Salt Stress

Previous studies have revealed positive effects of Ag-NPs under salinity stress (Latef et al. 2017). In *Lycopersicon esculentum,* Ag-NPs enhanced germination under salinity stress (Almutairi 2016) (Table 6.4). Application of different concentrations of Ag-NPs to *Triticum aestivum* increased fresh and dry weight, total chlorophyll content, soluble sugar content and antioxidant enzymes under salt stress (Mohamed et al. 2017) (Table 6.4). Application of Ag-NPs lowered MDA and H_2O_2 content in wheat (Mohamed et al. 2017) (Table 6.4). In *Lathyrus sativus*, application of 0–10 mg/L Ag-NPs enhanced seed germination, shoot and root length and proline content under salt stress (Hojjat 2019) (Table 6.4). Exogenous application of 0–10 mm/L Ag-NPs increased germination percentage, shoot and root length, and seed vigor in *Thymus vulgaris* and *T. daenensis* under salinity stress (Ghavam 2018) (Table 6.4). Exogenous application of 2 mg/L Ag-NPs enhanced root length, and seedling fresh and dry weight in tomato (Almutairi 2016) (Table 6.4). In *Foeniculum vulgare*, application of 20 mg/kg Ag-NPs improved germination percentage and vigor (Ekhtiyari et al. 2011) (Table 6.4). In tomato, 20 mg/L Ag-NPs increased percentage plant survival at different levels of salinity (Younes and Nassef 2015) (Table 6.4). Moreover, varied concentrations of Ag-NPs improved average fruit diameter and weight, number of branches and plant height (Younes and Nassef 2015) (Table 6.4). In *Cuminum cyminum*, 20 mg/kg Ag-NPs enhanced germination percentage, fresh weight as well as length of rootlets under salinity stress (Ekhtiyari and Moraghebi 2011) (Table 6.4). A dosage of 40 mg/kg Ag-NPs to *Ocimum basilicum* led to enhancement in germination percentage and improved resistance to salinity (Darvishzadeh 2015) (Table 6.5). In fenugreek, low concentrations of Ag-NPs enhanced germination percentage under different concentrations of NaCl (5, 10, 15 and 20 dS/m) (Hojjat and Kamyab 2017). Inhibition in root length of *Arabidopsis thaliana* under salt stress was alleviated by supplementation of Ag-NPs (Qian et al. 2013) (Table 6.3). An enhancement in total soluble sugars and proline content was reported (Mohamed et al. 2017) (Table 6.4). Ag-NPs and salt stress in combination increased activity of catalase (CAT) and peroxidase (POX) in wheat plants (Mohamed et al. 2017) (Table 6.4). In leaves of *Lycopersicon esculentum*, 75 mg/L Ag-NPs enhanced CAT and POX activity under salinity stress (Mehrian et al. 2015) (Table 6.5). In castor seedlings, Ag-NPs promoted the activities of SOD and POX under salt stress (Yasur and Rani 2013) (Table 6.5).

6.3.2 Flooding Stress

In saffron, application of 40 and 80 mg/L Ag-NPs increase root number under flooding stress. However, a marked reduction in root number was observed at 120 mg/L Ag-NPs (Rezvani et al. 2012) (Table 6.5). Root length was significantly increased at 40 mg/L and inhibited at 80 and 120 mg/L Ag-NPs (Rezvani et al. 2012) (Table 6.5). Root fresh and dry weight was significantly increased at 40 and 80 mg/L Ag-NPs whereas a marked reduction was reported at 120 mg/L Ag-NPs (Rezvani et al. 2012) (Table 6.5). Leaf bud number was significantly increased at 40, 80 and 120 mg/L Ag-NPs whereas leaf bud length showed a significant reduction at above-mentioned concentrations (Rezvani et al. 2012) (Table 6.5). In soybean, 2 mg/L Ag-NPs of various sizes (2, 15 and 50–80 nm) enhanced seedling weight under flooding (Mustafa et al. 2015) (Table 6.5).

Table 6.5 Effects of exogenous application of Ag-NPs in different plant species

	Dose-dependent effect of Ag-NPs		
Plant species	Positive effects	Negative effects	References
Ocimum basilicum	40 mg/kg Ag-NPs enhance germination percentage and improved resistance to salinity		Darvishzadeh (2015)
Lycopersicon esculentum	75 mg/L Ag-NPs enhanced CAT and POX activity under salinity stress		Mehrian et al. (2015)
Riccinus communis	Ag-NPs promoted the activities of SOD and POX under salt stress		Yasur and Rani (2013)
Crocus sativus	40 and 80 mg/L Ag-NPs increase root length; 40, 80 and 120 mg/L Ag-NPs increase leaf bud number; 40, 80 and 120 mg/L Ag-NPs increase root fresh and dry weight; 80 and 120 mg/L Ag-NPs increase leaf bud dry weight; 40 and 80 mg/L Ag-NPs increase root number under flooding stress	120 mg/L Ag-NPs decrease root length; 40 mg/L Ag-NPs decrease leaf bud dry weight; 40, 80 and 120 mg/L Ag-NPs decrease leaf bud dry weight; a marked reduction in root number was observed at 120 mg/L Ag-NPs	Rezvani et al. (2012)
Glycine max	2 mg/L Ag-NPs of various sizes (2,15 and 5080 nm) enhance seedling weight under flooding stress		Mustafa et al. (2015)
Triticum aestivum	50 and 75 mg/L Ag-NPs increase plant fresh weight; 25–100 mg/L Ag-NPs increase plant dry weight; 25–100 mg/L Ag-NPs increase root and shoot length; 25–100 mg/L Ag-NPs increase leaf number; 25–75 mg/L Ag-NPs increase leaf area; 25–100 mg/L Ag-NPs increase leaf fresh and dry weight	100 mg/L Ag-NPs decrease plant fresh weight, leaf area and leaf number	Iqbal et al. (2019)

6.3.3 *Heat Stress*

In wheat, application of Ag-NPs from 25–75 mg/L increased fresh and dry weight of plants under heat stress. However, a marked decline was observed at 100 mg/L (Iqbal et al. 2019) (Table 6.5). Root and shoot length also shows a significant increase at 25, 50 and 75 mg/L Ag-NPs whereas inhibited at 100 mg/L (Iqbal et al.

2019) (Table 6.5). A significant increase in leaf area, leaf number, leaf fresh weight and leaf dry weight was observed at 25, 50 and 75 mg/L Ag-NPs (Iqbal et al. 2019) (Table 6.5).

6.4 Conclusions

After summarizing the data, it is clear that Ag-NPs impart both positive and negative effects on plants. Additionally, Ag-NPs can improve growth, chlorophyll content, photosynthetic efficiency and antioxidative defense systems under abiotic stress. Despite being an active area of research, the effects of Ag-NPs on plants are still far from being conclusively studied. A well-established and coordinated research program is needed that clearly explains the optimal doses of nanoparticles to use for different crops under different environmental conditions. Undoubtedly, a rich area for future investigation remains.

References

Abd-Alla MH, Nafady NA, Khalaf DM (2016) Assessment of silver nanoparticles contamination on faba bean-Rhizobium leguminosarum bv. viciae-Glomus aggregatum symbiosis: implications for induction of autophagy process in root nodule. Agric Ecosyst Environ 218:163–177

Al-Huqail AA, Hatata MM, Al-Huqail AA, Ibrahim MM (2018) Preparation, characterization of silver phyto nanoparticles and their impact on growth potential of Lupinus termis L. seedlings. Saudi J Biol Sci 25(2):313–319

Almutairi ZM (2016) Influence of silver nano-particles on the salt resistance of tomato (Solanum lycopersicum) during germination. Int J Agric Biol 18(2):449–457

Almutairi ZM, Alharbi A (2015) Effect of silver nanoparticles on seed germination of crop plants. J Adv Agric 4(1):283–288

Cvjetko P, Zovko M, Štefanić PP, Biba R, Tkalec M, Domijan AM, Balen B (2018) Phytotoxic effects of silver nanoparticles in tobacco plants. Environ Sci Pollut Res 25(6):5590–5602

Darvishzadeh F (2015) Effects of silver nanoparticles on salinity tolerance in basil plant (Ocimum basilicum L.) during germination in vitro. New Cell Mol Biotechnol J 5(20):63–70

Dimkpa CO, McLean JE, Martineau N, Britt DW, Haverkamp R, Anderson AJ (2013) Silver nanoparticles disrupt wheat (Triticum aestivum L.) growth in a sand matrix. Environ Sci Technol 47(2):1082–1090

Durán N, Marcato PD, De Souza GI, Alves OL, Esposito E (2007) Antibacterial effect of silver nanoparticles produced by fungal process on textile fabrics and their effluent treatment. J Biomed Nanotechnol 3(2):203–208

Ekhtiyari R, Moraghebi F (2011) The study of the effects of nano silver technology on salinity tolerance of cumin seed (Cuminum cyminum L.). Plant Ecosyst 7(25):99–107

Ekhtiyari R, Mohebbi H, Mansouri M (2011) The study of the effects of nano silver technology on salinity tolerance of (Foeniculum vulgare mill.). Plant Ecosyst 7(27):55–62

El-Temsah YS, Joner EJ (2012) Impact of Fe and Ag nanoparticles on seed germination and differences in bioavailability during exposure in aqueous suspension and soil. Environ Toxicol 27(1):42–49

García-Sánchez S, Bernales I, Cristobal S (2015) Early response to nanoparticles in the Arabidopsis transcriptome compromises plant defence and root-hair development through salicylic acid signalling. BMC Genomics 16:1–16
Geisler-Lee J, Wang Q, Yao Y, Zhang W, Geisler M, Li K, Ma X (2012) Phytotoxicity, accumulation and transport of silver nanoparticles by Arabidopsis thaliana. Nanotoxicology 7(3):323–337
Geisler-Lee J, Brooks M, Gerfen J, Wang Q, Fotis C, Sparer A, Geisler M (2014) Reproductive toxicity and life history study of silver nanoparticle effect, uptake and transport in Arabidopsis thaliana. Nanomaterials 4(2):301–318
Ghavam M (2018) Effect of silver nanoparticles on seed germination and seedling growth in Thymus vulgaris L. and Thymus daenensis Celak under salinity stress. J Rangeland Sci 8(1):93–100
Hojjat SS (2019) Effect of interaction between Ag nanoparticles and salinity on germination stages of Lathyrus sativus L. J Environ Soil Sci 2(2):186–191
Hojjat SS, Hojjat H (2015) Effect of nano silver on seed germination and seedling growth in fenugreek seed. Int J Food Eng 1(2):106–110
Hojjat SS, Kamyab M (2017) The effect of silver nanoparticle on Fenugreek seed germination under salinity levels. Russ Agric Sci 43(1):61–65
Homaee MB, Ehsanpour AA (2016) Silver nanoparticles and silver ions: oxidative stress responses and toxicity in potato (Solanum tuberosum L) grown in vitro. Hortic Environ Biotechnol 57(6):544–553
Iqbal M, Raja NI, Hussain M, Ejaz M, Yasmeen F (2019) Effect of silver nanoparticles on growth of wheat under heat stress. Iranian J Sci Technol Trans A Sci 43(2):387–395
Jasim B, Thomas R, Mathew J, Radhakrishnan EK (2017) Plant growth and diosgenin enhancement effect of silver nanoparticles in Fenugreek (Trigonella foenum-graecum L.). Saudi Pharm J 25(3):443–447
Latef AAHA, Alhmad MFA, Abdelfattah KE (2017) The possible roles of priming with ZnO nanoparticles in mitigation of salinity stress in lupine (Lupinus termis) plants. J Plant Growth Regul 36(1):60–70
Li CC, Dang F, Li M, Zhu M, Zhong H, Hintelmann H, Zhou DM (2017) Effects of exposure pathways on the accumulation and phytotoxicity of silver nanoparticles in soybean and rice. Nanotoxicology 11(5):699–709
Liang L, Tang H, Deng Z, Liu Y, Chen X, Wang H (2018) Ag nanoparticles inhibit the growth of the bryophyte, Physcomitrella patens. Ecotoxicol Environ Saf 164:739–748
Mehrian SK, Heidari R, Rahmani F (2015) Effect of silver nanoparticles on free amino acids content and antioxidant defense system of tomato plants. Indian J Plant Physiol 20(3):257–263
Mehta CM, Srivastava R, Arora S, Sharma AK (2016) Impact assessment of silver nanoparticles on plant growth and soil bacterial diversity. 3 Biotech 6(2):254
Mirzajani F, Askari H, Hamzelou S, Farzaneh M, Ghassempour A (2013) Effect of silver nanoparticles on Oryza sativa L. and its rhizosphere bacteria. Ecotoxicol Environ Saf 88:48–54
Mohamed AKS, Qayyum MF, Abdel-Hadi AM, Rehman RA, Ali S, Rizwan M (2017) Interactive effect of salinity and silver nanoparticles on photosynthetic and biochemical parameters of wheat. Arch Agron Soil Sci 63(12):1736–1747
Mueller NC, Nowack B (2008) Exposure modeling of engineered nanoparticles in the environment. Environ Sci Technol 42(12):4447–4453
Mustafa G, Sakata K, Hossain Z, Komatsu S (2015) Proteomic study on the effects of silver nanoparticles on soybean under flooding stress. J Proteome 122:100–118
Nair PMG, Chung IM (2014) Physiological and molecular level effects of silver nanoparticles exposure in rice (Oryza sativa L.) seedlings. Chemosphere 112:105–113
Olchowik J, Bzdyk R, Studnicki M, Bederska-Błaszczyk M, Urban A, Aleksandrowicz-Trzcińska M (2017) The effect of silver and copper nanoparticles on the condition of english oak (Quercus robur L.) seedlings in a container nursery experiment. Forests 8(9):310

Panda KK, Achary VMM, Krishnaveni R, Padhi BK, Sarangi SN, Sahu SN, Panda BB (2011) In vitro biosynthesis and genotoxicity bioassay of silver nanoparticles using plants. Toxicol In Vitro 25(5):1097–1105
Pandey C, Khan E, Mishra A, Sardar M, Gupta M (2014) Silver nanoparticles and its effect on seed germination and physiology in Brassica juncea L.(Indian mustard) plant. Adv Sci Lett 20(7–8):1673–1676
Panyala NR, Peña-Méndez EM, Havel J (2008) Silver or silver nanoparticles: a hazardous threat to the environment and human health? J Appl Biomed (De Gruyter Open) 6(3):117–129
Parveen A, Rao S (2015) Effect of nanosilver on seed germination and seedling growth in Pennisetum glaucum. J Clust Sci 26(3):693–701
Pokhrel LR, Dubey B (2013) Evaluation of developmental responses of two crop plants exposed to silver and zinc oxide nanoparticles. Sci Total Environ 452:321–332
Qian H, Peng X, Han X, Ren J, Sun L, Fu Z (2013) Comparison of the toxicity of silver nanoparticles and silver ions on the growth of terrestrial plant model Arabidopsis thaliana. J Environ Sci 25(9):1947–1956
Rani PU, Yasur J, Loke KS, Dutta D (2016) Effect of synthetic and biosynthesized silver nanoparticles on growth, physiology and oxidative stress of water hyacinth: Eichhornia crassipes (Mart) Solms. Acta Physiol Plant 38(2):58
Rastogi A, Zivcak M, Sytar O, Kalaji HM, He X, Mbarki S, Brestic M (2017) Impact of metal and metal oxide nanoparticles on plant: a critical review. Front Chem 5:78
Rezvani N, Sorooshzadeh A, Farhadi N (2012) Effect of nano-silver on growth of saffron in flooding stress. World Acad Sci Eng Technol 6(1):517–522
Salama HM (2012) Effects of silver nanoparticles in some crop plants, common bean (Phaseolus vulgaris L.) and corn (Zea mays L.). Int J Biotechnol Res 3(10):190–197
Savithramma N, Rao ML, Rukmini K, Devi PS (2011) Antimicrobial activity of silver nanoparticles synthesized by using medicinal plants. Int J ChemTech Res 3(3):1394–1402
Savithramma N, Ankanna S, Bhumi G (2012) Effect of nanoparticles on seed germination and seedling growth of Boswellia ovalifoliolata an endemic and endangered medicinal tree taxon. Nano Vision 2(1):2
Sharma P, Bhatt D, Zaidi MGH, Saradhi PP, Khanna PK, Arora S (2012) Silver nanoparticle-mediated enhancement in growth and antioxidant status of Brassica juncea. Appl Biochem Biotechnol 167(8):2225–2233
Sosan A, Svistunenko D, Straltsova D, Tsiurkina K, Smolich I, Lawson T, Colbeck I (2016) Engineered silver nanoparticles are sensed at the plasma membrane and dramatically modify the physiology of Arabidopsis thaliana plants. Plant J 85(2):245–257
Thuesombat P, Hannongbua S, Akasit S, Chadchawan S (2014) Effect of silver nanoparticles on rice (Oryza sativa L. cv. KDML 105) seed germination and seedling growth. Ecotoxicol Environ Saf 104:302–309
Tomacheski D, Pittol M, Simões DN, Ribeiro VF, Santana RMC (2017) Impact of silver ions and silver nanoparticles on the plant growth and soil microorganisms. Glob J Environ Sci Manag 3(4):341–350
Tripathi DK, Singh S, Singh S, Srivastava PK, Singh VP, Singh S, Chauhan DK (2017) Nitric oxide alleviates silver nanoparticles (AgNps)-induced phytotoxicity in Pisum sativum seedlings. Plant Physiol Biochem 110:167–177
Vannini C, Domingo G, Onelli E, Prinsi B, Marsoni M, Espen L, Bracale M (2013) Morphological and proteomic responses of Eruca sativa exposed to silver nanoparticles or silver nitrate. PLoS One 8(7):e68752
Vinković T, Štolfa Čamagajevac I, Tkalec M, Goessler W, Domazet Jurašin D, Vinković Vrček I (2018) Does plant growing condition affects biodistribution and biological effects of silver nanoparticles? Span J Agric Res 16:1–13
Wang J, Koo Y, Alexander A, Yang Y, Westerhof S, Zhang Q, Alvarez PJ (2013) Phytostimulation of poplars and Arabidopsis exposed to silver nanoparticles and Ag+ at sublethal concentrations. Environ Sci Technol 47(10):5442–5449

Wang P, Menzies NW, Lombi E, Sekine R, Blamey FPC, Hernandez-Soriano MC, Kopittke PM (2015) Silver sulfide nanoparticles (Ag2S-NPs) are taken up by plants and are phytotoxic. Nanotoxicology 9(8):1041–1049

Yang J, Cao W, Rui Y (2017) Interactions between nanoparticles and plants: phytotoxicity and defense mechanisms. J Plant Interact 12(1):158–169

Yang J, Jiang F, Ma C, Rui Y, Rui M, Adeel M, Xing B (2018) Alteration of crop yield and quality of wheat upon exposure to silver nanoparticles in a life cycle study. J Agric Food Chem 66(11):2589–2597

Yasur J, Rani PU (2013) Environmental effects of nanosilver: impact on castor seed germination, seedling growth, and plant physiology. Environ Sci Pollut Res 20(12):8636–8648

Yin L, Colman BP, McGill BM, Wright JP, Bernhardt ES (2012) Effects of silver nanoparticle exposure on germination and early growth of eleven wetland plants. PLoS One 7(10):e47674

Younes NA, Nassef DM (2015) Effect of silver nanoparticles on salt tolerancy of tomato transplants (Solanum lycopersicom L. Mill.). *Assiut*. J Agric Sci 46:76–85

Zou X, Li P, Lou J, Zhang H (2017) Surface coating-modulated toxic responses to silver nanoparticles in Wolffia globosa. Aquat Toxicol 189:150–158

Chapter 7
Silicon Nanoparticles and Plants: Current Knowledge and Future Perspectives

Husna Siddiqui, Khan Bilal Mukhtar Ahmed, Fareen Sami, and Shamsul Hayat

Abstract The use of nanotechnology in agriculture is increasing at a phenomenal rate. It is, therefore, necessary to appreciate and elucidate the role of nanoparticles (NPs) in plant growth and development. Silicon is regarded as a 'quasi-essential' element for plants and regulates a range of physiological processes including germination, vegetative growth, photosynthesis and stress tolerance. It is, therefore, of importance to assess the effects of silicon nanoparticles (SNPs) on these physiological processes, as SNPs are considered more efficient than their bulk particles due to their small size and high surface area and reactivity. The present chapter deals with the role of SNPs in plant growth, photosynthesis and stress tolerance. Additionally, potential toxic effects of NPs are presented.

Keywords Antioxidants · Drought · Germination · Growth · Heavy metals · Nanotechnology · OsNAC protein · *OsHMA3* · Oxidative stress · Photosynthesis · Protein · Salinity · Silicon

7.1 Introduction

Meeting the demands of ever-increasing world population requires a concurrent increase in crop yield. Use of fertilizer, herbicides and pesticides to enhance crop production per unit area is a common strategy adopted by farmers. Synthetic fertilizers have proven beneficial in improving crop productivity; however, they may also cause nutrient imbalances and are often costly. Heavy use of chemical fertilizers may be detrimental to local environments, for example by leaching and runoff to nearby water bodies. The development of an alternative eco-friendly approach that

---

H. Siddiqui (✉) · K. B. M. Ahmed · F. Sami · S. Hayat
Plant Physiology & Biochemistry Section, Department of Botany, Faculty of Life Sciences, Aligarh Muslim University, Aligarh, Uttar Pradesh, India

S. Hayat et al. (eds.), *Sustainable Agriculture Reviews 41*, Sustainable Agriculture Reviews 41, https://doi.org/10.1007/978-3-030-33996-8_7

could address mineral deficiencies, and plant growth and development is necessary. One such approach is to employ nanotechnology. The synthesis and use of nanoparticles is of substantial interest in recent research, due to the remarkable progress of nano-science and nanotechnology and the broad applications of nanomaterials (Huan and Shu-Qing 2014).

Nanotechnology is an emerging tool that offers remarkable applications to biotechnology, agriculture, and other disciplines. In agriculture, nanotechnology has the potential to increase food quality, global food production, plant protection, detection of plant and animal diseases, monitoring of plant growth and reduction of waste (Gruere et al. 2011; Frewer et al. 2011; Bagchi et al. 2012; Prasad et al. 2014; Biswal et al. 2012; Ditta 2012; Sonkaria et al. 2012). Due to their small size, high surface area to weight ratio, and different shapes, nanoparticles (NPs) exhibit markedly different properties than their bulk counterparts (Roduner 2006). The high surface to volume ratio increases their reactivity including biochemical activity (Dubchak et al. 2010). By virtue of their small size, they have the potential to cross cell walls and plasma membranes to facilitate effective absorption (Monica and Cremonini 2009). Therefore, NPs can be used to increase the supply of elements to plants.

Silicon (Si), the second-most abundant element found in the earth's crust, is a metalloid considered beneficial to plants (Epstein 1994; Luyckx et al. 2017; Siddiqui et al. 2018). Silicon is absorbed in the form of mono-silicic acid by plants and is transported across the plant via different transporter governed by genes such as *LSi1*, *LSi2*, and *LSi6* (Rao and Susmitha 2017). Silicon is known to be deposited on the epidermal wall and vascular tissue of the stem, leaf and sheath in plants, especially monocots (Ma and Yamaji 2006; Currie and Perry 2007; Parveen and Ashraf 2010). Silicon is also known to regulate numerous physiological activities in various flora (Bao-Shan et al. 2004). However, Si nanoparticles (Si-NPs) present different physio-chemical properties as compared to bulk formulations (O'Farrell et al. 2006). Si-NP-mediated response is a function of the size, shape, physio-chemical properties, and method of application of the NPs (Rastogi et al. 2017).

7.2 Silicon Nanoparticle-Mediated Regulation of Physiological Processes

7.2.1 Germination

The use of NPs for improving seed germination and seedling growth is well documented (Zheng et al. 2005; Shah and Belozerova 2009; Siddiqui and Al-Whaibi 2014; Sabaghnia and Janmohammadi 2014). In tomato (*Lycopersicon esculentum*), Si-NPs improve seed germination, mean germination time, seed germination index, seed vigor index, seedling fresh weight and dry weight when supplied at low concentrations (Siddiqui and Al-Whaibi 2014). In maize, Si-NP-mediated increases in seed germination are attributed to improved availability of nutrients to seeds (Suriyaprabha et al. 2012). In Changbai larch (*Larixolgensis*) seedlings, Si-NPs improved seedling growth and quality including mean height, root collar diameter, main root length, and number of lateral roots of seedlings. The Si-NPs also induced synthesis of chlorophyll (Bao-shan et al. 2004). Seed germination traits including percent germination and germination rate, length, fresh and dry mass of root and shoot of *Thymus kotschyanus* seedlings increased after application of nano-Si (Khalaki et al. 2016). Pre-chilling seeds with Si-NPs break dormancy and promote seed germination, and increase dry weight of seedling roots and shoots in tall wheatgrass (*Thinopyrum intermedium*) (Azimi et al. 2014). Moreover, vigor index and mean germination time increased in the presence of Si-NPs. Seed priming as well as seed soaking in Si-NPs increase seedling root and shoot length along with seedling biomass and vigor index of *Helianthus annuus* (Janmohammadi and Sabaghnia 2015). Germination and growth of soybean (*Glycine max* L.) was improved when a mixture of nano-size silicon dioxide was incorporated into the growth medium. The ability to absorb water and fertilizer increased due to enhanced nitrate reductase (NR) activity and stimulation of antioxidant systems (Lu et al. 2002). Promotive effects of Si-NPs on percent seed germination, length and dry weight of root and shoot has been reported in rice (*Oryza sativa*) seedlings (Adhikari et al. 2013). SiO_2 NPs improved the activity of carbonic anhydrase and synthesis of photosynthetic pigments that resulted in elevated photosynthetic rate of seedlings (Siddiqui et al. 2014; Xie et al. 2012). It is clear, therefore, that nano-SiO_2 has a significant impact on seed germination potential.

7.2.2 Growth

Fitriani and Haryanti (2016) reported that different concentrations of Si-NPs, when used as fertilizer, promoted plant height, leaf number, and root length of *Solanum lycopersicum*. Nano-silica fertilizer promoted net assimilation rate (NAR), leaf area index (LAI), relative growth rate (RGR) and yield of soybean plants but did not affect height, number of leaves or stem diameters of plants (Suciaty et al. 2018). It is suggested that accumulation of Si in leaves is beneficial in maintaining leaves upright and stretching leaf surfaces to capture maximum sunlight, thus optimizing photosynthesis. Increased photosynthesis will enhance photosynthate accumulation, which can be stored as a reserve in vegetative organs such as roots, leaves and stems (Putri et al. 2017). Si-NPs at low doses increased shoot and root fresh and dry weight of *Triticum aestivum* (Karimi and Mohsenzadeh 2016). Treating fenugreek seeds with 50 mg/L Si-NPs increased shoot length by 66% and seedling length by 55% compared to the control treatment (Ivani et al. 2018). Foliar spray of Si-NPs (0, 15, 30, 60 and 120 mg/L) increased the growth, yield and chemical composition of cucumber (*Cucumis sativus*) (Yassen et al. 2017). Growth parameters studied included plant height, fresh and dry weight of leaves/plant, number of leaves and fruits/plant, mean weight of fruit, fruit length and yield. Of all concentrations of

Si-NPs, 60 mg/L proved the most effective dose for increasing biomass (Yassen et al. 2017). A Si-NP-mediated increase in growth could be an outcome of elevated levels of free amino acids, protein, total nitrogen, phosphorus and potassium (Li et al. 2012).

7.2.3 Photosynthesis

Plants convert solar energy to chemical energy via photosynthesis. Only 2–4% of available radiation energy converted by plants is used in plant growth and development (Kirschbaum 2011). A prime focus of research at present is to improve photosynthetic efficiency of plants by gene manipulation and other techniques. Nanotechnology is also capable of improving the function of photosynthetic machinery. As Si bulk particles are known to improve photosynthetic efficiency (Siddiqui et al. 2018), Si-NPs have gained significant attention by researchers. Unfortunately, however, data related to the effects of NPs on the photosynthetic apparatus remain scarce.

The light absorbed by plants could be more efficiently utilized when treated with Si-NPs; consequent increases in content of photosynthetic pigments may occur as a natural response. In *Zea mays*, Si-NPs positively affected contents of chlorophyll (chl) a, b and carotenoids at all concentrations used (400, 2000 and 4000 mg/L); however, 400 mg/L resulted in maximal increase in all photosynthetic pigments in relation to the control (Sharifi et al. 2016). SiO_2 NPs improved photosynthesis, PSII activity, photochemical efficiency, electron transport rate, photochemical quenching, carbonic anhydrase activity, net photosynthetic rate, transpiration rate, stomatal conductance, and synthesis of photosynthetic pigments in *Indocalamus barbatus* and *Cucurbita pepo* (Siddiqui et al. 2014; Xie et al. 2012).

7.2.4 Cellular Redox Status during Abiotic Stress

Adverse climatic factors such as drought, heat, freezing, and soil contamination by salt and heavy metals comprise key growth stressors that significantly limit productivity and quality of crop species worldwide (Kumar 2013). Abiotic stress alters morphology, physiology and biochemistry of plants, ultimately reducing growth and productivity (Kumar 2013). Emerging information and novel approaches must continuously be applied in a timely and effective manner by both the research and applied agricultural communities. Nanotechnology is a promising strategy to cope with the above adverse scenarios and must be exploited for alleviation of abiotic stresses.

7.2.4.1 Salt Stress

Soil salinity is one of the primary environmental constraints worldwide, particularly in arid and semiarid regions, which limit plant growth and productivity in agricultural systems (Flowers 2004; Koca et al. 2007). The Salinity Laboratory of the U.S. Department of Agriculture (USDA) defines a saline soil as having an electrical conductivity of 4 dS/m or greater. Salinity stress in soils occurs due to natural accumulation of salts, poor irrigation practices and over-fertilization.

The inhibitory effect of salinity on growth, elongation and division of cells is attributed to alteration in water status of plants which ultimately causes cell death (Munns 2002). Salinity imparts both osmotic (cell hydration) and toxic (ion accumulation) impacts on plants (Desai et al. 2011). Salinity stress induces synthesis of abscisic acid (ABA) which is transported to guard cells and causes stomatal closure; photosynthesis becomes limited, with consequent photo-inhibition and oxidative stress. These effects lead to immediate inhibition of cell expansion, evident as inhibited plant growth and development, and accelerated senescence (Chinnusamy et al. 2006).

Salinity limits photosynthesis and carbon uptake for dark reactions, causing a reduction in the electron transport during photosynthesis, and channelization of photon energy towards processes involved in reactive oxygen species (ROS) generation (Hichem and Mounir 2009). Singlet oxygen, superoxide radical (O_2^-), hydroxyl radical (OH), and hydrogen peroxide (H_2O_2) are some of the major ROS generated in plants (Zushi et al. 2009) that can cause peroxidation of membrane lipids and consequent destruction to proteins, DNA, and lipids (Pompelli et al. 2010; Sgherri et al. 2000; Vardharajula et al. 2012). Over-generation of ROS causes oxidative damage to cellular components, leading to cell death (Noctor and Foyer 1998).

Si-NPs improved water use efficiency and relative water content thereby increasing turgor pressure (Rawson et al. 1988). One gram of Si-NPs, diameter 7 nm, has an absorption surface of approximately 400 m^2; hence, Si-NP application affects water translocation that ultimately improves water use efficiency (Wang and Naser 1994).

Salinity stress increases osmotic potential and increases Na^+ ion accumulation which results in stunted growth. However, Si-NPs reduce Na toxicity by reducing Na absorption, resulting in restoration of vigorous growth (Raven 1983). Application of Si-NPs increased shoot fresh and dry weight of basil (*Ocimum basilicum*) under saline conditions (Kalteh et al. 2018). Salinity results in marked reduction in germination percentage, length of roots and shoots, seedling fresh weight and dry weight. However, application of Si-NPs in salt-stressed lentil allowed for growth attributes to return to optimal levels (Sabaghnia and Janmohammadi 2014). An increase in seed germination under saline conditions might be due to absorption and utilization of Si-NPs by seeds (Suriyaprabha et al. 2012). Content of nitrogen, phosphorus and potassium (NPK) in leaves decreased under saline conditions; however, application of different concentrations of Si-NPs (15, 30, 60 and 120 mg L^{-1}) restored the NPK content (Yassen et al. 2017). Si-NPs significantly mitigated symptoms of salinity stress in *Capsicum annum* L. and increased the values of various growth (plant height, leaves fresh and dry weight) and physiological parameters (chl and leaf NPK content) (Tantawy et al. 2015).

7.2.4.2 Water Stress

Water is necessary for plant survival – it is required for maintaining cell turgor pressure, transport of nutrients, and as a reactant in numerous processes. During drought conditions plants weaken, resulting in restricted growth and high mortality (Martinez-Vilata and Pinol 2002; Bigler et al. 2007; Mahajan and Tuteja 2005). Upon encountering drought conditions plants opt for either avoidance or tolerance; these strategies include morphological and/or physiological adjustments (Bassett 2013).

Photosynthesis and stomatal conductance of hawthorn (*Crataegus* sp.) increased during drought stress in the presence of Si-NPs (Ashkavand et al. 2015; Pei et al. 2010). Leaf water potential acts as a prime indicator for estimating degree of plant stress during water deficits (McCutchan and Shackel 1992). Xylem water potential decreased in seedlings suffering from drought stress; however, application of Si-NPs mitigated the effect of drought on xylem water potential (Ashkavand et al. 2015).

Relative water content (RWC), which operates complementary to xylem water potential, is used to evaluate the water status of plants (Zarafshar et al. 2014). RWC is affected by water availability but was not influenced by Si-NPs (Ashkavand et al. 2015). During water stress, both cell membrane permeability and selectivity change, where the former increases and latter decreases, which is evident from increased electrolyte leakage (Blokhina et al. 2003). We found a general increase of electrolyte leakage under severe drought stress, which suggests the occurrence of cell membrane damage (Campos et al. 2003).

Estimation of malondialdehyde (MDA) content is a useful indicator of oxidative damage to membrane lipids, which increases under severe drought conditions. Treating plants with Si-NPs decreased MDA content (Ozkur et al. 2009; Ashkavand et al. 2015). Decreased content of leaf pigment (chl and carotenoid) is an indicator of drought stress severity which was restored upon Si-NP application to plants (Egert and Tevini 2002; Ashkavand et al. 2015). Si-NP application increased chlorophyll content in *Zea mays* L. and *Larix olgensis* seedlings (Yuvakkumar et al. 2011; Bao-shan et al. 2004). The increase in pigment, growth and yield of plants upon Si-NP supplementation under drought conditions might be due to an increase in nitrogen, potassium and silicon content in plant tissue (Alsaeedi et al. 2019). Controlled-release fertilizer (CRF) containing Si-NPs in the core is proficient in gradually releasing nutrients and withholding substantial quantities of water in soil; this could help plants tolerate drought conditions (Mushtaq et al. 2018).

7.2.4.3 Heavy Metals

Cadmium (Cd) is one of the most toxic, persistent and bio-accumulative heavy metals (Yousaf et al. 2016; Shoeva and Khlestkina 2018). It is harmful even at low concentrations. Cadmium enters soil in industrial and mining waste, via sewage sludge release to water bodies and soil, and atmospheric deposition (Rizwan et al. 2018). Cadmium reduces photosynthesis, growth and yield of plants (Hayat et al. 2010). Si-NP application to plants in the presence of Cd has been found to enhance biomass and growth (Hussain et al. 2019). Si-NP-mediated increases in plant growth might be due to its alleviative role under heavy metal stress (Tripathi et al. 2015). NPs promote nutrient availability to plants and hence could serve as the driver behind NP-mediated enhancement of growth (Liu and Lal 2015).

Silicon enhances protein content and protects cell membranes from injury (Nazaralian et al. 2017; Merwad et al. 2018). Moreover, Si accumulates in leaf apoplasts and acts as a barrier to protect plants from stress, thus promoting growth (Silva et al. 2017). Si-NPs protect the photosynthetic apparatus from Cd toxicity, which is possibly due to Si-mediated enhancement of accumulation of nutrients and water leading to opening of the xylem and cell wall thickening (Hussain et al. 2019; Asgari et al. 2018; Gao et al. 2018). In the presence of Cd, Si-NPs decreased ROS content and increased antioxidant enzyme activities to protect cell membrane integrity, plausibly due to restricted Cd entry into plants (Hussain et al. 2019). Cui et al. (2017) reported that survival of rice cells under Cd toxicity was observed to be dependent on size of Si-NPs. NPs up-regulate gene expression for Si uptake (OsLsi1) and Cd transport to vacuoles (OsHMA3), and down-regulate the genes responsible for Cd uptake (OsLCT1 and OsNRAMP5).

Arsenic (As) is a toxic metalloid that adversely affects plant growth by decreasing photosynthesis and increasing ROS generation (Tripathi et al. 2016). However, application of Si-NPs prevents damage and restores the photosynthetic machinery. Si-NPs also enhance the activities of antioxidant enzymes to counter ROS generation (Tripathi et al. 2016).

Si-NPs improved growth of pea seedlings experiencing chromium (Cr) toxicity. Chromium accumulates in plant cells and accelerates the accumulation of ROS that alters photosynthetic activity as well as nutrient uptake, ultimately reducing plant growth (Gangwar and Singh 2011). Tripathi et al. (2015) demonstrated that Si-NP addition to Cr-stressed plants ameliorates Cr-induced toxicity symptoms related to chlorophyll florescence, pigment content, and protein and nutrient status of the plant, resulting in improved growth. The improvement in physiological parameters is accompanied by a marked reduction in Cr accumulation in plant parts. The enhanced stress tolerance, mediated by the Si-NPs, could be attributed to the ability of Si to up-regulate the expression of osNAC proteins that are responsible for up-regulation of genes for stress tolerance, proline synthesis, soluble sugar biosynthesis and redox homeostasis (Manivannan and Ahn 2017).

7.3 Si Nanoparticle-Generated Phytotoxicity

The use of NPs in agriculture has become an area of great interest for agronomists and plant biology researchers (Haghighi and Pessarakli 2013; Pourkhaloee et al. 2011). At present, results demonstrating phytotoxicity of NPs are contradictory (Dietz and Herth 2011; Miralles et al. 2012; Judy and Bertsch 2014). Some studies show no toxicity symptoms of Si-NPs in plants while others reveal negative effects on plants (Bao-shan et al. 2004; Siddiqui and Al-Whaibi 2014). Lu et al. (2002) reported that Si-NPs did not impart any toxic effects to *Glycine max*. Bao-Shan et al. (2004) observed that growth and quality of *Larix* seedlings increased in the presence of Si-NPs. Si-NP-mediated toxicity was reported by Lee et al. (2010) on *Arabidopsis thaliana*; however, the toxicity was not as strong as that of Zn and Fe-NPs. High doses of Si-NPs decreased fresh and dry weight, and volume of roots in tomato when compared with control plants (Haghighi and Pessarakli 2013).

There exist very few reports where Si-NPs were proved to have a negative impact on plants (Slomberg and Schoenfisch 2012). High levels of Si-NPs are known to induce cytotoxic effects in meristematic cells of *Allium cepa* (Silva and Monteiro 2017) where cytotoxicity was confirmed by the decline of the mitotic index. Si-NPs interfere with regular mitosis, preventing cells from entering prophase and obstructing the mitotic cycle during interphase. This effect constrains synthesis of DNA/protein which ultimately results in reduced root growth and germination rate. This Si-NP-mediated effect may be regarded as mitodepressive. Likewise, Lee et al. (2010) observed that 42.8 nm Si-NPs promoted root elongation in *Arabidopsis thaliana* at low concentration, but proved toxic at higher concentrations. The phytotoxicity of Si-NPs was assessed as a function of particle size (14, 50, and 200 nm) with 50 and 200 nm Si-NPs displaying phytotoxicity expressed in the form of chlorosis due to inadequate synthesis of chlorophyll (Slomberg and Schoenfisch 2012).

It has been suggested that Si-NP-generated toxicity arises due to the ability of the negatively charged Si-NPs (size 50 and 200 nm) to adsorb nutrients, making them unavailable for plant uptake (Sollins et al. 1988). Growth retardation and chlorosis are major development problems in plants which result from deficiencies of macro- and micronutrients like nitrogen, phosphorus, potassium, iron, zinc, and manganese (Van Patten 2004).

Particle size and surface area are considered crucial from a toxicological perspective: reduction in particle size increases surface area and the proportion of atoms or molecules (Nel et al. 2006). This size-dependent feature of NPs affects interfacial reactivity and the ability to pass through physiological barriers. The uptake and phytotoxicity of NPs depend upon particle size, where smaller particles accumulate to higher levels and prove more toxic as compared to their bulk particles (Larue et al. 2012; Slomberg and Schoenfisch 2012; Judy et al. 2012). However, it is uncertain whether this variation in toxicity results from changes in the size-dependent specific surface area. For example, SiO_2 NPs measuring 12.5-nm exhibited more toxicity to *Pseudokirchneriella subcapitata* in comparison to the NPs having 27.0 nm, but once normalized by surface area, no significant difference exists (Van Hoecke et al. 2008).

7.4 Summary and Conclusions

Si-NPs are very small particles, and it is speculated that their entry into the plant cell via Si transporters (lsi1) is a relatively simple task. Si-NPs increase seed germination by enhancing nutrient availability (Fig. 7.1). They promote stomatal conductance and water uptake that leads to elevation of photosynthesis. Si-NPs also increase uptake of Si, which increases expression of OsNAC proteins that are known to regulate stress, and proline and sugar accumulation (Fig. 7.1).

During salinity stress Si-NPs block the uptake of NaCl to protect the plant from salt toxicity. Si-NPs also may protect plants from heavy metal toxicity. Si-NPs direct the translocation of Cd into vacuoles via expression of *OsHMA3* which is valuable in reducing cellular toxicity (Fig. 7.1). Moreover, Si-NPs enhance the antioxidant defence system by reducing ROS accumulation, thereby preventing oxidative damage in plants. Growth promotive and stress mitigating properties of Si-NPs must be further examined and exploited in different horticultural crops.

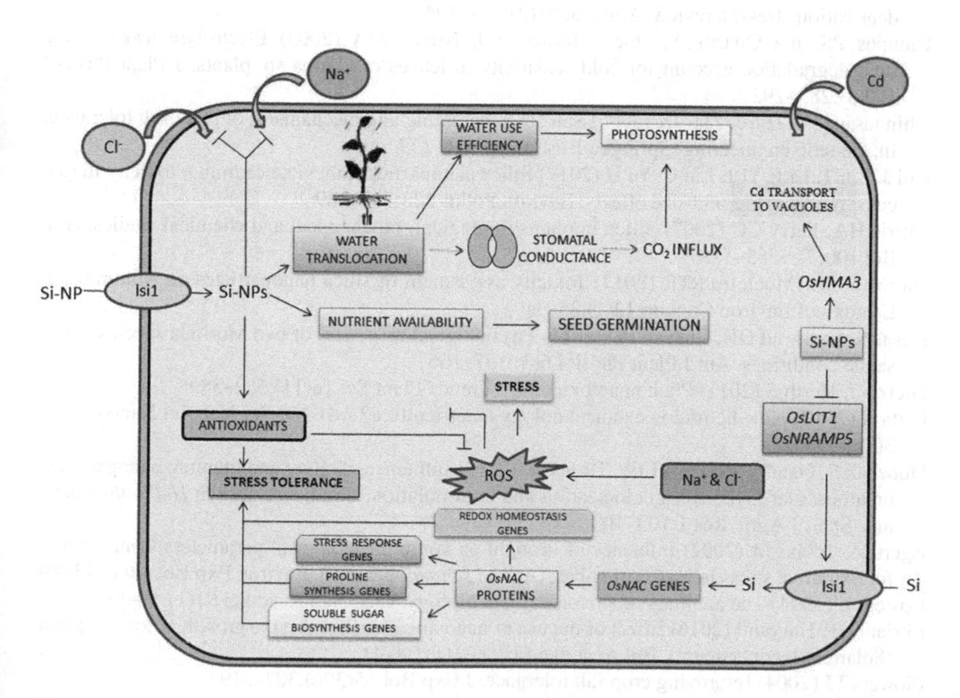

Fig. 7.1 Role of Si nanoparticles on the physiology of plants

References

Adhikari T, Kundu S, Rao AS (2013) Impact of SiO_2 and Mo nano particles on seed germination of rice (*Oryza sativa* L.). Int J Agric Food Sci Technol 4(8):809–816

Alsaeedi A, El-Ramady H, Alshaal T, El-Garawany M, Elhawat N, Al-Otaibi A (2019) Silica nanoparticles boost growth and productivity of cucumber under water deficit and salinity stresses by balancing nutrients uptake. Plant Physiol Biochem 139:1–10

Asgari F, Majd A, Jonoubi P, Najafi F (2018) Effects of silicon nanoparticles on molecular, chemical, structural and ultrastructural characteristics of oat (*Avena sativa* L.). Plant Physiol Biochem 127:152–160

Ashkavand P, Tabari M, Zarafshar M, Tomášková I, Struve D (2015) Effect of SiO_2 nanoparticles on drought resistance in hawthorn seedlings. For Res Pap 76(4):350–359

Azimi R, Borzelabad MJ, Feizi H, Azimi A (2014) Interaction of SiO_2 nanoparticles with seed prechilling on germination and early seedling growth of tall wheatgrass (*Agropyron elongatum* L.). Pol J Chem Technol 16(3):25–29

Bagchi M, Moriyama H, Shahidi F (2012) Bio-nanotechnology: a revolution in food, biomedical and health sciences. Wiley, New York

Bao-shan L, Chun-hui L, Li-jun F, Shu-chun Q, Min Y (2004) Effect of TMS (nanostructured silicon dioxide) on growth of Changbai larch seedlings. J For Res 15(2):138–140

Bassett CL (2013) Water use and drought response in cultivated and wild apples. Abiotic stress-plant responses and applications in agriculture, pp 249–275

Bigler C, Gavin DG, Gunning C, Veblen TT (2007) Drought induces lagged tree mortality in a subalpine forest in the Rocky Mountains. Oikos 116(12):1983–1994

Biswal SK, Nayak AK, Parida UK, Nayak PL (2012) Applications of nanotechnology in agriculture and food sciences. Int J Sci Innov Discov 2(1):21–36

Blokhina O, Virolainen E, Fagerstedt KV (2003) Antioxidants, oxidative damage and oxygen deprivation stress: a review. Ann Bot 91(2):179–194

Campos PS, nia Quartin V, chicho Ramalho J, Nunes MA (2003) Electrolyte leakage and lipid degradation account for cold sensitivity in leaves of Coffea sp. plants. J Plant Physiol 160(3):283–292

Chinnusamy V, Zhu J, Zhu JK (2006) Salt stress signaling and mechanisms of plant salt tolerance. In: Genetic engineering. Springer, Boston, pp 141–177

Cui J, Liu T, Li F, Yi J, Liu C, Yu H (2017) Silica nanoparticles alleviate cadmium toxicity in rice cells: mechanisms and size effects. Environ Pollut 228:363–369

Currie HA, Perry CC (2007) Silica in plants: biological, biochemical and chemical studies. Ann Bot 100(7):1383–1389

Da Silva GH, Monteiro RTR (2017) Toxicity assessment of silica nanoparticles on Allium cepa. Ecotoxicol Environ Contam 12(1):25–31

Desai N, Gaikwad DK, Chavan PD (2011) Physiological responses of two Morinda species under saline conditions. Am J Plant Physiol 6(3):157–166

Dietz KJ, Herth S (2011) Plant nanotoxicology. Trends Plant Sci 16(11):582–589

Ditta A (2012) How helpful is nanotechnology in agriculture? Adv Nat Sci Nanosci Nanotechnol 3(3):033002

Dubchak S, Ogar A, Mietelski JW, Turnau K (2010) Influence of silver and titanium nanoparticles on arbuscular mycorrhiza colonization and accumulation of radiocaesium in *Helianthus annuus*. Span J Agric Res 1:103–108

Egert M, Tevini M (2002) Influence of drought on some physiological parameters symptomatic for oxidative stress in leaves of chives (*Allium schoenoprasum*). Environ Exp Bot 48(1):43–49

Epstein E (1994) The anomaly of silicon in plant biology. Proc Natl Acad Sci 91(1):11–17

Fitriani HP, Haryanti (2016) Effect of the use of nanosilica fertilizer on the growth of tomato plant (Solanum lycopersicum). Bul Anat dan Fisiol 24(1):34–41

Flowers TJ (2004) Improving crop salt tolerance. J Exp Bot 55(396):307–319

Frewer LJ, Norde W, Fischer A, Kampers F (eds) (2011) Nanotechnology in the agri-food sector: implications for the future. Wiley, Weinheim

Gangwar S, Singh VP (2011) Indole acetic acid differently changes growth and nitrogen metabolism in *Pisum sativum* L. seedlings under chromium (VI) phytotoxicity: implication of oxidative stress. Sci Hortic 129(2):321–328

Gao M, Zhou J, Liu H, Zhang W, Hu Y, Liang J, Zhou J (2018) Foliar spraying with silicon and selenium reduces cadmium uptake and mitigates cadmium toxicity in rice. Sci Total Environ 631:1100–1108

Gruère G, Narrod C, Abbott L (2011) Agricultural, food, and water nanotechnologies for the poor. International Food Policy Research Institute, Washington, DC

Haghighi M, Pessarakli M (2013) Influence of silicon and nano-silicon on salinity tolerance of cherry tomatoes (*Solanum lycopersicum* L.) at early growth stage. Sci Hortic 161:111–117

Hayat S, Hasan SA, Hayat Q, Ahmad A (2010) Brassinosteroids protect *Lycopersicon esculentum* from cadmium toxicity applied as shotgun approach. Protoplasma 239(1–4):3–14

Hichem H, Mounir D (2009) Differential responses of two maize (*Zea mays* L.) varieties to salt stress: changes on polyphenols composition of foliage and oxidative damages. Ind Crop Prod 30(1):144–151

Huan C, Shu-Qing S (2014) Silicon nanoparticles: preparation, properties, and applications. Chin Phys B 23(8):088102

Hussain A, Rizwan M, Ali Q, Ali S (2019) Seed priming with silicon nanoparticles improved the biomass and yield while reduced the oxidative stress and cadmium concentration in wheat grains. Environ Sci Pollut Res 26:1–10

Ivani R, Sanaei Nejad SH, Ghahraman B, Astaraei AR, Feizi H (2018) Role of bulk and Nanosized SiO2 to overcome salt stress during Fenugreek germination (*Trigonella foenum-graceum* L.). Plant Signal Behav 13(7):e1044190

Janmohammadi M, Sabaghnia N (2015) Effect of pre-sowing seed treatments with silicon nanoparticles on germinability of sunflower (*Helianthus annuus*). Botanica Lithuanica 21(1):13–21

Judy JD, Bertsch PM (2014) Bioavailability, toxicity, and fate of manufactured nanomaterials in terrestrial ecosystems. In: Advances in agronomy, vol 123. Academic, Cambridge, pp 1–64

Judy JD, Unrine JM, Rao W, Wirick S, Bertsch PM (2012) Bioavailability of gold nanomaterials to plants: importance of particle size and surface coating. Environ Sci Technol 46(15):8467–8474

Kalteh M, Alipour ZT, Ashraf S, Marashi Aliabadi M, Falah Nosratabadi A (2018) Effect of silica nanoparticles on basil (*Ocimum basilicum*) under salinity stress. J Chem Health Risks 4(3):49–55

Karimi J, Mohsenzadeh S (2016) Effects of silicon oxide nanoparticles on growth and physiology of wheat seedlings. Russ J Plant Physiol 63(1):119–123

Khalaki M, Ghorbani A, Moameri M (2016) Effects of silica and silver nanoparticles on seed germination traits of Thymus kotschyanus in laboratory conditions. J Rangeland Sci 6(3):221–231

Kirschbaum MU (2011) Does enhanced photosynthesis enhance growth? Lessons learned from CO_2 enrichment studies. Plant Physiol 155(1):117–124

Koca H, Bor M, Ozdemir F, Turkan I (2007) The effect of salt stress on lipid peroxidation, antioxidative enzymes and proline content of sesame cultivars. Environ Exp Bot 60:344–351

Kumar M (2013) Crop plants and abiotic stresses. J Biomol Res Ther 3:1

Larue C, Veronesi G, Flank AM, Surble S, Herlin-Boime N, Carrière M (2012) Comparative uptake and impact of TiO_2 nanoparticles in wheat and rapeseed. J Toxic Environ Health A 75(13–15):722–734

Lee CW, Mahendra S, Zodrow K, Li D, Tsai YC, Braam J, Alvarez PJ (2010) Developmental phytotoxicity of metal oxide nanoparticles to *Arabidopsis thaliana*. Environ Toxicol Chem Int J 29(3):669–675

Li B, Tao G, Xie Y, Cai X (2012) Physiological effects under the condition of spraying nano-SiO_2 onto the *Indocalamus barbatus* McClure leaves. J Nanjing For Univ (Natural Sciences Edition) 36(4):161–164

Liu R, Lal R (2015) Potentials of engineered nanoparticles as fertilizers for increasing agronomic productions. Sci Total Environ 514:131–139
Lu C, Zhang C, Wen J, Wu G, Tao M (2002) Research of the effect of nanometer materials on germination and growth enhancement of *Glycine max* and its mechanism. Soybean Sci 21(3):168–171
Luyckx M, Hausman JF, Lutts S, Guerriero G (2017) Silicon and plants: current knowledge and technological perspectives. Front Plant Sci 8:411
Ma JF, Yamaji N (2006) Silicon uptake and accumulation in higher plants. Trends Plant Sci 11(8):392–397
Mahajan S, Tuteja N (2005) Cold, salinity and drought stresses: an overview. Arch Biochem Biophys 444:139–158
Manivannan A, Ahn YK (2017) Silicon regulates potential genes involved in major physiological processes in plants to combat stress. Front Plant Sci 8:1346
Martínez-Vilalta J, Piñol J (2002) Drought-induced mortality and hydraulic architecture in pine populations of the NE Iberian Peninsula. For Ecol Manag 161(1–3):247–256
McCutchan H, Shackel KA (1992) Stem-water potential as a sensitive indicator of water stress in prune trees (*Prunus domestica* L. cv. French). J Am Soc Hortic Sci 117(4):607–611
Merwad ARM, Desoky ESM, Rady MM (2018) Response of water deficit-stressed *Vigna unguiculata* performances to silicon, proline or methionine foliar application. Sci Hortic 228:132–144
Miralles P, Church TL, Harris AT (2012) Toxicity, uptake, and translocation of engineered nanomaterials in vascular plants. Environ Sci Technol 46(17):9224–9239
Monica RC, Cremonini R (2009) Nanoparticles and higher plants. Caryologia 62(2):161–165
Munns R (2002) Comparative physiology of salt and water stress. Plant Cell Environ 25:239–250
Mushtaq A, Jamil N, Rizwan S, Mandokhel F, Riaz M, Hornyak GL, …, Shahwani MN (2018, September) Engineered silica nanoparticles and silica nanoparticles containing Controlled Release Fertilizer for drought and saline areas. In: IOP conference series: materials science and engineering, vol 414, no. 1, p 012029. IOP Publishing
Nazaralian S, Majd A, Irian S, Najafi F, Ghahremaninejad F, Landberg T, Greger M (2017) Comparison of silicon nanoparticles and silicate treatments in fenugreek. Plant Physiol Biochem 115:25–33
Nel A, Xia T, Mädler L, Li N (2006) Toxic potential of materials at the nanolevel. Science 311(5761):622–627
Noctor G, Foyer CH (1998) Ascorbate and glutathione: keeping active oxygen under control. Annu Rev Plant Physiol Plant Mol Biol 49(1):249–279
O'Farrell N, Houlton A, Horrocks BR (2006) Silicon nanoparticles: applications in cell biology and medicine. Int J Nanomedicine 1(4):451
Ozkur O, Ozdemir F, Bor M, Turkan I (2009) Physiochemical and antioxidant responses of the perennial xerophyte *Capparis ovata* Desf. to drought. Environ Exp Bot 66(3):487–492
Parveen N, Ashraf M (2010) Role of silicon in mitigating the adverse effects of salt stress on growth and photosynthetic attributes of two maize (*Zea mays* L.) cultivars grown hydroponically. Pak J Bot 42(3):1675–1684
Pei ZF, Ming DF, Liu D, Wan GL, Geng XX, Gong HJ, Zhou WJ (2010) Silicon improves the tolerance to water-deficit stress induced by polyethylene glycol in wheat (*Triticum aestivum* L.) seedlings. J Plant Growth Regul 29(1):106–115
Pompelli MF, Martins SC, Antunes WC, Chaves AR, DaMatta FM (2010) Photosynthesis and photoprotection in coffee leaves is affected by nitrogen and light availabilities in winter conditions. J Plant Physiol 167(13):1052–1060
Pourkhaloee A, Haghighi M, Saharkhiz MJ, Jouzi H, Doroodmand MM (2011) Investigation on the effects of carbon nanotubes (CNTs) on seed germination and seedling growth of salvia (*Salvia microsiphon*), pepper (*Capsicum annum*) and tall fescue (*Festuca arundinacea*). J Seed Technol 33:155–160
Prasad R, Kumar V, Prasad KS (2014) Nanotechnology in sustainable agriculture: present concerns and future aspects. Afr J Biotechnol 13(6):705–713

Putri FM, Suedy SWA, Darmanti S (2017) The effects of nano-silica fertilizer on the number of stomata, chlorophyll content and growth of black rice (*Oryza sativa L. Cv.* Japonica. Available online: http://www.ejournal.undip.ac.id/indek.php/baf/index

Rao GB, Susmitha P (2017) Silicon uptake, transportation and accumulation in Rice. J Pharmacogn Phytochem 6(6):290–293

Rastogi A, Zivcak M, Sytar O, Kalaji HM, He X, Mbarki S, Brestic M (2017) Impact of metal and metal oxide nanoparticles on plant: a critical review. Front Chem 5:78

Raven JA (1983) The transport and function of silicon in plants. Biol Rev 58(2):179–207

Rawson HM, Long MJ, Munns R (1988) Growth and development in NaCl-treated plants. I. Leaf Na^+ and Cl^-concentrations do not determine gas exchange of leaf blades in barley. Funct Plant Biol 15(4):519–527

Rizwan M, Ali S, Zia ur Rehman M, Rinklebe J, Tsang DC, Bashir A et al (2018) Cadmium phytoremediation potential of *Brassica* crop species: a review. Sci Total Environ 631:1175–1191

Roduner E (2006) Size matters: why nanomaterials are different. Chem Soc Rev 35(7):583–592

Sabaghnia N, Janmohammadi M (2014) Graphic analysis of nano-silicon by salinity stress interaction on germination properties of lentil using the biplot method. Agric For Poljoprivreda i Sumarstvo 60(3):29–40

Sgherri CLM, Maffei M, Navari-Izzo F (2000) Antioxidative enzymes in wheat subjected to increasing water deficit and rewatering. J Plant Physiol 157(3):273–279

Shah V, Belozerova I (2009) Influence of metal nanoparticles on the soil microbial community and germination of lettuce seeds. Water Air Soil Pollut 197(1–4):143–148

Sharifi RJ, Sharifirad M, Teixeira DSJ (2016) Morphological, physiological and biochemical responses of crops (*Zea mays* L., *Phaseolus vulgaris* L.), medicinal plants (*Hyssopus officinalis* L., *Nigella sativa* L.), and weeds (*Amaranthus retroflexus* L., *Taraxacum officinale* FH Wigg) exposed to SiO_2 nanoparticles. J Agric Sci Technol 18:1027–1040

Shoeva OY, Khlestkina EK (2018) Anthocyanins participate in the protection of wheat seedlings against cadmium stress. Cereal Res Commun 46(2):242–252

Siddiqui MH, Al-Whaibi MH (2014) Role of nano-SiO_2 in germination of tomato (*Lycopersicum esculentum* seeds Mill.). Saudi J Biol Sci 21(1):13–17

Siddiqui MH, Al-Whaibi MH, Faisal M, Al Sahli AA (2014) Nano-silicon dioxide mitigates the adverse effects of salt stress on *Cucurbita pepo* L. Environ Toxicol Chem 33(11):2429–2437

Siddiqui H, Yusuf M, Faraz A, Faizan M, Sami F, Hayat S (2018) 24-Epibrassinolide supplemented with silicon enhances the photosynthetic efficiency of *Brassica juncea* under salt stress. S Afr J Bot 118:120–128

Silva AJ, Nascimento CWA, Gouveia-Neto AS (2017) Assessment of cadmium phytotoxicity alleviation by silicon using chlorophyll a fluorescence. Photosynthetica 55(4):648–654

Slomberg DL, Schoenfisch MH (2012) Silica nanoparticle phytotoxicity to Arabidopsis thaliana. Environ Sci Technol 46(18):10247–10254

Sollins P, Robertson GP, Uehara G (1988) Nutrient mobility in variable-and permanent-charge soils. Biogeochemistry 6(3):181–199

Sonkaria S, Ahn SH, Khare V (2012) Nanotechnology and its impact on food and nutrition: a review. Recent Pat Food Nutr Agric 4(1):8–18

Suciaty T, Purnomo D, Sakya AT (2018) The effect of nano-silica fertilizer concentration and rice hull ash doses on soybean (*Glycine max* (L.) Merrill) growth and yield. In: IOP conference series: earth and environmental science, vol 129, no. 1, p 012009. IOP Publishing

Suriyaprabha R, Karunakaran G, Yuvakkumar R, Prabu P, Rajendran V, Kannan N (2012) Growth and physiological responses of maize (*Zea mays* L.) to porous silica nanoparticles in soil. J Nanopart Res 14(12):1294

Tantawy AS, Salama YAM, El-Nemr MA, Abdel-Mawgoud AMR (2015) Nano silicon application improves salinity tolerance of sweet pepper plants. Int J ChemTech Res 8(10):11–17

Tripathi DK, Singh VP, Prasad SM, Chauhan DK, Dubey NK (2015) Silicon nanoparticles (SiNp) alleviate chromium (VI) phytotoxicity in *Pisum sativum* (L.) seedlings. Plant Physiol Biochem 96:189–198

Tripathi DK, Singh S, Singh VP, Prasad SM, Chauhan DK, Dubey NK (2016) Silicon nanoparticles more efficiently alleviate arsenate toxicity than silicon in maize cultiver and hybrid differing in arsenate tolerance. Front Environ Sci 4:46
Van Hoecke K, De Schamphelaere KA, Van der Meeren P, Lcucas S, Janssen CR (2008) Ecotoxicity of silica nanoparticles to the green alga Pseudokirchneriella subcapitata: importance of surface area. Environ Toxicol Chem Int J 27(9):1948–1957
Van Patten GF (2004) Hydroponic basics. Van Patten Publishing, Portland
Vardharajula S, Ali SZ, Tiwari PM, Eroğlu E, Vig K, Dennis VA, Singh SR (2012) Functionalized carbon nanotubes: biomedical applications. Int J Nanomedicine 7:5361
Wang J, Naser N (1994) Improved performance of carbon paste amperometric biosensors through the incorporation of fumed silica. Electroanalysis 6(7):571–575
Xie Y, Li B, Tao G, Zhang Q, Zhang C (2012) Effects of nano-silicon dioxide on photosynthetic fluorescence characteristics of *Indocalamus barbatus* McClure. J Nanjing For Univ (Natural Sciences Edition) 36(2):59–63
Yassen A, Abdallah E, Gaballah M, Zaghloul S (2017) Role of Silicon dioxide nano fertilizer in mitigating salt stress on growth, yield and chemical composition of Cucumber (*Cucumis sativus* L.). Int J Agric Res 22:130–135
Yousaf B, Liu G, Wang R, Zia-ur-Rehman M, Rizwan MS, Imtiaz M et al (2016) Investigating the potential influence of biochar and traditional organic amendments on the bioavailability and transfer of cd in the soil–plant system. Environ Earth Sci 75(5):374
Yuvakkumar R, Elango V, Rajendran V, Kannan NS, Prabu P (2011) Influence of nanosilica powder on the growth of maize crop (*Zea mays* L.). Int J Green Nanotechnol 3(3):180–190
Zarafshar M, Akbarinia M, Askari H, Hosseini SM, Rahaie M, Struve D, Striker GG (2014) Morphological, physiological and biochemical responses to soil water deficit in seedlings of three populations of wild pear (*Pyrus boisseriana*). Université de Liège Gembloux Agro-Bio Tech; Biotechnologie, Agronomie, Société Et Environnement 18(3):353–366
Zheng L, Hong F, Lu S, Liu C (2005) Effect of nano-TiO_2 on strength of naturally aged seeds and growth of spinach. Biol Trace Elem Res 104(1):83–91
Zushi K, Matsuzoe N, Kitano M (2009) Developmental and tissue-specific changes in oxidative parameters and antioxidant systems in tomato fruits grown under salt stress. Sci Hortic 122(3):362–368

Chapter 8
Interaction Between Copperoxide Nanoparticles and Plants: Uptake, Accumulation and Phytotoxicity

Abreeq Fatima, Shikha Singh, and Sheo Mohan Prasad

Abstract A natural question arose when scientists and engineers began formulating and using nanoparticles (NPs): "Why are they so interesting? Why are studies of these extremely small entities are so fascinating, and why are they so challenging to handle as well as to synthesize?" The unique property possessed by all nanoparticles is where the answer lies. The term *nano* is adapted from the Greek word 'dwarf' and denotes 10^{-9} when used as a prefix. The use of nanoparticles (NPs) extends their potential into agricultural soils and indeed the formulations of NPs may be developed to improve nutrient and quality of crops. The rapid development of synthesized nanoparticles combined with their potential risks to public health and the environment has raised considerable concerns. A significant aspect regarding risk assessment of NPs is understanding the interaction between plants and NPs. Plants, which are fundamental components of all ecosystems, play an important role in fate and transport of NPs in the environment through uptake and bioaccumulation. The degree of accumulation of nanoparticles by plants depends on physicochemical characteristics such as shape, size, agglomeration state, chemical composition and others. Since, copper is an essential micronutrient for plants and play important role in the activation of several enzymes such as cytochrome c oxidase, superoxide dismutase, ascorbate oxidase, amine oxidase etc. and as electron transport carriers in plants i.e. plastocyanin (Sekine R, Marzouk ER, Khaksar M, Scheckel KG, Stegemeier JP, Lowry GV, Donner E, Lombi E, J Environ Qual, 46(6):1198–1205, 2017). This chapter discusses the nature of copperoxide nanoparticles (CuO NPs), their uptake and translocation mechanisms, and their toxic effects on different plant species at both physiological and cellular levels. This chapter also addresses tolerance mechanisms generated by plants and a critical assessment of the necessity for further research.

Keywords Copper oxide nanoparticle (CuO NPs) · Phytotoxicity · Tolerance

A. Fatima · S. Singh · S. M. Prasad (✉)
Ranjan Plant Physiology and Biochemistry Laboratory, Department of Botany, University of Allahabad, Allahabad, India

S. Hayat et al. (eds.), *Sustainable Agriculture Reviews 41*, Sustainable Agriculture Reviews 41, https://doi.org/10.1007/978-3-030-33996-8_8

8.1 Introduction

Nanotechnology is an interdisciplinary area of science which has encountered immense progress due to its vast applications in recent decades. Various public health, industrial, consumer and environmental challenges have been mitigated due to a boom in nanotechnologies and nanomaterials development. The term 'nanotechnology' is defined as the study of manipulating matter at the atomic and molecular scales. This is an exciting field of research with growing interest in its application for biological and environmental safety. In recent years, the scientific community has gained substantial interest in nanometer-sized materials with unique physical, chemical, and biological properties. Particles having a size less than 100 nm in diameter are called nanoparticles; they are recognized for possessing different size-dependent properties compared to their original bulk material. Nanoparticles have been widely used in environmental applications and have shown promising performance in pollutant removal and toxicity mitigation.

Based on the published literature, there are essentially four types of nanoparticles:

(a) *Metallic nanoparticles* (within the range of 1–100 nm, having a large surface-area-to-volume ratio as compared to bulk materials)
(b) *Polymeric nanoparticles* (within the range of 10–1000 nm)
(c) *Carbon-based nanostructures* (small dimensions, high chemical stability, high thermal conductivity and low resistivity)
(d) *Metal oxide nanoparticles*

Currently, several types of metal oxide nanoparticles are reported to play important roles in physics, chemistry, materials and medical sciences. Several types of metal oxide nanoparticles exist such as ZnO, CuO, TiO_2, MgO, NiO, ZrO_2, and others. Metal oxide nanoparticles exhibit unique chemical and physical properties due to their high density structural geometries and limited size which impart them the characteristics of semiconductor.

Copper is widely used in agriculture and in food, chemical, medical, cosmetics and textile industries for coatings, in environmental remediation and wastewater treatment, in fungicides, fuel additives, paints, plastics, and electronics (Rafique et al. 2017).CuO nanoparticles posses photovoltaic and photoconductive properties. These CuO nanoparticles can improve fluid viscosity and thermal conductivity, these novel properties make them a potentially useful energy-saving material that can improve the effect of energy conversion. Because of these beneficial properties, CuO nanoparticles have drawn the attention of scientists for use as an essential component in future nano-devices.

Copper is an essential micronutrient which is incorporated in many proteins and enzymes, thus playing a significant role in plant health and nutrition; it is involved in various physiological processes (Chibber et al. 2013). Copper is available in two oxidation states, i.e., Cu^{+} and Cu^{2+}. This allows it to function both as a reducing and an oxidizing agent in biochemical reactions. At the same time, however this property makes Cu potentially toxic as Cu ions may catalyze production of free radicals (Hänsch and Mendel 2009; Ivask et al. 2010).

Several metal oxide nanoparticles act as effective photocatalysts in UV or visible light. Nano copper oxide (CuO) offers unique photocatalytic properties in sunlight. Sundaramurthy and Parthiban (2015) reported that CuO NPs efficiently degrade methylene blue under solar irradiation. Nano CuO is used in a wide range of applications such as antioxidants, heterogeneous catalysts, sensing and thermoelectric materials, ceramics, imaging agents, superconducting materials, anti-microbials and many more (Yallappa et al. 2013; Yamamoto 2001; Faheem et al. 2016; Keller et al. 2017). The low cost, high surface reactivity and specific high surface area of this material qualifies CuO NPs as a cost-effective catalyst for numerous chemical reactions (White et al. 2006; Yurderi et al. 2015).

Despite its high application potential, various disadvantages occur with use of CuO NPs. One double-edged feature of CuO NPs is their biocidal activity. On onc hand, CuO NPs are effectively used in anti-fouling paint, wood preservatives, sterile surface coatings, water filters, textiles and bandages (Almeida et al. 2007; Evans et al. 2008; Ahmad et al. 2012; Perreault et al. 2012; Ben-Sasson et al. 2014; Dankovich and Smith 2014). On the other hand, the biocidal activity of CuO NPs could be inadvertently harmful to human health and the environment (Karlsson et al. 2008). Therefore, careful monitoring of the toxic potential of these particles is necessary to evaluate their risk for utilization.

Increasing applications and use of NPs have a direct correlation with their release to the environment. The effects of NPs have been documented in a wide variety of biota including microorganisms (Pelletier et al. 2010; Dimkpa et al. 2011), protozoa (Mortimer et al. 2010), invertebrates (Valant et al. 2012), and vertebrates (Federici et al. 2007). However, interactions of NPs with plants and related organisms such as algae, have not been fully studied. The general consequences of NPs exposure to plant cells remain unclear (Zhang et al. 2012). This lack of data leads to an incomplete understanding of how nanoparticles are transferred and accumulated in food chains (Kahru and Dubourguier 2010).

This chapter summarizes current understanding and the future possibilities of interactions between plants and CuO nanoparticles.

8.2 Synthesis of Copper Oxide Nanoparticles (CuO NPs)

During past two decades, synthesis of metallic nanoparticles has drawn considerable attention in academics and applications in nanotechnology because of their unusual properties and potential applications in optical, electronic, catalytic, and magnetic materials. A number of technologies are available for manufacturing metallic nanoparticles. There are two basic approaches *i.e.,* top-down and bottom-up. The top-down approach enables manufacture of particles in the nanorange by the conversion of larger particles using cutting, grinding and etching techniques;

whereas in the bottom-up approach, small particles are converted into larger ones which occur in the nano range.

With respect to other metal nanoparticles, synthesis of copper NPs is more challenging because of their reactivity in water and air. A variety of methods for copper nanoparticle synthesis using chemical, physical and biological procedures have been studied. Copper nanoparticles are manufactured using both bottom-up and top-down methods (Iravani et al. 2014). All physical (top-down) methods used for synthesis of CuO nanoparticles usually require high temperatures, operation under vacuum and expensive equipment, which makes this technique uneconomical (Umer et al. 2012). In contrast, chemical methods (bottom-up) are eco-friendly, relatively simple, low in cost, high-yielding, and can be carried out under ambient conditions with simple laboratory equipment.

8.2.1 Physical Methods of Copper Nanoparticlesynthesis

Physical methods for synthesis of nanoparticles include laser pulse ablation (Mafuné et al. 2000), gamma irradiation (Long et al. 2007), electron irradiation, pulsed wire discharge (Tanori and Pileni 1997), mechanical irradiation and others. An extensive variety of nanoparticles are produced via physical methods with few modifications for different metals; however, the major drawbacks of these methods are yield of low-quality product and requirements of costly equipment.

Pulse Laser Ablation/Deposition It is a commonly used technique for preparation of copper nanoparticles in colloidal form. This technique takes place in a vacuum chamber in the presence of an inert gas. A high-power pulsed laser beam (mostly Second Harmonic Generation (ND: YAG) type) is focused inside a vacuum chamber to strike a target in the material whereas plasma is created which is converted to a colloidal suspension of nanoparticles.

Mechanical/Ball Milling Method It is a solid state processing technique, for the production of super alloys. This method consists of balls and a mill chamber and runs on mechanical energy. The ball mill have a stainless steel container and many small iron, hardened steel, silicon carbide, or tungsten carbide balls are made to rotate inside a mill and powdered material is taken inside the steel container. This powder is converted into nanosize. A magnet is placed outside the container to provide the pulling force to the material and this magnetic force increases the milling energy when milling container or chamber rotates the metal balls.

Pulsed Wire Discharge Method (PWD) The metal (copper) wire is evaporated by a pulsed current to produce a vapor which is then cooled by an ambient gas to form nanoparticles. This method has the potential of a high production rate and high energy efficiency.

8.2.2 Chemical Method of Copper Nanoparticles Synthesis

Many chemical methods are available for synthesis of nanoparticles; some, like chemical reduction (Nikhil et al. 2000), colloidal techniques (Panigrah et al. 2006), electrochemical and hydrothermal synthesis, sol gel, and microwave irradiation are among the primary techniques. The chemical reduction of copper salts is the easiest, simplest and most commonly used method for copper NP synthesis. The production of nano-sized metal copper particles with good control of morphology and size using chemical reduction of copper salts can be achieved.

Electrochemical Method This process is very simple and effective and a broad range of reduced ions can be used in this method. This process has no limitation with respect to the shape and size of the sample. In this method electricity is used as the driving force for the synthesis of nanoparticles. There are two methods: one in which constant current is applied to electrodes whereas in the second, working of the electrode is controlled. In this process deposition of the layer can be done either by direct current (DC) or accelerating current (AC) mode depending on the requirement. This is due to the fact that the wires have grown atom (Kalska-Szostko 2011). The main disadvantage of electrochemically synthesized nanoparticles is the toxicity resulting from the use of hazardous substances such as organic solvents and reducing agents.

A variety of chemical and physical procedures are available for synthesis of metallic nanoparticles; however, they are fraught with problems including use of toxic solvents, generation of hazardous by-products, and high energy consumption. Thus, there is a need to develop environmentally benign procedures for synthesis of metallic nanoparticles. Chemists, physicists and material scientists have shown much interest in the development of innovative methods for the synthesis of nanomaterials.

Synthesis by the biological route is a promising approach which will diversify the area of application of metal nanoparticles with less toxicity. Biological synthesis of nanoparticles usually employs microbes or green plants. In recent years plants, algae, fungi, bacteria, and viruses have been used for production of low-cost, energy-efficient, and non-toxic metallic nanoparticles. In plant-assisted synthesis alkaloids, terpenoids, polyphenols, proteins, carbohydrates, sugars etc. are reported to have a key function in chemically reducing as well as stabilizing metal ions.

8.2.3 Biosynthesis of Copper Nanoparticles

Both physical and chemical methods for nanoparticle synthesis are expensive and potentially hazardous; therefore, improved synthesis methods are needed. One such method is organic synthesis, involving variousreducing and capping agents including bacteria, fungi, actinomycetes, yeast and plants (Krumov et al. 2009; Abdul

Hameed and Al-Samarrai 2012; Marcia et al. 2013). Sparse literature is available on biosynthesis of CuO NPs as compared to chemical synthesis (Rahman et al. 2009; Gunalan et al. 2012; Honary et al. 2012). Studies by several researchers have shown that plantscan bescreened for preparation of biological extracts for copper nanoparticle synthesis. Plant samples are washed with distilled water and shade-dried for 2 weeks, then milled using a household blender (Ramesh et al. 2018; Nasrollahzadeh et al. 2018).

Synthesis of nanoparticles using plants involves simple preparation protocols and fewer toxic reagents with a broad variability of metabolites. Such nanomaterials are stable and of variable sizes and shapes. Many researchers have exploited plant extracts, exudates, gums and other components for synthesis of Cu NPs (Iravani 2011; Vellora et al. 2013). Copper nanoparticles have been synthesized using leaf extracts of *Brassica juncea*, *Helianthus annuus*, *Lantana camara, Medicago sativa,* and *Tridaxprocumbens* (Cioffi et al. 2005; Majumder 2012; Umer at al. 2012). *Magnolia* leaf extract and stem latex of *Euphorbia nivulia* have been used for synthesis of Cu NPs ranging in size from 40 to 100 nm (Lee et al. 2011). Vellora et al. (2013) reported the synthesis of Cu NPs from gum karaya (a natural hydrocolloid) which is used as both: reducing as well as capping agent in nanoparticle synthesis. Chemical constituents such as alkaloids, glycosides, tannins, and aromatic compounds of *Ocimum sanctum* have also been used as stabilizers for synthesis of Cu NPs (Kulkarni and Kulkarni 2013; Kulkarni and Muddapur 2014).

The use of fungi to synthesize nanoparticles has been reported. Many fungi have been examined and many are suitable candidates, as they secrete enzymes in large quantitiesand are easy to handle in the laboratory (Sahayaraj and Rajesh 2011). Honary et al. (2012) reported the extracellular synthesis of Cu NPs using *Penicilliumaurantiogriseum, Penicilliumcitrinum*and*Penicilliumwaksmanii*. Also, Majumder (2012) reported the synthesis of Cu NPs using *Fusarium oxysporum* at ambient temperature.

8.3 Uptake, Translocation and Accumulation of Engineered CuO Nanoparticles

The importance of uptake and accumulation of nanomaterials by plants is increasingly recognized by researchers. The shape, size, chemical stability, and functionalization of NPs play a pivotal role in influencing uptake, translocation, and accumulation. Researchers have determined a linear relationship showing that high concentrations of Cu-NPs in the development media results in high accumulation of Cu-NPs in plant tissues (Kasana et al. 2017). Copper nanoparticles are taken up and accumulated in bean and wheat plants, a responsive relationship exists between bioaccumulated Cu NPs in plant tissue and in growth media (Lee et al. 2008).

CuONPs are absorbed from soil by plant roots and then converted to simpler forms. The plant cell wall acts as a restriction site which inhibits entry of the nanoparticle into the cell. The size of pores in the cell wall is a determining factor for internalization of NPs. Pore sizes ranging between 5 and 20 nm provide the plant with a sieving property (Fleischer et al. 1999). It is sometimes observed that formation of new cell wall pores occurs during reproduction, or enlargement of previously existing pores under the influence of NPs. This phenomenon makes the cell wall more permeable and enhances uptake of nanoparticle several-folds (Wessels 1993; Ovecka et al. 2005).

Nanoparticles may also be functionalized before entering the plant. The internalization of nanomaterials is therefore selective and occurs via channels in the plasma membrane. The active functionalized sites in the cell wall includes carboxylate, phosphate, hydroxyl amine, sulfhydryl, and imidazole functional groups (Vinopal et al. 2007) which interlink themselves and form complex biomolecules like cellulose, carbohydrates, and proteins (Knox 1995) and facilitate the selective uptake of nanoparticles.

Aquaporins, ion channels, or the organic chemicals of the environmental media which are membrane-embedded carrier proteins responsible for transport, bind with nano particles to facilitate their entrance into the plant. Other transportation routes include the complex formation of NPs with root exudates (Watanabe et al. 2008; Kurepa et al. 2010; Mishra et al. 2014).

Transport of these NPs across cell may occur either through symplastic or apoplastic pathway depending upon the availability of nanoparticle whereas cell to cell transportation occurs via plasmodesmata. The efficiency of uptake and transport of NPs is greater in some plants than in others, depending on the unique physiology of the plant species.

CuO-NPs are more actively accumulated than Cu^{+} (Nekrasova et al. 2011). When NPs are sprayed over leaf surfaces they permeate into tissue through stomatal openings (Eichert et al. 2008; Uzu et al. 2010). Such entry may obstruct the stomatal openings in photosynthetic components resulting in heating of foliar chambers which alter gas exchange and modify physiological and cellular functions (Da Silva et al. 2006).

8.4 Phytotoxicity of CuO Nanoparticles to Algae and Plant Seedlings

Based on the extensive use of nanomaterials, the risk of loss to the biosphere and subsequent accumulation in biota isobvious. Nanoparticles enter the environment via losses in wastewater, application to agricultural soil, and atmospheric deposition. Due to the slow movement of nanoparticles in soil they can accumulate and ultimately be taken up by plants and create toxic effects in plant (Arif et al. 2018).

Discharge of CuO nanoparticles to the environment may impart serious consequences on crop productivity. Negative impacts of nanoparticles have been reported in crop plants such as cucumber, lettuce, radish (Lin and Xing 2007), zucchini

(Stampoulis et al. 2009), and wheat (Du et al. 2011). CuONP toxicity chiefly depends on its solubility in water and temperature fluctuation. CuONPs pose minimal toxicity within the pH range 9–11. Toxicity curve of temperature can be deduced on the basis of solubility curve of CuO in water as done for pH curve (Chang et al. 2012). A one-unit shift in pH on either side may cause an extreme shift in toxicity; shifts in temperature also affect CuO NP toxicity. CuONPs cause toxicity upon crossing the cell membrane by releasing Cu ions inside the cell (Karlsson et al. 2008; Anjum et al. 2015). Inside the cytoplasm, NPs bind differently in different organelles and interact with metabolic pathways in both positive as well as negative ways (Jia et al. 2005).

These metal and their nanoparticles could impair and affect overall growth and development of plants as they influence timing of senescence, flowering, fruiting, abscission, dormancy and many other physiological processes (Gardea-Torresdey et al. 2004; Thul and Sarangi 2015). Cabiscol et al. (2010) reported that ROS causes lipid peroxidation. The CuO-NPs, once accumulated in the roots and shoots of plants, may transfer electrons to molecular oxygen and possibly lead to the formation of superoxide radicals (SOR) and hydrogen peroxide (H_2O_2) leading to membrane damage and lipid peroxidation (Dietz and Herth 2011; Shaw and Hossain 2013). The ROS are mainly responsible for changes in the fluidity and permeability of cell membranes and consequently the acquisition kinetics of nutrients. Scientists have studied the toxicity of NPs in plants such as *Cicer arietinum, Brassica nigra, Arabidopsis thaliana*, *Pisum sativum*, *Zea mays*, and green alga *Picochlorum* sp. (Wang et al. 2016; Zafar et al. 2017). Recently a study was conducted on the toxicogenic effects of CuO NPs on *Arabidopsis thaliana* using microarray analysis. The results suggest that when the Cu^{2+} ions from CuO NPs are released they contribute to partial toxicity during CuO NP exposure (Tang et al. 2016). A diagrammatic representation of uptake and translocation of CuO nanoparticles in plants and their metabolic consequences is given in Fig. 8.1.

With the germination of seed also begins the physiological processes within a plant. The embryo within the seed is protected by the seed coat; once ruptured, the radicle emerges which may come into direct contact with a metal (Wierzbicka and Obidzinska 1998). Nanoparticles can cause reduction in seed germination and suppress plant elongation, and can even cause plant death. Nanoparticles influence seedgermination and growthas a function of the characteristics of the nanomaterials (concentration/size/category/stability), the plant seed (size/species), growth medium, plant growth stage and type of NP coating material, if any.

CuO-NPs have been found to significantly reduce germination rate of *Hordeum sativum* by 23% with respect to control (Rajput et al. 2018). Zuverza-Menan et al. (2015) reported that 80 mg kg^{-1}nano-CuO, micro-CuO, and Cu ions, reduced seed germination of *Cucumis sativum* by 50%. The NPs reduced rateof germination, biomass, root and shoot length. CuO NPs inhibited seed germination of *Cucumis sativus* by 23.3% at a dose of 600 mg L^{-1}compared to control (Moon et al. 2014). The results suggest that Cu concentration increases considerably in roots and in aboveground tissue after treatment withCuONPs. An increased concentration of CuO was also observed in *Origanum vulgare* with application of CuO NPs in roots as well as

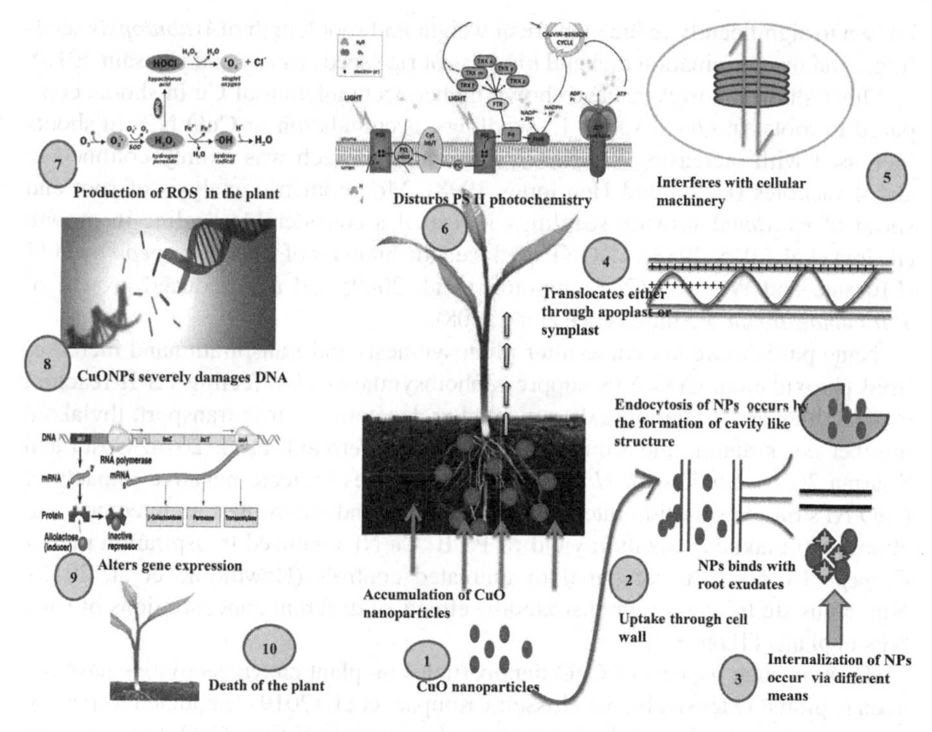

Fig. 8.1 Diagrammatic representation of the uptake and translocation of CuO nanoparticles in plants and the potential effects on plant metabolism

shoots (Du et al. 2018). These metal NPs accumulated more actively in roots than shoots (Costa and Sharma 2016).

Since roots are the primary target of CuONPs, functional and structural disorders generally appear in them as opposed to above-ground tissue (Shaw et al. 2014).CuO NPs shortenedprimary and lateral roots of *Brassicajuncea* L.seedlings (Nair and Chung 2015) and decreased root growth of*M. sativa* (Hong et al. 2015). Atha et al. (2012) reported that CuO NPs markedly inhibitedgrowth of *Raphanussativus*, *Loliumperenne* and *Loliumrigidum*. Singh and Kumar (2016) observed that irrigation water containing Cu NPs reducedshoot and root length of*Solanum oleracea*. Addition of 1000 mg L^{-1}CuO NPsto*Oryza sativa*, var. Jyoti significantly decreased shoot and root length. Exposure of *Allium cepa* seedlings to 80 mg L^{-1} CuO NPs damaged the root cap andmeristematic zone and reducedgrowth of the root tip (Deng et al. 2016). This, in turn, limitedtotal surface area for water uptake and affectedoverall plant growth.

Stampoulis et al. (2009) and Wang et al. (2012) reported that CuO-NPs suppressed root elongation in zucchini and maize. In *Arabidopsis thaliana,* components primarily affected were germination percentage, root elongation, biomass and leaf number (Lee et al. 2010). When wheat plants were grown in a sand matrix, CuONPs inhibited growth and changed root structure (Dimkpa et al. 2012). CuO NPs are known to significantly reduce the fresh weight and root length of*Arabidopsis* seedlings, and the germination rate and biomass of rice seeds (Shaw and Hossain 2013).

Other studies, however, have shown higher accumulation of Cu in shoots compared to roots. In *Oryza sativa* L. seedlings, accumulation of CuO NPs in shoots increased with increasing Cu NPs concentration, which was mainly confined to shoot vacuoles (Lidon and Henriques 1998). Morphometric analysis of root and shoot of *Hordeum sativum* seedlings indicated a considerable decline in growth (Rajput et al. 2015, 2018). CuONPs reduced the biomass of *Cucurbita pepo* by 90% (Musante and White 2012; Stampoulis et al. 2009) and also reduced growth of *Triticumaestivum* seedlings (Lee et al. 2008).

Nano particlesare known to alter photosynthesis and transpirationand increaselipid peroxidation. CuO NPs suppress photosynthesis by affecting PS II reaction centers; they are furthermore documented to decrease electron transport, thylakoid number per granum, and stomatal conductance (Perreault et al. 2014; Costa and Sharma 2016). Studies on *Hordeum sativum* leaves indicate negative impacts of CuO NPs on transpiration rate, stomatal aperture and chloroplast architecture, and changes in maximal quantum yield of PS II. Cu NPs reduced transpiration rate of *C. pepo* by 60–70% compared to untreated controls (Hawthorne et al. 2012). Numerous studies have demonstratedthe effects of different concentrations of CuO NPs in plants (Table 8.1).

Themolecular aspects of CuO nanoparticles on plant enzymeactivities have not been explored extensively, but Hosseini-Koupaei et al. (2019) conducted a spectroscopic characterization of the interaction of proteinase K with CuO nanoparticles. Data revealed that increasing concentrations of CuO nanoparticles decrease the activity as well as thermostability of proteinase K. They also found that microenvironments of the tryptophan residues decreased under increased concentrations of CuO nanoparticles. Results also revealed that the characteristics of CuONPs depend on temperature which greatly affects thermodynamic stability and binding affinity of proteinase K. CuO nanoparticles increase the stability of the native folded state of proteinase K at room temperature; conversely, it stabilizes the unfolded state at 310–333 K.

Accumulated nanoparticles may also hinder the growth of microflora such as algae in paddy fields either growing naturally or provided as a biofertilizer. However, NPs produce toxic effects only when present above a certain concentration (Rico et al. 2011).

8.5 Tolerance Mechanisms of Plants Against CuO Nanoparticle Toxicity

As documented above, CuO nanoparticles alter the growth and development of plants by increasing ROS production, thereby unbalancing homeostasis of essential elements (Du et al. 2017). It is reported that nanoparticles in excess quantities may

Table 8.1 Effect of different concentrations of CuO nanoparticles on selected crop plants

Crop	CuO Dose	Effect s on the plant	References
Elodea densa (Brazilian waterweed)	0.25 mg L^{-1}	Stimulated rate of photosynthesis; catalase and superoxide dismutase activity also increased by up to 1.5- to 2- times	Nekrasova et al. (2011)
Hordeum vulgare (Barley)	0.5, 1, 1.5 mM	Reduced shoot and root growth, increased hydrogen peroxide and lipid peroxidation, decreased GSH/GSSG ratio	Shaw et al. (2014)
Schoenoplectustabernaemontani (Bulrush)	0.5, 5.0, 50 mg L^{-1}	Marked accumulation of Cu in roots by up to 2.3, 3.8 and 2.7-times.Biomass of plants exposed to CuO NPs was reduced by 9% and 18% with respect to control	Zhang et al. (2014)
Cucumis sativus (Cucumber)	100–600 mg L^{-1}	Inhibited seed germination	Moon et al. (2014)
Coriandrum sativum (Coriander)	1000 mg L^{-1}	The rate of seed germination was inhibited due to decreased radical growth	Zuverza-Menan et al. (2015)
Lactuca sativa (Lettuce)	100–300 mg L^{-1}	Seed germination, vigor index, and fresh weight negatively affected; root length reduced up to 49%	Shams et al. (2018); Hong et al. (2015)
Lolium perenne (Perennial ryegrass)	10–1000 mg L^{-1}	Slowed down defense system and interrupted metabolic processes such as photosynthesis and respiration	Wang et al. (2015)
Coriandrum sativum (Cilantro)	0–80 mg L^{-1}	Germination rate and shoot elongation diminished.	Zuverza-Menan et al. (2015)
Lolium multiflorum (Annual ryegrass)	10–1000 mg L^{-1}	Enhanced production of ROS	Wang et al. (2015)
Oryza sativa (Rice)	5 mg L^{-1}	Overproduction of ROS was detected in root meristem thereby inhibits root growth	Wang et al. (2015)
Alium cepa (Onion root tip)	80 mg L^{-1}	Completely suppressed growth at 72 h exposure	Deng et al. (2016)
Raphanus raphanistrum (radish)	10, 100, 500, 1000 mg L^{-1}	Caused severe oxidative damage	Atha et al. (2012)
Solanum lycopersicum (tomato)	0–500 mg L^{-1}	Affected pigment content by decreasing total chlorophyll Content	Singh et al. (2017)

(continued)

Table 8.1 (continued)

Crop	CuO Dose	Effect s on the plant	References
Hordeum vulgare (Barley)	10,000 mg L^{-1}	Inhibited germination rate along with root and shoot Length	Shaw et al. (2014)
Lemna minor (duckweed)	50, 150, 300 μg L^{-1}	Frond number, frond area and biomass decreased after 7 days of exposure, ROS overproduction with subsequent increase in antioxidant activity	Yue et al. (2018)

cause severe toxicity and impair physiological processes and hamper the levels of cellular ROS (H_2O_2, $O^{\cdot -}{}_2$, ${}_2O^1$ and $^{\cdot}OH$). These ROS may act either as signaling or damaging molecules and play a vital role in maintaining the equilibrium between sites of production and scavenging.

Vacuoles in leaf cells are generally considered to be key metal sequestrations site in plants (MacFarlane and Burchett 2000; Clemens et al. 2002) which comprise a portion of the tolerance mechanism (Tong et al. 2004; Vymazal and Brezinova 2016).

ROS are involved in intercellular signaling cascades as second messengers in regulating plant responses in plant cells such as gravitropism (Joo et al. 2001), programmed cell death (Mittler 2002), stomatal closure (Kwak et al. 2003), root development and senescence. Plants have developed intrinsic mechanisms of antioxidant production to scavenge ROS and detoxify the system. The defense system consists of two types of non-enzymatic antioxidants including thiols, glutathione (GSH), phenolics, ascorbate (AA), and enzymatic ones such as catalase (CAT), superoxide dismutase (SOD), guaiacol peroxidase (GPOX), ascorbate peroxidase (APX), glutathione reductase (GR), monodehydro-ascorbate reductase, dehydroascorbate reductase (DHAR), glutathione S-transferases (GST) and glutathione peroxidase (GPX) (Singh et al. 2015). Shaw and Hossain 2013; Shaw et al. 2014 investigated the impact of CuO NPs on rice seedlings and found a consistent increase in APX activity for the scavenging of H_2O_2 in seedling leaves when exposed to 1.5 mMCuO NPs. It is reported that CuO induces GPX synthesis, and high concentrations of CuO NPs in soil also increase the metal content in plant leaves and affects physiological activity such as growth, photosynthesis and respiration.Further, production of RNS like nitric oxide helps in the removal of these ROS and also assists in the signaling under CuOtoxicity.

Dimkpa et al. (2012) reported higher CAT activity in wheat roots exposed to 500 mg/kg of nano-CuO, and Nair and Chung (2015) reported higher CAT activity in pea seedlings exposed to nano-CuO at both 100 and 200 mg/kg. This suggests that the inherent tolerance mechanism to combat phytotoxicity can be further enhanced by the use of phytohormones for better crop yield.

Various strategies are additionally employed to induce tolerance against CuO by the application of phytohormones in different plants such as pea in field studies. In a recent study, a bacterium isolated*Rhizobium leguminosarum* bv. viciae from root nodules of pea was shown to produce (IAA) (Tariq et al. 2014).These reports suggest that green pea plants associated with high IAA-producing bacteria have interaction with nano-CuO.

Exogenously applied IAA may counteract oxidation caused by CuO particles (Chaoui et al. 2004). It was found that CuO treatment significantly decreased CAT activity, and application of IAA at 10 μM resulted in a significant increase of CAT activity (~75%) (Buchanan et al. 2000). This evidence suggests that IAA plays an important role in increasing CAT activity and in protecting plants against oxidative stress induced by CuO particles.

8.6 Future Prospects

Immense advancements in the field of nano technological applications have been documented; however, the field is still in its infancy. Utilization of NPs in agricultural crop improvement programs is at the elementary phase. Each nanoparticle is characteristically different from all others in terms of shape, size, mode of action and other variables. So, in order to exploit the promised advantages of NPs, it is essential to further enhance our understanding regarding the interactions of particles and respective plant species.

Nanoparticles impose negative effects to plants in the form of phytotoxicity. Plants, however, possess inbuilt mechanisms for detoxification. These mechanisms are necessary for the removal of the toxins produced inside the plant cells. Volumes of emerging evidence suggest the impact of toxic effects of NPs. However, various reports suggest about surface modification of NPs to yield positive effects such as high crop yield. This characteristic could be utilized simultaneously to promote growth of edible crops and kill weeds or phytopathogens affecting crops. Size and concentration of NPs could be optimized to attain such desirable effects. Antimicrobial activity also makes NPs a strong tool as a biological control agent to help in crop management and improvement.

NP assimilation and subsequent accumulation in the food web is a major concern. This directly or indirectly affects growth and reproduction of plants. Additionally NPs may also cover cell surface of algae. It adversely affects food uptake by herbivores as well as nutrient exchange between microbial cells and their environment.

The NP-plant cell interaction modifies the gene as well as protein profiles of plant cells, ultimately leading to changes in biological pathways that bring about changes in growth and development of plants. Hence, experimentation should be conducted to generate data at the molecular level caused by uptake and translocation of NPs.

References

Abdul Hameed M, Al-Samarrai (2012) Nanoparticles as alternative to pesticides in management plant diseases-a review. Int J Sci Res Publ 2(*4*):1–4

Ahmad Z, Vargas-Reus MA, Bakhshi R, Ryan F, Ren GG, Oktar F, Allaker RP (2012) Antimicrobial properties of electrically formed elastomeric polyurethane-copper oxide nanocomposites for medical and dental applications. Methods Enzymol 509:87–99

Almeida E, Diamantino TC, de Sousa O (2007) Marine paints: the particular case of antifouling paints. Prog Org Coat 59:2–20

Anjum NA, Singh HP, Khan MI, Masood A, Per TS, Negi A, Batish DR, Khan NA, Duarte AC, Pereira E, Ahmad I (2015) Too much is bad-an appraisal of phytotoxic¬ity of elevated plant-beneficial heavy metal ions. Environ Sci Pollut Res 22:3361–3382

Arif N, Yadav V, Singh S, Tripathi DK, Dubey NK, Chauhan DK, Giorgetti L (2018) Interaction of Copper Oxide nanoparticles with plants: uptake, accumulation, and toxicity. nanomaterials in plants, algae, and microorganisms. Elsevier. https://doi.org/10.1016/B978-0-12-811487-2.00013-X

Atha DH, Wang H, Petersen EJ, Cleveland D, Holbrook R, Jaruga DP, Dizdaroglu M, Xing B, Nelson BC (2012) Copper oxide nanoparticle mediated DNA damage in terrestrial plant models. Environ Sci Technol 46:1819–1827

Ben-Sasson M, Zodrow KR, Genggeng Q, Kang Y, Giannelis EP, Elimelech M (2014) Surface functionalization of thin-film composite membranes with copper nanoparticles for antimicrobial surface properties. Environ Sci Technol 48:384–393

Buchanan BB, Gruissem W, Jones RL (2000) Biochemistry and molecular biology of plants, vol 40. American Society of Plant Physiologists, Rockville, p 1367

Cabiscol E, Tamarit J, Ros J (2010) Oxidative stress in bacteria and protein damage by reactive oxygen species. Int Microbiol 3(1):3–8

Chang YN, Zhang M, Xia L, Zhang J, Xing G (2012) The toxic effects and mechanisms of CuO and ZnO nanoparticles. Materials 5(12):2850–2871

Chaoui A, Jarrar B, El Ferjani E (2004) Effect of cadmium and copper on peroxidase, NADH oxidase and IAA oxidase activities in cell wall, soluble and microsomal membrane fractions of pea roots. J Plant Physiol 161(11):1225–1234

Chibber S, Ansari SA, Satar R (2013) New vision to CuO, ZnO, and TiO_2 nanoparticles: their outcome and effects. J Nanopart Res 15:1–13

Cioffi N, Ditaranto N, Torsi L, Picca RA, Giglio ED, Zambonin PG (2005) Synthesis, analytical characterization and bioactivity of Ag and Cu nanoparticles embedded in poly-vinyl-methyl-ketone films. Anal Bioanal Chem 382:1912

Clemens S, Palmgren MG, Kraemer U (2002) A long way ahead: understanding and engineering plant metal accumualtion. Trends Plant Sci 7:309–315

Costa MVJD, Sharma PK (2016) Effect of copper oxide nanoparticles on growth, morphology, photosynthesis, and antioxidant response in *Oryza sativa*. Photosynthetica 54:110–119

Da Silva LC, Oliva MA, Azevedo AA, De Araujo JM (2006) Responses of resting plant species to pollution from an iron pelletization factory. Water Air Soil Pollut 175:241–256

Dankovich TA, Smith JA (2014) Incorporation of copper nanoparticles into paper for point-of use water purification. Water Res 63:245–251

Deng F, Wang S, Xin H (2016) Toxicity of CuO nanoparticles to structure and metabolic activity of *Allium cepa* root tips. Bull Environ Contam Toxicol 97:702–708

Dietz KJ, Herth S (2011) Plant nanotoxicology. Trends Plant Sci 16(11):582–589

Dimkpa CO, Calder A, Britt DW, McLean JE, Anderson AJ (2011) Responses of a soil bacterium, *Pseudomonas chlororaphis* O6 to commercial metal oxide nanoparticles compared with responses to metal ions. Environ Pollut 159:1749–1756

Dimkpa CO, McLean JE, Latta DE, Manangón E, Britt DW, Johnson WP, Boyanov MI, Anderson AJ (2012) Cuo and ZnO nanoparticles: phytotoxicity, metal speciation, and induction of oxidative stress in sand-grown wheat. J Nanopart Res 14:1–15

Du W, Sun Y, Ji R, Zhu J, Wu J, Guo H (2011) TiO_2 and ZnO nanoparticles negatively affect wheat growth and soil enzyme activities in agricultural soil. J Environ Monit 13(4):822–828

Du W, Tan W, Peralta-Videa JR, Gardea-Torresdey JL, Ji R, Yin Y, Guo H (2017) Interaction of metal oxide nanoparticles with higher terrestrial plants: physiological and biochemical aspects. Plant Physiol Biochem 110:210–225

Du W, Tan W, Yin Y, Ji R, Peralta-Videa JR, Guo H, Gardea-Torresdey JL (2018) Differential effects of copper nanoparticles/microparticles in agronomic and physiological parameters of oregano (*Origanum vulgare*). Sci Tot Environ 618:306–312

Eichert T, Kurtz A, Steiner U, Goldbach HE (2008) Size exclusion limits and lateral heterogeneity of the stomatal foliar uptake pathway for aqueous solutes and water-suspended nanoparticles. Physiolgea Plantarum 134:151–160

Evans P, Matsunaga H, Kiguchi M (2008) Large-scale application of nanotechnology for wood protection. Nat Nanotechnol 3:577

Faheem I, Sammia S, Shakeel AK et al (2016) Green synthesis of copper oxide nanoparticles using *Abutilon indicum* leaf extract: antimicrobial, antioxidant and photocatalytic dye degradation activities. Trop J Pharm Res 16:743–753

Federici G, Shaw BJ, Handy RD (2007) Toxicity of titanium dioxide nanoparticles to rainbow trout (*Oncorhynchus mykiss*): gill injury, oxidative stress, and other physiological effects. Aquat Toxicol 84(4):415–430

Fleischer MA, Neill O, Ehwald R (1999) The pore size of non-graminaceous plant cell wall is rapidly decreased by borate ester cross-linking of the pectic polysaccharide rhamnogalacturon II. Plant Physiol 121:829–838

Gardea-Torresdey JL, Peralta-Videa JR, Montes M, De la Rosa G, Corral-Diaz B (2004) Bioaccumulation of cadmium, chromium and copper by *Convolvulus arvensis* L.: impact on plant growth and uptake of nutritional elements. Bioresour Technol 92(3):229–235

Gunalan S, Sivaraj R, Venckatesh R (2012) *Aloe barbadensis* Miller mediated green synthesis of mono-disperse copper oxide nanoparticles: optical properties. Spectrochim Acta A Mol Biomol Spectrosc 97:1140–1144

Hänsch R, Mendel RR (2009) Physiological functions of mineral micronutrients (Cu, Zn, Mn, Fe, Ni, Mo, B, Cl). Curr Opin Plant Biol 12:259–266

Hawthorne J, Musante C, Sinha SK, White JC (2012) Accumulation and phytotoxicity of engineered nanoparticles to *Cucurbita pepo*. Int J Phytoremediation 1:429–442

Honary S, Barabadi H, Fathabad EG, Naghibi F (2012) Green synthesis of copper oxide nanoparticles using *penicilliumaurantiogriseum*, *penicilliumcitrinum* and *penicilliumwakasmanii*. Dig J Nanomater Biostruct 7:999–1005

Hong J, Rico CM, Zhao L, Adeleye AS, Keller AA, Peralta-Videa JR, Gardea-Torresdey JL (2015) Toxic effects of copper-based nanoparticles or compounds to lettuce (*Lactuca sativa*) and alfalfa (*Medicago sativa*). Environ Sci Processes and Impacts 17:177–185

Hosseini – Koupaei M, Shareghi B, Saboury AA, Davar F, Sirotkin VA, Hosseini-Koupaei MH, Enteshari Z (2019) Catalytic activity, structure and stability of proteinase K in the presence of biosynthesized CuO nanoparticles. Int J Biol Macromol 122:732–744

Iravani S (2011) Green synthesis of metal nanoparticles using plants. Green Chem 13(10):2638–2650

Iravani S, Korbekandi H, Mirmohammadi SV, Zolfaghari B (2014) Synthesis of silver nanoparticles: chemical, physical and biological methods. Res Pharm Sci 9(6):385–406

Ivask A, Bondarenko O, Jepihhina N, Kahru A (2010) Profiling of the reactive oxygen species related ecotoxicity of CuO, ZnO, TiO_2, silver and fullerene nanoparticles using a set of recombinant luminescent *Escherichia coli* strains: differentiating the impact of particles and solubilised metals. Anal Bioanal Chem 398:701–716

Jia G, Wang H, Yan L, Wang X, Pei R, Yan T (2005) Cytotoxicity of carbon nanomaterials: single-wall nanotube, multi-wall nanotube, and fullerene. Environ Sci Technol 39:1378–1383

Joo JH, Bae YS, Lee JS (2001) Role of auxin-induced reactive oxygen species in root gravitropism. Plant Physiol 126:1055–1060

Kahru A, Dubourguier HC (2010) From ecotoxicology to nano ecotoxicology. Toxicology 269:105–119

Kalska-Szostko B (2011) Electrochemical methods in nanomaterials preparation. Recent Trend Electrochem Sci Technol. https://doi.org/10.5772/33662

Karlsson HL, Cronholm P, Gustafsson J, Moller L (2008) Copper oxide nanoparticles are highly toxic: a comparison between metal oxide nanoparticles and carbon nanotubes. Chem Res Toxicol 21(9):1726–1732

Kasana RC, Panwar NR, Kaul RK, Kumar P (2017) Biosynthesis and effects of copper nanoparticles on plants. Environ Chem Lett 15:233–240

Keller AA, Adeleye AS, Conway JR, Garner KL, Zhao L, Cherr GN et al (2017) Comparative environmental fate and toxicity of copper nanomaterials. NanoImpact 7:28–40

Knox JP (1995) The extra cellular-matrix in higher-plants. 4. Developmentally-regulated proteoglycans and glycoproteins of the plant-cell surface. FASEB J 9:1004–1012

Krumov N, Nochta IP, Oder SV, Gotcheva A, Posten AC (2009) Production of inorganic nanoparticles by microorganisms. Chem Eng Technol 32(7):1026–1035

Kulkarni V, Kulkarni P (2013) Green synthesis of copper nanoparticles using *Ocimum Sanctum* leaf extract. Int J Chem Stud 1(3):1–4

Kulkarni N, Muddapur U (2014) Biosynthesis of metal nanoparticles: a review. J Nanotechnol. https://doi.org/10.1155/2014/510246

Kurepa J, Paunesku T, Vogt S, Arora H, Rabatic BM, Lu J et al (2010) Uptake and distribution of ultrasmallanatase TiO_2 Alizarin red S nanoconjugates in *Arabidopsis thaliana*. Nano Lett 10:2296–2302

Kwak JM, Mori IC, Pei ZM, Leonhardt N, Torres MA, Dangl JL, Bloom RE, Bodde S, Jones JD, Schroeder JI (2003) NADPH oxidase AtrbohD and AtrbohF genes function in ROS-dependent ABA signaling in Arabidopsis. Eur Mol Biol Org J 22:2623–2633

Lee WM, An YJ, Yoon H, Kweon HS (2008) Toxicity and bioavailability of copper nanoparticles to the terrestrial plants mung bean (*Phaseolus radiatus*) and wheat (*Triticumaestivum*): plant agar test for water-insoluble nanoparticles. Environ Toxicol Chem 27:1915–1921

Lee CW, Mahendra S, Zodrow K, Li D, Tsai YC, Braam J, Alvarez PJ (2010) Developmental phytotoxicity of metal oxide nanoparticles to *Arabidopsis thaliana*. Environ Toxicol Chem 29:669–675

Lee HJ, Lee G, Jang NR, Yun JM, Song JY, Kim BS (2011) Biological synthesis of copper nanoparticles using plant extract. Nanotechnology 1:371–374

Lidon FC, Henriques FS (1998) Role of rice shoot vacuoles in copper toxicity regulation. Environ Exp Bot 39:197–202

Lin D, Xing B (2007) Phytotoxicity of nanoparticles: inhibition of seed germination and root growth. Environ Pollut 150(2):243–250

Long D, Wu G, Chen S (2007) Preparation of oligochitosan stabilized silver nanoparticles by gamma irradiation. Rad Phys Chem 76:1126–1131

MacFarlane GR, Burchett MD (2000) Cellular distribution of copper, lead and zinc in the grey mangrove, *Avicennia marina* (Forsk.) Vierh. Aquat Bot 68:45–59

Mafuné F, Kohno J, Takeda Y, Kondow T, Sawabe H (2000) Structure and stability of silver nanoparticles in aqueous solution produced by laser ablation. J Phys Chem B 104(35):8333–8337

Majumder DR (2012) Bioremediation: copper nanoparticles from electronic-waste. Int J Eng Sci Technol 4:4380–4389

Marcia R, Salvadori LF, Lepre AR, Oller do Nascimento CA (2013) Biosynthesis and uptake of copper nanoparticles by dead biomass of *Hypocrealixii*isolated from the metal mine in the Brazilian Amazon region. PLoS One 8(11):1–8

Mishra V, Mishra RK, Dikshit A, Pandey AC (2014) Interactions of nanoparticles with plants: an emerging prospective in the agriculture industry. Emerging Technologies and Management of Crop Stress Tolerance 1:159–180

Mittler R (2002) Oxidative stress, antioxidants and stress tolerance. Trends Plant Sci 7:405–410

Moon YS, Park ES, Kim TO, Lee HS, Lee SE (2014) SELDI-TOF MS-based discovery of a biomarker in *Cucumis sativus* seeds exposed to CuO nanoparticles. Environ Toxicol Pharmacol 38:922–931

Mortimer M, Kasemets K, Kahru A (2010) Toxicity of ZnO and CuO nanoparticles to ciliated protozoa *Tetrahymena thermophile*. Toxicology 269:182–189

Musante C, White JC (2012) Toxicity of silver and copper to *Cucurbita pepo*: differential effects of nano and bulk-size particles. Environ Toxicol 27(9):510–517

Nair PMG, Chung IM (2015) Study on the correlation between copper oxide nanoparticles induced growth suppression and enhanced lignification in Indian mustard (*Brassica juncea* L.). Ecotoxicol Environ Saf 113:302–313

Nasrollahzadeh M, Sajjadi M, Dasmeh HR, Sajadi SM (2018) Green synthesis of the Cu/sodium borosilicate nanocomposite and investigation of its catalytic activity. J Alloys Compd 763:1024–1034

Nekrasova GF, Ushakova OS, Ermakov AE, Uimin MA (2011) Effects of copper (II) ions and copper oxide nanoparticles on *Elodea densa* planch. Russ J Ecol 42:458–463

Nikhil J, Zhong LW, Tapan KS, Tarasankar P (2000) Seed mediated growth method to prepare cubic copper nanoparticles. Curr Sci 79:1367

Ovecka M, Lang I, Baluska F, Ismail A, Illes P, Lichtscheidl IK (2005) Endocytosis and vesicle trafficking during tip growth of root hairs. Protoplasma 226:39–54

Panigrah S, Kundu S, Ghosh SK, Nath S, Praharaj S, Soumen B, Pal T (2006) Cysteine functionalized copper organosol: synthesis, characterization and catalytic application. Polyhydron 25:1263

Pelletier DA, Suresh AK, Holton GA et al (2010) Effects of engineered cerium oxide nanoparticles on bacterial growth and viability. Appl Environ Microbiol 76(24):7981–7989

Perreault F, Oukarroum A, Melegari SP, Matias WG, Popovic R (2012) Polymer coating of copper oxide nanoparticles increases nanoparticles uptake and toxicity in the green alga *Chlamydomonasreinhardtii*. Chemosphere 87:1388–1394

Perreault F, Samadani M, Dewez D (2014) Effect of soluble copper released from copper oxide nanoparticles solubilisation on growth and photosynthetic processes of *Lemna gibba* L. Nanotoxicology 8:374–382

Rafique M, Shaikh AJ, Rasheed R, Tahir MB, Bakhat HF, Rafique MS, Rabbani F (2017) A review on synthesis, characterization and applications of copper nanoparticles using green method. Nanotechnology 12(04):1750043

Rahman A, Ismail A, Jumbianti D, Magdalena S, Sudrajat H (2009) Synthesis of copper oxide nanoparticles by using *Phormidium* cyanobacterium. Indones J Chem 9:355–360

Rajput VD, Chen Y, Ayup M (2015) Effects of high salinity on physiological and anatomical indices in the early stages of *Populuseuphratica* growth. Russ J Plant Physiol 62:229–236

Rajput V, Minkina T, Fedorenko A, Sushkova S, Mandzhieva S, Lysenko V, Duplii N, Fedorenko G, Dvadnenko K, Ghazaryan K (2018) Toxicity of copper oxide nanoparticles on spring barley (*Hordeumsativum distichum*). Sci Total Environ 645:1103–1113

Ramesh AV, Devi SR, Botsa SM, Basavaiah K (2018) Facile green synthesis of Fe_3O_4 nanoparticles using aqueous leaf extract of *Zanthoxylum armatum* DC. for efficient adsorption of methylene blue. J Asian Ceramic Soc 6:145–155

Rico CM, Majumdar S, Duarte-Gardea M, Peralta-Videa JR, Gardea-Torresdey JL (2011) Interaction of nanoparticles with edible plants and their possible implications in the food chain. J Agric Food Chem 59:3485–3498

Sahayaraj K, Rajesh S (2011) Bionanoparticles: synthesis and antimicrobial applications. In: Mendez-Vilas A (ed) Science against microbial pathogens: communicating current research and technological advances. Research Center, Badajoz, pp 228–244

Sekine R, Marzouk ER, Khaksar M, Scheckel KG, Stegemeier JP, Lowry GV, Donner E, Lombi E (2017) Aging of dissolved copper and copper-based nanoparticles in five different soils: short-term kinetics vs. long-term fate. J Environ Qual 46(6):1198–1205. https://doi.org/10.2134/jeq2016.12.0485

Shams M, Yildirim E, Agar G, Ercisli S, Dursun A, Ekinci M, Kul R (2018) Nitric oxide alleviates copper toxicity in germinating seed and seedling growth of *Lactuca sativa* L. Notulae Botanicae Horti Agrobotanici 46(1):167–172

Shaw AK, Hossain Z (2013) Impact of nano-CuO stress on rice (*Oryza sativa* L.) seedlings. Chemosphere 93:906–915

Shaw AK, Ghosh S, Kalaji HM, Bosa K, Brestic M, Zivcak M et al (2014) Nano-CuO stress induced modulation of antioxidative defense and photosynthetic performance of Syrian barley (*Hordeum vulgare* L.). Environ Exp Bot 102:37–47

Singh D, Kumar A (2016) Impact of irrigation using water containing CuO and ZnO nanoparticles on *Spinach oleracea* grown in soil media. Bull Environ Contam Toxicol 97:548–553

Singh VP, Singh S, Kumar J, Prasad SM (2015) Investigating the roles of ascorbate-glutathione cycle and thiol metabolism in arsenate tolerance in ridged *Luffa* seedlings. Protoplasma 252(5):1217–1229

Singh S, Vishwakarma K, Singh S, Sharma S, Dubey NK, Singh VK, Liu S, Tripathi DK, Chauhan DK (2017) Understanding the plant and nanoparticle interface at transcriptomic and proteomic level: a concentric overview. Plant Gene 11:265–272

Stampoulis D, Sinha SK, White JC (2009) Assay dependent phytotoxicity of nanoparticles to plants. Environ Sci Technol 43:9473–9479

Sundaramurthy N, Parthiban C (2015) Biosynthesis of copper oxide nanoparticles using *pyrus-pyrifolia* leaf extract and evolve the catalytic activity. Int Res J Eng Technol 2:332–338

Tanori J, Pileni MP (1997) Control of the shape of copper metallic particles by using a colloidal system as template. Langmuir 13(4):639–646

Tang Y, He R, Zhao J, Nie G, Xu L, Xing B (2016) Oxidative stress induced toxicity of CuO nanoparticles and related toxicogenomic responses in *Arabidopsis thaliana*. Environ Pollut 212:605–614

Tariq M, Hameed S, Yasmeen T, Zahid M, Zafar M (2014) Molecular characterization and identification of plant growth promoting endophytic bacteria isolated from the root nodules of pea (*Pisum sativum* L.). World J Microbiol Biotechnol 30(2):719–725

Thul ST, Sarangi BK (2015) Implications of nanotechnology on plant productivity and its rhizospheric environment. In: Nanotechnology and plant sciences. Springer, Cham, pp 37–53

Tong YP, Kneer R, Zhu YG (2004) Vacuolar compartmentalization: a second generation approach to engineering plants for phytoremediation. Trends Plant Sci 9:7–9

Umer S, Naveed S, Ramzan N, Rafique MS (2012) Selection of a suitable method for the synthesis of copper nanoparticles. Nanotechnology. https://doi.org/10.1142/S1793292012300058

Uzu G, Sobanska S, Sarret G, Muñoz M, Dumat C (2010) Foliar lead uptake by lettuce exposed to atmospheric fallouts. Environ Sci Technol 44(3):1036–1042

Valant J, Drobne D, Novak S (2012) Effect of ingested titanium dioxide nanoparticles on the digestive gland cell membrane of terrestrial isopods. Chemosphere 87(1):19–25

Vellora V, Padil T, Cernik M (2013) Green synthesis of copper oxide nanoparticles using gum karaya as a biotemplate and their antibacterial application. Int J Nanomedicine 8:889–898

Vinopal S, Ruml T, Kotrba P (2007) Biosorption of Cd21 and Zn21 by cell surface-engineered *Saccharomyces cerevisiae*. Int Biodeterior Biodegradation 60:96–102

Vymazal J, Brezinová T (2016) Accumulation of heavy metals in aboveground biomass of *Phragmitesaustralis* in horizontal flow constructed wetlands for wastewater treatment: a review. Chem Eng J 290:232–242

Wang Z, Xie X, Zhao J, Liu X, Feng W, White JC, Xing B (2012) Xylem and phloem-based transport of CuO nanoparticles in maize (*Zea mays* L.). Environ Sci Technol 46:4434–4441

Wang S, Liu H, Zhang Y, Xin H (2015) The effect of CuO nanoparticles on reactive oxygen species and cell cycle gene expression in roots of rice. Environ Toxicol Chem 34:554–561

Wang Z, Xu L, Zhao J, Wang X, White JC, Xing B (2016) CuO nanoparticle interaction with *Arabidopsis thaliana*: toxicity, parentprogeny transfer, and gene expression. Environ Sci Technol 50:6008–6016

Watanabe T, Misawa S, Hiradate S, Osaki M (2008) Root mucilage enhances aluminum accumulation in *Melastomamalabathricum*, an aluminum accumulator. Plant Signal Behav 3:603–605
Wessels JGH (1993) Wall growth, protein excretion and morphogenesis in fungi. New Phytol 123:397–413
White B, Yin M, Hall A, Le D, Stolbov S, Rahman T, Turro N, O'Brien S (2006) Complete CO oxidation over CuO nanoparticles supported on silica gel. Nano Lett 6:2095–2098
Wierzbicka M, Obidzinska J (1998) The uptake of lead on seed imbibition and germination in different plant species. Plant Sci 137:155–171
Yallappa S, Manjanna J, Sindhe MA et al (2013) Microwave assisted rapid synthesis and biological evaluation of stable copper nanoparticles using T. arjuna bark extract. Spectrochim Acta A Mol Biomol Spectrosc 110:108–115
Yamamoto O (2001) Influence of particle size on the antibacterial activity of zinc oxide. Int J Inorg Mater 3:643–646
Yue L, Zhao J, Yu X, Lv K, Wang Z, Xing B (2018) Interaction of CuO nanoparticles with duckweed (*Lemna minor*. L): uptake, distribution and ROS production sites. Environ Pollut. https://doi.org/10.1016/j.envpol.2018.09.013
Yurderi M, Bulut A, Ertas İE, Zahmakiran M, Kaya M (2015) Supported copper–copper oxide nanoparticles as active, stable and low-cost catalyst in the methanolysis of ammonia–borane for chemical hydrogen storage. Appl Catal B Environ 165:169–175
Zafar MS, Khurshid Z, Najeeb S, Zohaib S, Rehman IU (2017) Therapeutic applications of nanotechnology in dentistry. In: Andronescu E, Grumezescu AM (eds) Nanostructures for Oral Medicine. Elsevier
Zhang P, Ma Y, Zhang Z et al (2012) Comparative toxicity of nanoparticulate/bulk Yb2O3 and YbCl3 to cucumber (*Cucumis sativus*). Environ Sci Technol 46(3):1834–1841
Zhang D, Hua T, Xiao F, Chen C, Gersberg RM, Liu Y, Ng WJ, Tan SK (2014) Uptake and accumulation of CuO nanoparticles and CdS/ZnSquantum dot nanoparticles by *Schoenoplectus tabernaemontaniin* hydroponic mesocosms. Ecol Eng 70:114–123
Zuverza-Menan N, Medina-Velo IA, Barrios AC, Tan W, Peralta-Videa JR, Gardea-Torresdey JL (2015) Copper nanoparticles/compounds impact agronomic and physiological parameters in cilantro (*Coriandrum sativum*). Environ Sci: Processes Impacts 17:1783–1793

Chapter 9
Nanotechnological Advances with PGPR Applications

A. R. Nayana, Bicky Jerin Joseph, Ashitha Jose, and E. K. Radhakrishnan

Abstract Plant growth promoting rhizobacteria (PGPR) are soil bacteria which have the potential for direct and indirect effects on plant growth. These organisms may have the capability to limit or replace the use of chemical fertilizers and inputs of toxic chemicals. Exploring PGPR in agriculture is, thus, one of the more promising techniques for increasing agricultural production without harming ecosystems. At the same time, nano-technological applications are greatly imparting their influence in agriculture. When compared against conventional fertilizers, nano-fertilizers play an effective role in promoting plant growth as they are rapidly absorbed by plants. Hence, nano-materials such as nano-fibers, nano-fertilizers and nano-pesticides may produce revolutionary effects in the agricultural sector. PGPR together with nanomaterials may thus be a favorable strategy for managing plant growth and productivity. The application of nanomaterials like silver, titanium, zinc oxide, silica, gold and others with PGPR holds great promise. However, there can be both positive and negative impacts of engineered metal nanoparticles on rhizobacteria. Hence, engineered nanoparticle (ENPs) must be studied further to explore their use as ecofriendly agents for field application. In this chapter we describe the effects of nanofertilizers with PGPR as an innovative method for improving crop productivity.

Keywords Agriculture · Biofertilizer · Nanofertilizer · Nanoparticle · PGPR

9.1 Introduction

In this period of climate change and resource limitation, challenges to crop production in terms of abiotic stress, nutritional deficiency and disease are considerable. Managing these challenges with conventional agrochemicals is no longer practical as they will only produce significant negative impacts on both the environment and

---

A. R. Nayana · B. J. Joseph · A. Jose · E. K. Radhakrishnan (✉)
School of Biosciences, Mahatma Gandhi University, Kottayam, India
e-mail: radhakrishnanek@mgu.ac.in

S. Hayat et al. (eds.), *Sustainable Agriculture Reviews 41*, Sustainable Agriculture Reviews 41, https://doi.org/10.1007/978-3-030-33996-8_9

human health. Hence, to successfully counteract the adverse impacts of climate stress and lower yields, sustainable and innovative approaches are essential. Plant growth promoting rhizobacteria (PGPR) are heterogeneous root-associated beneficial bacteria which are known for their ability to enhance plant growth by either direct or indirect phytostimulatory mechanisms. Direct mechanisms involve those related to mobilization of important nutrients such as phosphorous, zinc, sulfur and iron, and for promoting non-symbiotic nitrogen fixation along with production of various phytohormones like indole acetic acid (Glick 2012). Indirectly, PGPR reduce the deleterious effects of phytopathogens and protect the plant against biotic and abiotic stress conditions (Beneduzi et al. 2012). However, the variability in performance of PGPR under varied climate, weather parameters and soil characteristics is a major difficulty to exploring its field efficacy (Timmusk et al. 2017).

PGPR formulations are applied as suspensions to seeds, root surfaces or directly to soil (Mendis et al. 2018). It is difficult for a single microbial inoculant to perform consistently under varying agro climatic conditions and stresses; therefore, recent trends in PGPR applications adopt multiple inocula. Microbial consortia have proven to have higher efficiency than application of a single species (Pandey et al. 2012). Their survival and colonization, however, are dependent on the physical, chemical and biological nature of the recipient environment. Declining microbial diversity and numbers within the consortia can result in inefficient colonization of the rhizosphere of the host plant.

Microbial consortia can be prepared in liquid, organic, inorganic, polymeric, and encapsulated formulations for wider use (Bashan et al. 2014). The carrier of the consortia can provide the necessary microenvironment to ensure survival of organisms and also act as a niche for security against soil predators. Peat, coal, clay, waste plant materials, vermiculite, and residues of azolla are commonly employed for PGPR applications (Maiyappan et al. 2011).

Maintenance of adequate growth conditions over time in terms of nutrition and climate are major hurdles in transferring the developed consortia from the lab to the field. Failure to maintain the desired environs can considerably affect microbial counts, which in turn can adversely affect field results. Hence, introducing innovative and effective methods for field delivery of PGPR is important.

Nanotechnology is, an emerging field that offers tremendous applications in all aspects of science including chemistry, biology, physics, materials science and engineering. The application of nanotechnology in the agricultural sector has gained immense attention due to its ability to enhance biotic and abiotic stress tolerance, disease detection and prevention along with refined nutrient absorption (Shalaby et al. 2016). Nanomaterials can improve nutrient utilization efficiency of plants when compared to conventional approaches. Nanoparticles (NPs) can boost plant metabolism by their defined physicochemical properties, and thereby enhance crop yield and provide nutrients to soil (Siddiqui et al. 2015).

Nanoparticles can generally be classified as carbon nanoparticles, metal nanoparticles, organic, and semiconductor nanoparticles (Buzea and Pacheco 2017). Among these, silver (Yin et al. 2012), titanium (Abdel Latef et al. 2018), zinc oxide (Laware and Raskar 2014), silica (Rastogi et al. 2019), carbon (Mohamed et al. 2018), boron

(Goudar et al. 2018; Shireen et al. 2018), gold (Shukla et al. 2015) and zeolite (Yılmaz et al. 2014) nanoparticles have been reported to have plant growth promoting effects. Nano-fertilizers are more effective than conventional chemical fertilizers as they do not cause problems with leaching and nutrient loss following application and only minimal amounts are required which thereby reduces risk of soil and water pollution.

Nanotechnology-based plant viral disease detection kits and nanobiosensors are gaining popularity by virtue of their improved efficacy in detection of various viral diseases (Chaudhary et al. 2018). Nanobiosensors can be used to detect even minute levels of fertilizers, herbicides, pesticides, insecticides, pathogens, moisture, and soil pH, thus supporting sustainable agriculture for enhanced crop production.

The rhizosphere is the zone surrounding the plant root and which contains abundant plant growth promoting microorganisms. Plants secrete various exudates into rhizosphere soil which attract microorganisms. The exudates and microbial communities also have the capability to produce various nano-size minerals in soil which have not yet been fully studied (Yu 2018). Reports are available on the plant growth promotion activity exhibited by nanoparticles in combination with various PGPR organisms. PGPR and nanotechnological applications can make the agriculture sector more powerful than conventional technologies used for crop improvement. By developing a conjugative approach of both NP and PGPR, there is immense potential to improve both yields and disease resistance of plants (Table 9.1; Figs. 9.1 and 9.2). Nanoparticles offering such potential are discussed in the following sections.

9.2 Titania Nanoparticles

Titania nanoparticles are widely used in cosmetics, agriculture, and the chemical and food industries. Titania in the form of titanium dioxide (TiO_2) nanoparticles has a positive effect on plants by its involvement in plant nitrogen metabolism and modulating ROS signaling (Abdel Latef et al. 2018). Application of low concentrations of nano-TiO_2 to roots or leaves can stimulate the crop, leading to the improved activities of certain enzymes (Lyu et al. 2017), enhanced chlorophyll content and thereby photosynthesis (Gao et al. 2008). This can also promote nutrient uptake (Larue et al. 2012), and improve stress tolerance (Karami and Sepehri 2018) and hence improve overall crop yield.

Titania nanoparticles impart an adhesive effect on bacteria as evident from the potent nanophase adhesion of *Pseudomonas fluorescens* 5RL and *Pseudomonas putida* TVA8 when compared to conventional titania (Park et al. 2008). Double inoculation of PGPR with sol gel-synthesized titania nanoparticles has been shown to enhance colonization of PGPR to about 25%. This occurs via formation of microniches around the root which are entirely different from the surrounding microbiome and thus allow beneficial bacteria to work as a functional unit leading to crop improvement (Timmusk et al. 2018). Establishment of significantly larger and

Table 9.1 Effects of selected nanoparticles with PGPR on plants

Nano particle	Method of synthesis	PGPR	Plant	Effect	References
Silica	Nanosilica synthesized from rice husk ash using alkaline extraction followed by acid precipitation	*Bacillus megaterium, Bacillus brevis, Pseudomonas fluorescens, Azotobacter vinelandii*	Maize (*Zea mays*)	Promoted seed germination percentage than conventional silica	Karunakaran et al. (2013)
	Nanosilica synthesized from rice husk using alkaline treatment followed by acid precipitation	–	Maize (*Zea mays*)	Increased bacterial biomass	Rangaraj et al. (2014)
Titania	Acidic hydrolysis of titanium ethoxide, $(Ti(OC_2H_5)_4)$ modified with triethanolamine, $N(C_2H_4OH)_3$	*B.thuringiensis AZP2, P.polymyxa A26*	Wheat (*Triticum aestivum cv. Stava*)	Enhanced performance of PGPR and their colonization	Timmusk et al. (2018)
	Sol-Gel approach (Captigel method and applying TiBALDH precursor)	*B.amyloliquefaciens subsp. plantarum UCMB5113*	Oilseed rape plants (*Brassica napus*)	Helped adhesion of beneficial bacteria to roots of oilseed rape; protected plant from the fungal pathogen *Alternaria brassicae*	Palmqvist et al. (2015)
Silver	Chemical method using sodium borohydride $(NaBH_4)$	*Pseudomonas fluorescence, Bacillus cereus*	Maize (*Zea mays*)	Augmented PGPR and induced increase in root area, root length and root-shoot ratio of maize irrigated with municipal wastewater	Khan and Bano (2015)

(continued)

Table 9.1 (continued)

Nano particle	Method of synthesis	PGPR	Plant	Effect	References
Gold	Biosynthesized by *Bacillus subtilis* SJ15	*P. monteilii*	Cowpea (*Vigna unguiculata*)	Enhancement of IAA production in *P. monteilii*; improved the plant probiotic effect	Panichikkal et al. (2019)
Zeolite	Commercially purchased	*Bacillus* sp.	Maize (*Zea mays*)	Increased plant height, leaf area, number of leaves, chlorophyll and total protein	Khati et al. (2018)
Zinc Oxide	–	*Brady rhizobium japonicum, Pseudomonas putida, Azospirillum lipoferum*	Soybean (*Glycine max* L.)	Increased plant height, number of nodules per plant, grain yield and grain weight	Khoramdel (2016)
Chitosan	–	*Bacillus* sp.	Maize (*Zea mays*)	Increased seed germination, plant height and leaf area	Khati et al. (2017)

thicker bacterial clumps than a self-produced biofilm matrix is possible through application of titania nanoparticles (Timmusk et al. 2018). Titania nanoparticles are reported to act at the nano interface between the beneficial plant growth promoting bacterium *Bacillus amyloliquefaciens* UCMB5113 and oilseed rape plants (*Brassica napus*) – titania nanoparticles increased adhesion of beneficial bacteria to the roots of oilseed rape and protected the plant from the fungal pathogen *Alternaria brassicae* (Palmqvist et al. 2015). Analysis by SEM, EDS, CLSM, SDS-PAGE and fluorescence measurements confirmed the nanoparticle-mediated colonization which eventually enhanced bacterial biomass (Palmqvist et al. 2015). Titania nanoparticles provide an effective platform for PGPR colonization on the plant and hence can be used as a potential agent for PGPR development for sustainable agriculture.

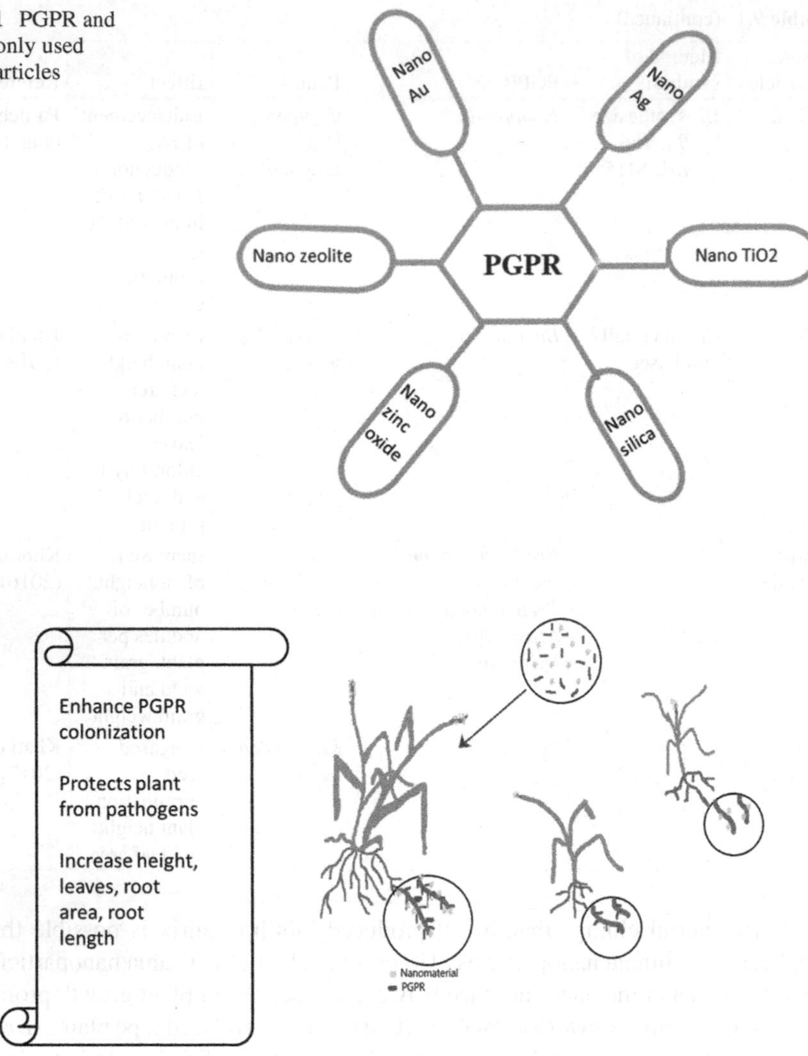

Fig 9.1 PGPR and commonly used nanoparticles

Fig 9.2 Combined action of nanomaterials with PGPR

9.3 Silica Nanoparticles

Silicon and silica nanoparticles are well established plant growth promoters. Silica nanoparticles have the capability to enter the plant and influence metabolic activities either directly or indirectly (Rastogi et al. 2019). Application of silica nanoparticles can lead to the formation of a film on the epidermal cell wall following absorption, which can provide additional structural leaf colour to the plant (Strout et al. 2013) and increase resistance to fungal, bacterial, and nematodal infections (Rastogi et al. 2019). The nano-silica layer can also reduce plant transpiration, which thereby makes plants more resistant to drought, high temperature, and humidity (Ashkavand et al. 2015). These modifications allow for improved plant growth and yield under adverse environmental conditions.

Nano-silica oxide has a significant impact on seed germination (Siddiqui and Al-whaibi 2014). Silica nanoparticles are also used as an effective nano-pesticide (Aa et al. 2016), nanoherbicide, and nanofertilizer in sustainable agriculture. Studies by Karunakaran et al. (2013) showed nanosilica to have a favorable effect on both beneficial bacterial populations and nutrient value of soil. Nanosilica synthesized using rice husks was proven to enhance the microbial population greater than that of sodium silicate, which inhibited plant growth promoting rhizobacteria. This effect may be due to the hydration properties of the nanosilica surface, which could facilitate attraction to the microbial surface (Gordienko and Kurdish 2007). In a study by Rangaraj et al. (2014) nanosilica treatment induced the populations of phosphate solubilizing bacteria and nitrogen fixers, where silica indirectly acted as a substrate for the bacteria.

Silica and PGPR individually are reported to be potent agents for crop improvement (Suriyaprabha et al. 2012; Ramprasad et al. 2015). Hence, the combination of these inputs has the potential to replace traditional fertilizers and to be used as an efficient agent for biofertilizer development.

9.4 Silver Nanoparticles

The wide acceptance of silver nanoparticles (AgNPs) is attributed to its well-known antibacterial activity (Radhakrishnan 2017). AgNPs are used in many applications from medical devices (Gherasim et al. 2018) to sports socks and washing machines, to deter microbial growth. Utilization of this nanoparticle occurs in almost all fields but especially in medical, dentistry, clothing, catalysis, mirrors, optics, photography, electronics, and the food industry (Dargo et al. 2017). Silver nanoparticles are also gaining attention and acceptance in the agricultural sector (Babu et al. 2014), but their ecotoxicological properties and underlying risks must be considered (Mao et al. 2018).

Treatment of fenugreek seedlings with biosynthesized silver nanoparticles was found to have beneficial impact on growth parameters such as number of leaves, root length, shoot length, and wet weight (Jasim et al. 2017).

Biologically synthesized silver nanoparticles are less toxic when compared with chemically synthesized AgNPs (Sharma et al. 2015). Silver nanoparticles are effective in facilitating the penetration of water and nutrients through the seed coat. AgNPs accelerated seed germination and seedling growth in *Boswellia ovalifoliolata* (Savithramma et al. 2012). In a study by Khan and Bano (2015), silver nano particles augmented PGPR activity by increasing root area, root length and root-shoot ratio of maize. Phytohormones are commonly applied to promote plant growth, and silver nanoparticles are now known to be an efficient tool to enhance phytohormones to a greater extent. Upon treatment with AgNPs, PGPR were reported to induce abscisic acid (ABA) levels by 34%, indole acetic acid (IAA) to 55% and gibberlic acid (GA) to 82% (Khan and Bano 2015).

Silver nanoparticles can have significant impact on the diversity of soil bacteria even when applied at minute levels. The fungal endophyte mediated synthesis of plant secondary metabolites such as taxol, podophyllotoxin, polyketides, terpenes and peptides (Mishra and Sarma 2018) can be enhanced effectively through a wide range of elicitors (Stierle and Stierle 2016). Among these, nanoelicitors comprising silver nanoparticles are found to be highly efficient (Jasim et al. 2017). Thus, exploring the potential of silver nanoparticles as a nanoelicitor for endophytic fungi can enhance the yield of desired secondary metabolites manyfold.

9.5 Gold Nanoparticles

Gold nanoparticles are among the most commonly synthesized and studied metal nanoparticles (Rashid et al. 2014). Synthesis includes chemical, biological and physical methods (Herizchi et al. 2014; Shah et al. 2014). Among these, biological methods are the most environmentally friendly and most commonly employed (Raghuvanshi et al. 2017), utilizing both plant extracts (Aljabali et al. 2018) and microorganisms (Roshmi et al. 2015). Plant growth promoting *Bacillus thuringiensis* strain PG-4 has been described as an effective agent for biosynthesis of gold nanoparticles (Raghuvanshi et al. 2017).

Considering its negligible toxicity (Rashid et al. 2014), gold nanoparticle-based formulations have become a huge attraction to the agricultural sector (Pestovsky and Martínez-antonio 2017). There are reports of enhancement of growth and yield of *Brassica juncea* (Arora and Sharma 2012) and *Arabidopsis thaliana* (Kumar et al. 2013) upon treatment with gold nanoparticles.

Enhanced growth of the PGPR *Pseudomonas* and *Bacillus* in the presence of 6.25 μg/mL gold nanoparticles was reported by Shukla et al. (2015). In addition, IAA production by *P. monteilii* was enhanced when treated with gold nanoparticles (Panichikkal et al. 2019). IAA is a major phytohormone that plays important roles in vascular tissue differentiation, root initiation, flowering, fruit ripening, leaf senescence and the abscission of leaves and fruits thus leading to overall plant growth (Basheer 2017). The exact mechanism behind enhancement of growth and IAA production by PGPR in the presence of gold nanoparticles has not yet been determined. A suggested mechanism involves penetration of the cell surface by nanoparticle aggregates without inducing any further toxic effect (Shukla et al. 2015). This attachment of nanoparticles can alter the shape and size of the bacterial cell and might accelerate its growth (Phenrat et al. 2009).

9.6 Nanozeolites

Zeolite, a crystalline aluminosilicate, is one of the most common minerals present in sedimentary rocks. Among the 40 naturally-occurring zeolites, the most well-known are clinoptilolite, erionite, chabazite, heulandite, mordenite, stilbite, and phillipsite. The maintenance of moisture content and pH of soil has been carried out effectively for years by Japanese farmers with the aid of zeolites (Ramesh and Reddy 2011). Zeolites have pores and channels within their crystal structure along with a high cation exchange capacity (CEC) which is beneficial in agriculture (Mahesh et al. 2018). They make up a class of excellent nutrient carriers, enhancing soil nutrient levels, which in turn increase crop yield and also nutrient utilization efficiency. Other common applications include use as a carrier of slow-release fertilizers, insecticides, fungicides, and herbicides, and also as a trap for heavy metals (Sangeetha and Baskar 2016).

The nanozeolites with few side effects, make them a powerful tool in agriculture as compared to bulk zeolite material (Khati et al. 2018). The ability of nanozeolites to enhance the organic carbon content of the soil and to stabilize micro and macro-aggregates are superior to that of bulk zeolite (Mirzaei et al. 2015).

Composites of zeolites possess a greater water retention capacity, water absorbance, equilibrium water content and swelling ratio when compared to nanozeolites used alone (Lateef et al. 2016). This may encourage utilization of nanozeolite-based composites as an environmentally-friendly fertilizer. Thus, crop productivity can be improved by a combined application of nanozeolite and PGPR.

9.7 Nano Zinc Oxide

Zinc is an essential micronutrient required by all organisms including plants, animals and humans. The human genome encodes approximately 3000 zinc-containing proteins having structural and functional roles (Process 2013). Zinc acts as a catalytic and structural cofactor in all classes of enzymes, namely oxidoreductases, transferases, hydrolases, lyases, isomerases and ligases (Mccall et al. 2000). These enzymes play a pivotal role in cellular regulation where they act either as an extracellular stimuli or intracellular messengers (Maret 2017).

Zinc is a key factor in photosynthesis (Wang et al. 2009), protein synthesis (Obata and Umebayashi 1988), phytohormone synthesis (e.g. auxine, ABA) (Atici et al. 2005), seedling vigor (Boonchuay et al. 2013), sugar formation (Mousavi 2011), membrane function and defense against disease and abiotic stress (e.g., drought) (Ma et al. 2017; Dang et al. 2008). Zinc deficiency can result in severe yield loss with plant death in acute cases (Hafeez 2014). Zinc deficiency is often observed in fields despite its abundance (Sharma et al. 2013). Owing to its relative insolubility, application of zinc fertilizer is the sole method to overcome this issue.

Both zinc sulfate and EDTA-Zn chelate are commonly used to meet this deficiency (Cakmak and Kutman 2018).

Nano fertilizer formulations have enhanced performance compared to traditional fertilizers, as they release the required nutrients in controlled manners which, along with their small size and large surface area, promote its activity. Zinc oxide nanoparticles at concentrations of 20 and 30 μg ml^{-1} reduced the flowering period in onion and ensured better growth of plants along with production of healthy seeds (Laware and Raskar 2014). Zinc oxide nanoparticles can, however, exert negative impacts on plants. Treatment of zinc oxide nanoparticles on tomato plants negatively affected both plant growth and chlorophyll content which, in turn, affected photosynthetic rate (Wang et al. 2018). A supernatant of ZnO nanoparticle suspensions containing Zn^{2+} was not found to affect growth of tomato; hence, toxicity is attributed to ZnO nanoparticles.

ZnO NPs are reported to enhance the plant defense response by increasing transcription of genes related to antioxidant enzymes (Wang et al. 2018). Numerous zinc-solubilizing bacteria have been reported, which are capable of solubilizing zinc in soil (Kamran et al. 2017). Mumtaz et al. (2017) showed that inoculation of zinc solubilizing bacterial isolates to result in improved growth of maize. Zinc solubilizers also of have the capability to produce zinc nanoparticles. Sultana et al. (2019) reported the production of zinc nanoparticles by zinc solubilizing strains of *Pseudomonas, Bacillus* and *Azospirillum*.

Seed inoculation with PGPR along with nano zinc oxide, significantly increased plant height, number of nodules per plant, grain yield and grain weight (Khoramdel 2016). ZnO nanoparticles exhibit a dose-dependent enhancement in siderophore production of bacteria (Haris and Ahmad 2017). An environmentally-friendly dose of Zn nanoparticles along with PGPR could very well revolutionize the agriculture sector.

9.8 Nano Carbon

Use of carbon nanomaterials in agriculture has both negative and positive feedbacks (Mukherjee et al. 2016). They are widely used for plant growth promotion, plant protection, plant transformation and for nanodiagnostics in the agricultural sector (Al-whaibi and Mohammad 2017). Nano Fullerenes (nC_{60}) are reported to have little impact on the soil microbial community (Tong et al. 2007). It is possible; however, that application of PGPR along with suitable nano-carbon or fullerenes can have a future application.

9.9 Nano Boron

Boron, a trace element, plays an important role in plant growth such as cell division, elongation, nitrogen and carbohydrate metabolism, sugar transport, cytoskeletal proteins, ion fluxes (H^+, K^+, PO_4^{3-}, Rb^+, Ca^{2+}) across membranes and phenol metabolism and transport (Shireen et al. 2018). Maziah et al. (2010) reported that inoculation of plant growth promoting rhizobacteria, *Bacillus sphaericus* UPMB10 in modified MS medium containing boron improved the growth and root biomass of banana plantlets compared to control. Nanoboron nitride fertilization increased seed and stalk yield of sunflower (*Helianthus annuus* L) (Goudar et al. 2018). PGPR mediated boron uptake in boron-limited conditions would be a great achievement for the field of agriculture. Use of boron nanomaterials in PGPR formulations may offer promising applications.

9.10 Nano Chitosan

Chitosan, a natural biomaterial, can be formed by the deacetylation of chitin. In plants, chitosan induces biotic and abiotic stress tolerance and control of plant diseases, and promotes growth (Malerba and Cerana 2016). Nano chitosan in combination with plant growth promoting rhizobacteria (*Bacillus spp*) in maize has been reported to enhance plant height, leaf area, seed germination and increased organic acid production in response to stress tolerance (Khati et al. 2017). Thus, the use of natural biopolymers in PGPR formulation can potentially have an immense impact in PGPR inoculant technology.

9.11 Advances of Nanotechnology with PGPR

Nanotechnology has revolutionized the agricultural sector with its wide range of applications. This includes application of more efficient and targeted use of inputs, thereby increasing nutrient uptake by plants, disease control and its detection, storage and packaging (Prasad et al. 2017). An emerging trend of bio-nano-encapsulation using plant beneficial microorganisms and nano particles has paved a unique way of development in current agriculture scenarios.

Understanding potential toxicity of nanoparticles together with microbes is crucial for optimizing their application. In such cases the use of nanoparticles derived from biopolymers such as proteins, carbohydrates can offer a significant role.

9.11.1 Nano Encapsulation of PGPR

The use of nanomaterials for the delivery of macro- and micro-nutrients to plants is trending (Jampílek and Kráľová 2017). Nanoencapsulation also has the capability to protect crops from disease-causing organisms, insects, and other pests (Castañeda et al. 2014). De Gregorio et al. (2017) showed that immobilization of *Pantoea agglomerans* and *Burkholderia caribensis* in nanofibers did not alter the viability or beneficial properties of either rhizobacteria. The immobilized PGPR were associated with increased seed germination, and length and dry weight of soybean roots.

Maintaining viable counts of bacteria is a major concern when dealing with PGPR application in the field. Nanofibers and other nanomaterials can be a promising ecofriendly agent for inoculum development in which PGPR exerts its beneficial plant growth promoting traits and nanofibers protect bacteria and seeds from local abiotic stresses (Fig. 9.3).

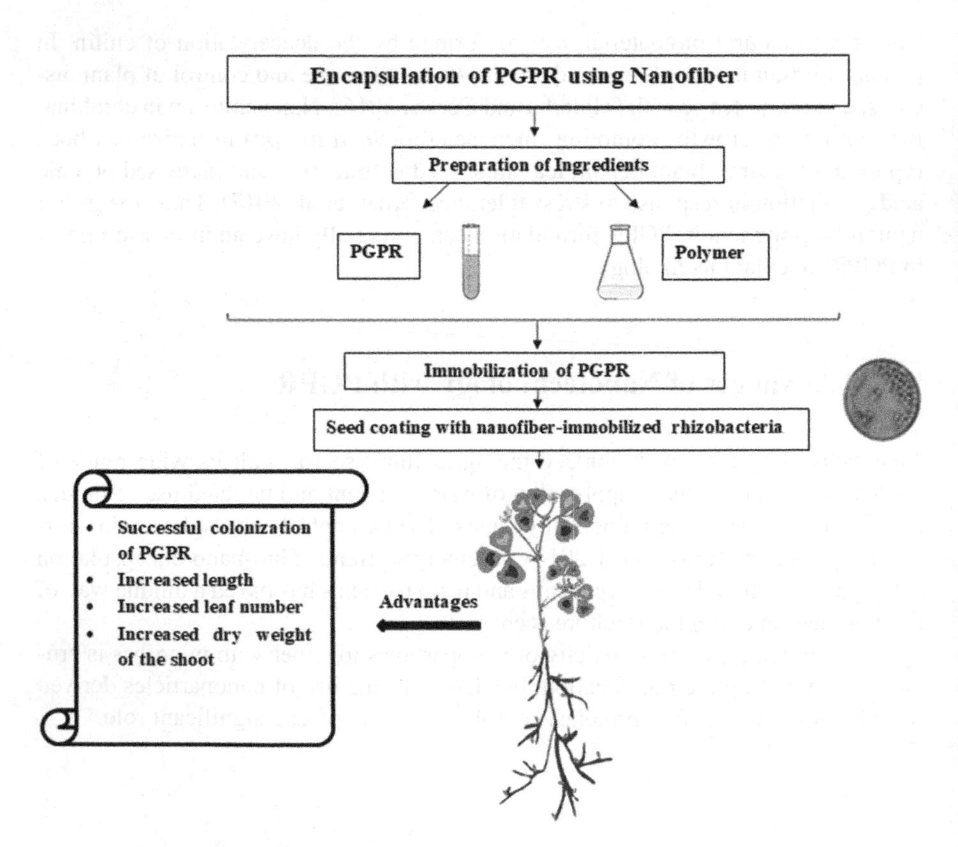

Fig 9.3 Benefits of encapsulation of PGPR using nanofiber

9.11.2 *Nanobiofertilizers with PGPR*

Nano-fertilizers have the potential to replace traditional fertilizers. They can increase nutrient use efficiency, reduce soil toxicity, minimize the potential negative effects associated with over dosage and reduce frequency of application (Qureshi et al. 2018). Slow-release nanofertilizers which release nutrients and plant growth promoting rhizobacteria were formulated by Mala et al. (2017); blending of nanofertilizer with neem cake, a byproduct of neem (*Azadirachta indica*) and PGPR was carried out. This slow release fertilizer system can accelerate the enzyme action during germination and increase seed vigour index (Mala et al. 2017). Use of PGPR facilitates the efficient use of NPK fertilizer and, hence, combined action can enhance overall crop yields.

9.12 Conclusions

Nanoparticles combined with PGPR have the potential to create a promising future for the upcoming age of agriculture. Nanoparticle interactions with PGPR can promote root colonization, phytohormone and secondary metabolite synthesis, and overall enhancement of the performance of rhizobacteria. Nano encapsulation of PGPR and nanobiofertilizer with PGPR are also trending. These eco-friendly compounds offer immense potential to replace traditional fertilizer practices. PGPR nanotechnology can be exploited as a low-input, sustainable and environmentally-friendly technology for management of plant stress and disease management. Future innovative technologies in this area will revitalize the agricultural sector.

References

Aa E et al (2016) The silica-nano particles treatment of squash foliage and survival and development of Spodoptera littoralis (Bosid.) larvae. J Entomol Zool Stud 4(1):175–180

Abdel Latef AAH et al (2018) Titanium dioxide nanoparticles improve growth and enhance tolerance of broad bean plants under saline soil conditions. Land Degrad Dev. John Wiley & Sons, Ltd, 29(4):1065–1073. https://doi.org/10.1002/ldr.2780

Aljabali AAA et al (2018) Synthesis of gold nanoparticles using leaf extract of Ziziphus zizyphus and their antimicrobial activity. Nanomaterials:1–15. https://doi.org/10.3390/nano8030174

Al-whaibi M, Mohammad F (2017) Role of nanoparticles in plants chapter 2 role of nanoparticles in plants. Nanotechnol Plant Sci (March). https://doi.org/10.1007/978-3-319-14502-0

Arora S, Sharma P (2012) Gold-nanoparticle induced enhancement in growth and seed yield of Brassica juncea. Plant Growth Regul:303–310. https://doi.org/10.1007/s10725-011-9649-z

Ashkavand P et al (2015) Effect of SiO2 nanoparticles on drought resistance in hawthorn seedlings. De Gruyter 76(4):350–359. https://doi.org/10.1515/frp-2015-0034

Atici Ö, Ağar G, Battal P (2005) Changes in phytohormone contents in chickpea seeds germinating under lead or zinc stress. Biol Plant 49(2):215–222. https://doi.org/10.1007/s10535-005-5222-9

Babu MY et al (2014) Application of biosynthesized silver nanoparticles in agricultural and marine application of biosynthesized silver nanoparticles in agricultural and marine Pest control. Curr Nanosci (May). https://doi.org/10.2174/157341371130966010З
Bashan Y, de-Bashan LE, Prabhu SR, Hernandez JP (2014) Advances in plant growth-promoting bacterial inoculant technology: formulations and practical perspectives (1998–2013). Plant Soil 378(1–2):1–33
Basheer J (2017) Plant growth promoting potential of endophytic bacteria isolated from Piper nigrum plant growth promoting potential of endophytic bacteria isolated from Piper nigrum. Plant Growth Regul (September 2013). https://doi.org/10.1007/s10725-013-9802-y
Beneduzi A, Ambrosini A, Passaglia LMP (2012) Plant growth-promoting rhizobacteria (PGPR): their potential as antagonists and biocontrol agents. Genet Mol Biol 4(35):1044–1051
Boonchuay P et al (2013) Effect of different foliar zinc application at different growth stages on seed zinc concentration and its impact on seedling vigor in rice. Soil Sci Plant Nutr 59(2):180–188. https://doi.org/10.1080/00380768.2013.763382
Buzea C, Pacheco I (2017) Nanomaterials and their classification. In: EMR/ESR/EPR spectroscopy for characterization of nanomaterials. Springer, New Delhi, pp 3–45
Cakmak I, Kutman UB (2018) Agronomic biofortification of cereals with zinc: a review. Eur J Soil Sci 69(1):172–180. https://doi.org/10.1111/ejss.12437
Castañeda L et al (2014) Innovative Rice seed coating (Oryza Sativa) with polymer nanofibres and microparticles using the electrospinning method. J Res Updates Polym Sci 3(1):33–39. https://doi.org/10.6000/1929-5995.2014.03.01.5
Chaudhary V, Jangra S, Yadav NR (2018) Nanotechnology based approaches for detection and delivery of microRNA in healthcare and crop protection. J Nanobiotechnol 16(1):40
Dang H et al (2008) Dominant chloramphenicol-resistant bacteria and resistance genes in coastal marine waters of Jiaozhou Bay, China. World J Microbiol Biotechnol 24(2):209–217. https://doi.org/10.1007/s11274-007-9458-8
Dargo H, Ayaliew A, Kassa H (2017) Synthesis paradigm and applications of silver nanoparticles (AgNPs), a review. Sustain Mater Technol Elsevier 13(August):18–23. https://doi.org/10.1016/j.susmat.2017.08.001
De Gregorio PR, Michavila G, Muller LR, de Souza Borges C, Pomares MF, de Sa ELS, Pereira C, Vincent PA (2017) Beneficial rhizobacteria immobilized in nanofibers for potential application as soybean seed bioinoculants. PLoS One 12(5):e0176930
Gao F et al (2008) Was improvement of spinach growth by nano-TiO(2) treatment related to the changes of Rubisco activase? Biometals. https://doi.org/10.1007/s10534-007-9110-y
Gherasim O, Ficai A, Andronescu E (2018) Biomedical applications of silver nanoparticles : an up-to-date overview. Nanomaterials:1–25. https://doi.org/10.3390/nano8090681
Glick BR (2012) Plant growth-promoting bacteria: mechanisms and applications. Scientifica 2012
Gordienko AS, Kurdish I (2007) Surface electrical properties of Bacillus subtilis cells and the effect of interaction with silicon dioxide particles. Biophysics. https://doi.org/10.1134/S0006350907020121
Goudar KM et al (2018) Response of sunflower (Helianthus annuus L.) to nano boron nitride fertilization. Int J Chem Stud 6(5):2624–2630
Hafeez B (2014) Role of zinc in plant nutrition- a review. Am J Exp Agric 3(2):374–391. https://doi.org/10.9734/ajea/2013/2746
Haris Z, Ahmad I (2017) Impact of metal oxide nanoparticles on beneficial soil microorganisms and their secondary metabolites. Int J Life-Sci Sci Res 3(April):1020–1030. https://doi.org/10.21276/ijlssr.2017.3.3.10
Herizchi R et al (2014) Current methods for synthesis of gold nanoparticles. Artif Cells Nanomed Biotechnol. https://doi.org/10.3109/21691401.2014.971807
Jampílek J, Kráľová K (2017) Nanomaterials for delivery of nutrients and growth-promoting compounds to plants. In: Nanotechnology. Springer, Singapore, pp 177–226
Jasim B et al (2017) Plant growth and diosgenin enhancement effect of silver nanoparticles in Fenugreek (Trigonella foenum-graecum L.). Saudi Pharm J King Saud Univ 25(3):443–447. https://doi.org/10.1016/j.jsps.2016.09.012.

Kamran S et al (2017) Contribution of Zinc solubilizing bacteria in growth promotion and zinc content of Wheat. Front Microbiol 8(December). https://doi.org/10.3389/fmicb.2017.02593

Karami A, Sepehri A (2018) Nano titanium dioxide and nitric oxide alleviate salt induced changes in seedling growth, physiological and photosynthesis attributes of barley. Zemdirbyste-Agric 105(2):123–132. https://doi.org/10.13080/z-a.2018.105.016

Karunakaran G et al (2013) Effect of nanosilica and silicon sources on plant growth promoting rhizobacteria, soil nutrients and maize seed germination. IET Nanobiotechnol (April):1–8. https://doi.org/10.1049/iet-nbt.2012.0048

Khan N, Bano A (2015) Role of plant growth promoting Rhizobacteria and Ag-nano particle in the bioremediation of heavy metals and maize growth under municipal wastewater irrigation. Int J Phytoremediation 6514(October). https://doi.org/10.1080/15226514.2015.1064352.

Khati P et al (2017) Nanochitosan supports growth of Zea mays and also maintains soil health following growth. 3 Biotech. Springer Berlin Heidelberg 7(1):1–9. https://doi.org/10.1007/s13205-017-0668-y.

Khati P et al (2018) Effect of nanozeolite and plant growth promoting rhizobacteria on maize. 3 Biotech. Springer Berlin Heidelberg (February) https://doi.org/10.1007/s13205-018-1142-1

Khoramdel RSSS (2016) Effects of Nano-zinc oxide and seed inoculation by plant growth promoting Rhizobacteria (PGPR) on yield, yield components and grain filling period of soybean (Glycine max L.). Iran J Field Crop Res 13(4):738–753

Kumar V et al (2013) Science of the total environment gold nanoparticle exposure induces growth and yield enhancement in Arabidopsis thaliana. Sci Tot Environ The Elsevier B.V:461–462, 462–468. https://doi.org/10.1016/j.scitotenv.2013.05.018

Larue C et al (2012) Comparative uptake and impact of TiO 2 nanoparticles in Wheat and Rapeseed. J Toxicol Environ Health Part A. https://doi.org/10.1080/15287394.2012.689800

Lateef A et al (2016) Microporous and mesoporous materials synthesis and characterization of zeolite based nano e composite : an environment friendly slow release fertilizer. Microporous Mesoporous Mater. Elsevier Ltd 232:174–183. https://doi.org/10.1016/j.micromeso.2016.06.020.

Laware SL, Raskar S (2014) Influence of Zinc oxide nanoparticles on growth, flowering and seed productivity in onion. Int J Appl Microbiol 3(7):874–881

Lyu S et al (2017) Titanium as a beneficial element for crop production. Front Plant Sci 8(April):1–19. https://doi.org/10.3389/fpls.2017.00597

Ma D et al (2017) Physiological responses and yield of wheat plants in Zinc-mediated alleviation of drought stress. Front Plant Sci 8(May):1–12. https://doi.org/10.3389/fpls.2017.00860

Maiyappan S et al (2011) Isolation, evaluation and formulation of selected microbial consortia for sustainable agriculture. J Biofertil Biopestic 2(2):2–7. https://doi.org/10.4172/2155-6202.1000109

Mahesh M et al (2018) Zeolite farming : a sustainable agricultural prospective. Int J Appl Sci 7(5):2912–2924

Mala R et al (2017) Evaluation of Nano structured slow release fertilizer on the soil fertility, yield and nutritional profile of Vigna radiata. Recent Pat Nanotechnol 11(1):50–62. https://doi.org/10.2174/1872210510666160727093554

Malerba M, Cerana R (2016) Chitosan effects on plant systems. Int J Mol Sci 17(7):1–15. https://doi.org/10.3390/ijms17070996

Mao B et al (2018) Silver nanoparticles have lethal and sublethal adverse effects on development and longevity by inducing ROS-mediated stress responses. Sci Rep. Springer US (500):1–16. https://doi.org/10.1038/s41598-018-20728-z.

Maret W (2017) Zinc in cellular regulation : the nature and significance of "Zinc signals". Int J Mol Sci. https://doi.org/10.3390/ijms18112285

Maziah M, Zuraida AR, Halimi MS, Zulkifli HS, Sreeramanan S (2010) Influence of boron on the growth and biochemical changes in plant growth promoting rhizobacteria (PGPR) inoculated banana plantlets. World J Microbiol Biotechnol 26(5):933–944

Mccall KA, Huang C, Fierke CA (2000) Zinc and health : current status and future directions. Function and mechanism of zinc Metalloenzymes 1. Am Soc Nutr Sci 29(5):1437S–1446S

Mendis HC, Thomas VP, Schwientek P, Salamzade R, Chien JT, Waidyarathne P, Kloepper J, De La Fuente L (2018) Strain-specific quantification of root colonization by plant growth promoting rhizobacteria Bacillus firmus I-1582 and Bacillus amyloliquefaciens QST713 in non-sterile soil and field conditions. PLoS One 13(2):e0193119

Mirzaei M, Ali A, Safari A (2015) Aggregation stability and organic carbon fraction in a soil amended with some plant residues, nanozeolite, and natural zeolite. Int J Recycling Organic Waste Agric:11–22. https://doi.org/10.1007/s40093-014-0080-0

Mishra R, Sarma VV (2018) Secondary metabolite production by endophytic Fungi: the gene clusters, nature, and expression in. In: Jha S (ed) Endophytes and secondary metabolites. Springer, Dordrecht, pp 1–16. https://doi.org/10.1007/978-3-319-76900-4_20-1.

Mohamed MA, Hashim AF, Alghuthaymi MA, Abd-Elsalam KA (2018) Nano-carbon: plant growth promotion and protection. In: Nanobiotechnology applications in plant protection. Springer, Cham, pp 155–188

Mousavi SR (2011) Zinc in crop production and interaction with phosphorus. Aust J Basic Appl Sci 5(9):1503–1509. Available at: https://www.scopus.com/inward/record.uri?eid=2-s2.0-84155179110&partnerID=40&md5=17b9048c5e857bbd085b53ac7d590ab4

Mukherjee A et al (2016) Carbon nanomaterials in agriculture: a critical review. Front Plant Sci 7(February):1–16. https://doi.org/10.3389/fpls.2016.00172

Mumtaz MZ et al (2017) Zinc solubilizing Bacillus spp. potential candidates for biofortification in maize. Microbiol Res. Elsevier, 202(June):51–60. https://doi.org/10.1016/j.micres.2017.06.001.

Obata H, Umebayashi M (1988) Effect of zinc deficiency on protein synthesis in cultured tobacco plant cells. Soil Sci Plant Nutr 34(3):351–357. https://doi.org/10.1080/00380768.1988.10415691

Palmqvist NGM et al (2015) Nano titania aided clustering and adhesion of beneficial bacteria to plant roots to enhance crop growth and stress management. Nature Publishing Group. Nature Publishing Group (December 2014):1–12. https://doi.org/10.1038/srep10146.

Pandey P, Bisht S, Sood A, Aeron A, Sharma GD, Maheshwari DK (2012) Consortium of plant-growth-promoting bacteria: future perspective in agriculture. In: Bacteria in agrobiology: plant probiotics. Springer, Berlin/Heidelberg, pp 185–200

Panichikkal J, Thomas R, John JC (2019) Biogenic gold nanoparticle supplementation to plant beneficial pseudomonas monteilii was found to enhance its plant probiotic effect. Curr Microbiol. Springer US 0(0):0. https://doi.org/10.1007/s00284-019-01649-0.

Park MR et al (2008) Influence of nanophase titania topography on bacterial attachment and metabolism. Int J Nanomedicine 3(4):497–504

Pestovsky YS, Martínez-antonio A (2017) The use of nanoparticles and Nanoformulations in agriculture. J Nanosci Nanotechnol 17(12):8699–8730. https://doi.org/10.1166/jnn.2017.15041

Phenrat T et al (2009) Partial oxidation (" Aging ") and surface modification decrease the toxicity of Nanosized Zerovalent. Environ Sci Technol. https://doi.org/10.1021/es801955n

Prasad R, Kumar M, Kumar V (2017) Nanomaterials for delivery of nutrients and growth-promoting compounds to plants. Nanotechnol Agric Paradigm. https://doi.org/10.1007/978-981-10-4573-8

Process DG (2013) Detonation gun process Zinc Biochemistry: from a single Zinc enzyme to a key element of life. Am Soc Nutr Adv Nutr (1):82–91. https://doi.org/10.3945/an.112.003038.82

Qureshi A, Singh DK, Dwivedi S (2018) Nano-fertilizers: a novel way for enhancing nutrient use efficiency and crop productivity. Int J Curr Microbiol App Sci 7(2):3325–3335. https://doi.org/10.20546/ijcmas.2018.702.398

Radhakrishnan EK (2017) Microbially and phytofabricated AgNPs with different mode of bactericidal action were identified to have comparable potential for surface fabrication of central venous catheters to combat Staphylococcus aureus biofilm. J Photochem Photobiol B Biol Elsevier B.V. https://doi.org/10.1016/j.jphotobiol.2017.04.036

Raghuvanshi P et al (2017) Biosynthesis and characterization of gold nanoparticles from plant growth promoting rhizobacteria. Int J Chem Stud 5(5):525–532

Ramesh K, Reddy DD (2011) Zeolites and their potential uses in agriculture. Adv Agron 113. https://doi.org/10.1016/B978-0-12-386473-4.00004-X
Ramprasad D, Sahoo D, Sreedhar B (2015) Plant growth promoting Rhizobacteria – an overview. Eur J Biotechnol Biosci 2(2):30–34
Rangaraj S et al (2014) Effect of silica nanoparticles on microbial biomass and silica availability in maize rhizosphere. Biotechnol Appl Biochem. https://doi.org/10.1002/bab.1191
Rashid R, Murtaza G, Zahra A (2014) Gold nanoparticles: synthesis and applications in drug. Trop J Pharm Res 13(July):1169–1177
Rastogi A et al (2019) Application of silicon nanoparticles in agriculture. 3 Biotech. Springer International Publishing 9(3):1–11. https://doi.org/10.1007/s13205-019-1626-7.
Roshmi T et al (2015) Effect of biofabricated gold nanoparticle-based antibiotic conjugates on minimum inhibitory concentration of bacterial isolates of clinical origin. Gold Bull:63–71. https://doi.org/10.1007/s13404-015-0162-4
Sangeetha C, Baskar P (2016) Zeolite and its potential uses in agriculture : a critical review. Agric Rev 37(2):101–108. https://doi.org/10.18805/ar.v0iof.9627.
Savithramma N, Ankanna S, Bhumi G (2012) Effect of nanoparticles on seed germination and seedling growth of Boswellia ovalifoliolata an endemic and endangered medicinal tree taxon. Nano Vision 2(1):2
Shah M et al (2014) Gold nanoparticles: various methods of synthesis and antibacterial applications. Fron Biosci (Landmark Ed) 1(1):1320–1344
Shalaby TA, Bayoumi Y, Abdalla N, Taha H, Alshaal T, Shehata S, Amer M, Domokos-Szabolcsy É, El-Ramady H (2016) Nanoparticles, soils, plants and sustainable agriculture. In: Nanoscience in food and agriculture 1. Springer, Cham, pp 283–312
Sharma A, Patni B, Shankhdhar D (2013) Zinc – an indispensable micronutrient. Physiol Mol Biol Plants 19(March):11–20. https://doi.org/10.1007/s12298-012-0139-1.
Sharma D, Kanchi S, Bisetty K (2015) Biogenic synthesis of nanoparticles : a review. Arab J Chem. https://doi.org/10.1016/j.arabjc.2015.11.002
Shireen F et al (2018) Boron : functions and approaches to enhance its availability in plants for sustainable agriculture. Int J Mol Sci:95–98. https://doi.org/10.3390/ijms19071856
Shukla SK et al (2015) Prediction and validation of gold nanoparticles (GNPs) on plant growth promoting rhizobacteria (PGPR): a step toward development of. De Gruyter 4(5):439–448. https://doi.org/10.1515/ntrev-2015-0036
Siddiqui MH, Al-whaibi MH (2014) Role of nano-SiO_2 in germination of tomato (Lycopersicum esculentum seeds mill.). Saudi J Biol Sci. King Saud Univ 21(1):13–17. https://doi.org/10.1016/j.sjbs.2013.04.005.
Siddiqui MH, Al-Whaibi MH, Mohammad F (2015) Nanotechnology and plant sciences: nanoparticles and their impact on plants. Nanotechnol Plant Sci Nanopart Impact Plants:1–303. https://doi.org/10.1007/978-3-319-14502-0.
Stierle AA, Stierle DB (2016) Bioactive secondary metabolites produced by the fungal endophytes of conifers. HHS Public Access 10(10):1671–1682
Strout G et al (2013) Silica nanoparticles aid in structural leaf coloration in the Malaysian tropical rainforest understorey herb Mapania caudata. Ann Bot:1141–1148. https://doi.org/10.1093/aob/mct172
Sultana U, Desai S, Reddy G, TNKVV P (2019) Zinc solubilizing plant growth promoting microbes produce zinc nanoparticles. bioRxiv:602219
Suriyaprabha R et al (2012) Growth and physiological responses of maize (Zea mays L.) to porous silica nanoparticles in soil. J Nanopart Res 14(12). https://doi.org/10.1007/S11051-012-1294-6
Timmusk S et al (2017) Perspectives and challenges of microbial application for crop improvement. Front Plant Sci 8(February):1–10. https://doi.org/10.3389/fpls.2017.00049
Timmusk S, Seisenbaeva G, Behers L (2018) Titania (TiO_2) nanoparticles enhance the performance of growth-promoting rhizobacteria. Sci Rep. Springer US:1–13. https://doi.org/10.1038/s41598-017-18939-x
Tong Z, Bischoff M, Nies L, Applegate B, Turco RF (2007) Impact of fullerene (C60) on a soil microbial community. Environ Sci Technol 41(8):2985–2991

Wang H, Liu RL, Jin JY (2009) Effects of zinc and soil moisture on photosynthetic rate and chlorophyll fluorescence parameters of maize. Biol Plant 53(1):191–194. https://doi.org/10.1007/s10535-009-0033-z

Wang XP et al (2018) Effects of zinc oxide nanoparticles on the growth, photosynthetic traits, and antioxidative enzymes in tomato plants. Biol Plant 62(4):801–808. https://doi.org/10.1007/s10535-018-0813-4

Yılmaz E, Sönmez İ, Demir H (2014) Effects of zeolite on seedling quality and nutrient contents of cucumber plant (Cucumis sativus L. cv. Mostar F1) grown in different mixtures of growing media. Commun Soil Sci Plant Anal 45(21):2767–2777

Yin L, Colman BP, McGill BM, Wright JP, Bernhardt ES (2012) Effects of silver nanoparticle exposure on germination and early growth of eleven wetland plants. PLoS One 7(10):e47674. https://doi.org/10.1371/journal.pone.0047674

Yu G (2018) Root exudates and microbial communities drive mineral dissolution and the formation of Nano-size minerals in soils: implications for soil carbon storage. Root Biol 52(May). https://doi.org/10.1007/978-3-319-75910-4.

Chapter 10
Interaction of Engineered Nanomaterials with Soil Microbiome and Plants: Their Impact on Plant and Soil Health

Shams Tabrez Khan

Abstract A large numbe of nanomaterials-based products are being commercially engineered and produced. Many of these engineered nanomaterials (ENMs) are disposed in soil in significant quantities. Furthermore, nanomaterials are being specially tailored for use in agriculture as nano-fertilizers, nano-pesticides, and nano-based biosensors. The behavior of ENMs in soil and their persistence depends on their chemical nature and soil characteristics. Furthermore, nanoparticles like silver and zinc oxide possess well-known antimicrobial activities. The presence and persistence of these nanomaterials in soil can alter the quality of the soil microbiome, thus influencing key microbial processes like mineralization, nitrogen fixation and plant growth promoting activities. It is, therefore, extremely important to understand how nanomaterials influence the soil microbiome and associated chemical and biochemical processes. Such investigations will provide necessary information for eventual regulation of the appropriate use of nanomaterials for sustainable agriculture and increased agricultural productivity. This chapter discusses some of these issues.

Keywords ENMs · Plant microbiome · Soil microbiome

10.1 Introduction

The industrial production of engineered nanomaterials (ENMs) is rapidly increasing with its commercial and domestic use consequently, their quantities reaching various environments including soil are also increasing (Keller et al. 2013). ENMs may be released to the environment during production, or during fabrication of ENM-containing products, during the use of such products or via disposal following use. ENMs may be added to soil directly or may experience various transformations

S. T. Khan (✉)
Department of Agricultural Microbiology, Faculty of Agricultural Sciences, Aligarh Muslim University, Aligarh, Uttar Pradesh, India

S. Hayat et al. (eds.), *Sustainable Agriculture Reviews 41*, Sustainable Agriculture Reviews 41, https://doi.org/10.1007/978-3-030-33996-8_10

before reaching soil. Therefore, it is important to understand the types of nanomaterials that are released to the environment especially to the soil and their release routes. Another key concern is to evaluate ENM fates in soil and how soil affects various properties of ENMs. For example, soil pH may modify certain physical and chemical properties of ENMs. Finally, it is both urgent and important to understand how nanomaterials affect the overall health of plants, the plant microbiome, soil and the soil microbiome. This chapter discusses the effect of ENMs on soil and plant microbiomes, as both greatly affect the health of the plant as well as soil fertility (Berendsen et al. 2012; Sergaki et al. 2018).

How ENMs influence soil and plant microbiomes depends on various factors. Some nanomaterials may be more microbicidal than others. ENM dose, size, chemical nature and various other properties influence their potential microbicidal activity. Nanomaterials that are microbicidal at high concentration may promote the growth of microorganisms at lower concentrations (Khan et al. 2018). Some nanomaterials may have selective activity against only a select group of microorganisms. Therefore, the ENMs in soil may influence important geochemicsl processes mediated by microorganisms. Similarly, toxicity of ENMs to plants also depends on a number of factors and their effect on plants can not be generalized. This chapter discusses these factors and the nanomaterials of immediate concern to soil fertility, plant health and the plant microbiome.

10.2 Plant and Soil Microbiomes

Recent studies on microbiomes from various habitats has made it clear that microorganisms are ubiquitous, abundant, and irreplaceable due to the diverse and vital roles they play in nature. One habitat that harbors the highest density of microorganisms is soil. It is established that one gram of soil may contain up to 10^8 bacterial cells, thus contributing greatly to soil biomass. The biomass of bacteria and fungi in the soil is 10^2–10^4 times higher than that of other microorganisms such as archaea, protists, and viruses. The quantity of soil microbial biomass rivals biomass occurring above-ground (˃1000 kg/hectare) (Fierer 2017). Abundance, diversity, and type of microorganisms present in soil depend on soil type, available nutrients, oxygen availability and climatic conditions. Soil represents a diverse environment which varies greatly with location; microbial communities may vary even within a distance of a few millimeters. In a review on soil microbiomes, it has was stated that most of the bacterial and archaeal species found in different soils are rare, and only a few microbial types occur abundantly in all studied samples (Fierer 2017). Based on data from 66 soil samples it was determined that, according to their abundance, the predominant fungi in soil belong to Agaricomycetes (Basidiomycota), Archaeorhizomycetes (Ascomycota), Zygomycota, Sordariomycetes (Ascomycota), Leotiomycetes (Ascomycota), Dothideomycetes (Ascomycota), Eurotiomycetes (Ascomycota), *Glomeromycota* and *Chytridiomycota*. The predominant bacteria in these samples belong to *Acidobacteria*, *Verrucomicrobia*, *Bacteroidetes*,

Proteobacteria (*Alphaproteobacteria*, *Gammaproteobacteria*, *Deltaproteobacteria*, and *Betaproteobacteria*), *Planctomycetes* and *Actinobacteria*. The archaea can be arranged in the following order of abundance: *Thaumarchaeota* (Crenarchaeota), marine benthic group archaea (MBGA; *Crenarchaeota*), *Thermoplasmata* (*Euryarchaeota*), *Parvarchaeota*, and *Euryarchaeota* (unclassified groups) (Janssen 2006). These microorganisms play important roles in soil including nutrient fixation (carbon and nitrogen), nutrient solubilization (phosphate and zinc), mineralization, and loss of nutrients from soil through processes like methane production and denitrification (Jacoby et al. 2017; Prakash et al. 2015). Therefore, the role of microorganisms in nutrient cycling is irreplaceable.

The plant microbiome can simply be defined as the community of microorganisms that live on plants, including the root surface, inside the root and on other surfaces such as leaves, stems, flowers, etc. These habitats are referred to as the phyllosphere, rhizoplane, rhizosphere, and endosphere. The plant microbiome is a complex system; interactions between the plant and its microbiome are affected by a number of factors including type of plant, age, health and nutrients secreted by the plant, the physical environment, the initial microbial load in soil and on the plants, and other factors. It has been demonstrated that plants are capable of selecting specific microorganisms for colonization of the rhizosphere (Lugtenberg and Kamilova 2009). Even within a plant under identical conditions, microorganisms present in the phyllosphere, rhizoplane, rhizosphere and endosphere may vary greatly. Research on the plant microbiome has focused on: (a) the extensive interplay among different microorganisms including bacteria, archaea, fungi, and protists; (b) plant-specific microbiomes to the level of cultivar; (c) the vertical transmission of core microbiomes; (d) the function of endophytes; and (e) unexpected functions and metabolic interactions (Berg et al. 2015). Some well-known examples of plant-microbe relationships are legume-rhizobia interactions and mycorrhizal associations, wherein the microbial partner provides nutrients for its host plant.

In addition to providing nutrients, microbiomes perform other support functions for plants (Hunter 2016). Microbial communities are so important that plants recruit specific microbial communities by the secretion of plant exudates (Berendsen et al. 2018). It has been demonstrated that flowering time and biomass yield of *A. thaliana* is controlled by its characteristic microbiome (Panke-Buisse et al. 2014). In another study on *A. thaliana* it was demonstrated that when plants are exposed to downy mildew disease they assemble protective microbial communities within their rhizosphere (Berendsen et al. 2018). The successful use of a microbial community for protecting *Nicotiana attenuata* against wilt disease has been demonstrated (Santhanam et al. 2015). The rhizosphere microbiome of *Phaseolus vulgaris* was shown to host a high population of the Bacteroidetes group of bacteria (Pérez-Jaramillo et al. 2017). While studying the domestication of *Phaseolus vulgaris* it was observed that domestication of plants changed the plant microbiome (Perez-Jaramillo et al. 2016): a decrease in the population of the Bacteroides group of bacteria, and an increase in populations of Actinobacteria and Proteobacteria was observed. Domestication also was found to adversely affect populations of symbiotic nitrogen fixers and mycorrhiza. Detailed studies on plant microbiomes will help in developing sustainable agriculture, minimizing dependency on fertilizers and pesticides while increasing the yield and nutrient content of crops.

Since different plants host distinct microbiomes that shift with varying conditions, it is necessary to understand the plant-specific microbiome which possibly can be customized for specific needs such as improved growth, protection against disease and better quality product (Fitzpatrick et al. 2018). Knowledge of microbiome will therefore make agriculture more extensive and less dependent on agrochemicals. In addition, it is of immense importance to understand how pollutants in soil may harm the microbiomes of plants and soil. Unparalleled urbanization in countries like China has resulted in extensive and serious soil pollution, it is estimated that 16% of Chinese soil, including 19.4% of farmland, is contaminated. An estimated 80% of these pollutants are toxic inorganic compounds (Yang et al. 2014). Nanomaterials are emerging as potential soil pollutant as they are being widely used and ultimately ends up in soil. According to some estimates, tons of nanomaterials are annually being produced and ending up in landfills and soil (Keller et al. 2013). Owing to their known antimicrobial activities, nanomaterials are expected to upset plant and soil microbiomes (Khan et al. 2016). This chapter discusses, in detail, nanomaterial pollution in soil and its consequences on plant and soil microbiomes and health.

10.3 Nanomaterials – A Brief Introduction

Nanomaterials are generally defined as materials that have at least one of their dimensions measuring less than 100 nm (Kreyling et al. 2010). Nanomaterials may be naturally occurring (e.g., volcanic ash) or engineered (TiO_2 and carbon nanotubes). Engineered nanomaterials are used in a number of commercial products. According to an estimate by the project on emerging nanotechnologies (PEN), more than 1824 products containing nanomaterials are already in the market (Berube et al. 2010). Fig. 10.1 shows some nanomaterial-based products. The extensive use of nanomaterials in industry is a consequence of their unique characteristics. Among various properties, the most important is their extremely small size and high volume to specific surface area ratio (VSSA). It has also been proposed that materials having a VSSA $\geq$60 m^2/cm^3 should be defined as nanomaterials (Kreyling et al. 2010). Nanomaterials represent an enormous class of compounds that are grouped based on size and other properties such as chemical nature, dimensionality, shape, and size. On the basis of chemical nature, nanomaterials are classified as organic, for example nanosized lipid micelles, and inorganic, such as silver nanoparticles. Nanomaterials can contain one repeating constituent unit or more than one type of repeating unit; the latter is termed a nanocomposite. Nanomaterials can be classified as 0D, 1D, 2D or 3D based on their dimensionality (Sun et al. 2014b). Nanomaterials having all dimensions in nanoscale like quantum dots and nanodispersions are termed 0D nanomaterials; materials having two dimensions in the nanoscale like nanotubes and nanowires, are referred to as 1D nanomaterials. 2D nanomaterials are

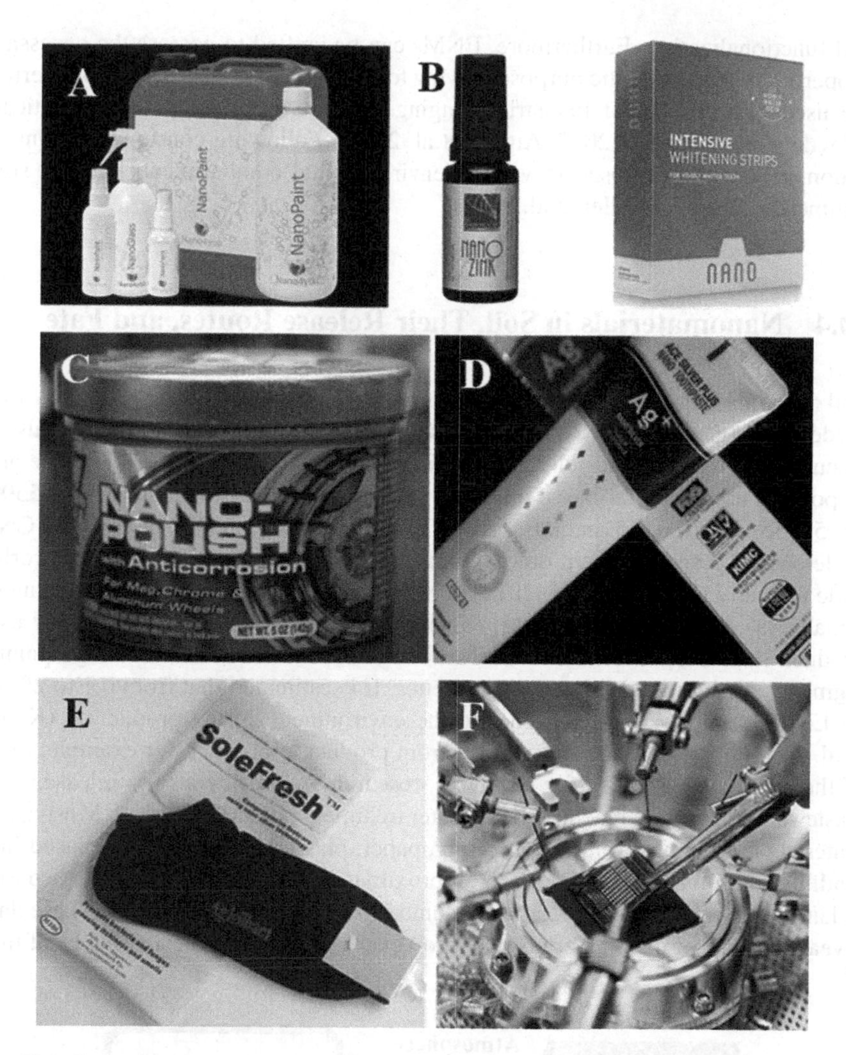

Fig. 10.1 Examples of some nanomaterial-containing products that may eventually release ENMs to soil: (**a**) paint; (**b**) teeth whitener; (**c**) automotive metal polish (**d**) Sun cream, (**e**) fabrics and (**f**) electronics

defined as nanomaterials with only one dimension in the nanometric size range, like nanofilms, while 3D nanomaterials are mesoporous structures and nanoclusters. Nanomaterials occur in various shapes such as nanotubes, nanocubes, nanosheets, nanospheres, and nanoflowers.

The small size of nanomaterials and their high surface area to volume ratio comprise the most unique properties of ENMs. Due to their small size ENMs are capable of penetration into living tissue including the blood-brain barrier in mammals and in various plant tissues (Khodakovskaya et al. 2011). The high volume to surface area ratio makes ENMs lighter and provides a greater surface for interaction and functionalization. Furthermore, ENMs can be crafted to possess the necessary properties to suit a specific purpose. Owing to their unique properties nanomaterials are used in a number of industries ranging from aeronautics to pharmaceuticals (Novikov and Voronina 2017; Aitken et al. 2006). Following consumption tons of nanomaterials are disposed in various environmental compartments such as soil, sediment and water (Keller et al. 2013).

10.4 Nanomaterials in Soil, Their Release Routes, and Fate

Soil contains a number of naturally-occurring nanomaterials such as clay, iron oxides and organic matter (Klaine et al. 2008). Several engineered nanomaterials, as discussed above, are also released to soil and other environments via the use and disposal of nanomaterial-based products. It is estimated that about 3000, 550, 5500, 55, 55, 55, 300, 0.6, 55 and 0.6 tons of TiO_2, ZnO, SiO_2, FeO_x, AlO_x, CeO_x, CNT Fullerenes, Ag and quantum dot nanomaterials, respectively, are produced worldwide annually (Piccinno et al. 2011). ENMs are released intentionally or unintentionally (a) during manufacturing; (b) during use; and (c) via disposal after use. Products that contribute the most ENMs to the environment include coatings, paints, pigments, electronics, optics, and cosmetics. It is estimated that from 0.1 to 2% of all ENMs produced are released into the environment during production (Keller et al. 2013). Release during use varies from product to product; for example, most of the nanomaterials used for academic, research and cosmetics are released into wastewater ultimately reaching wastewater treatment plants (Fig. 10.2). Most nanomaterials used in electronics, packaging, paper, plastics, and board are placed into landfills and soil (Keller et al. 2013). Approximately 63–91% of ENMs are disposed in landfills, while the second largest volume (8–28%) is disposed in soil. These data reveal that tons of nanomaterials reach soil every year. The highest volumes of tita-

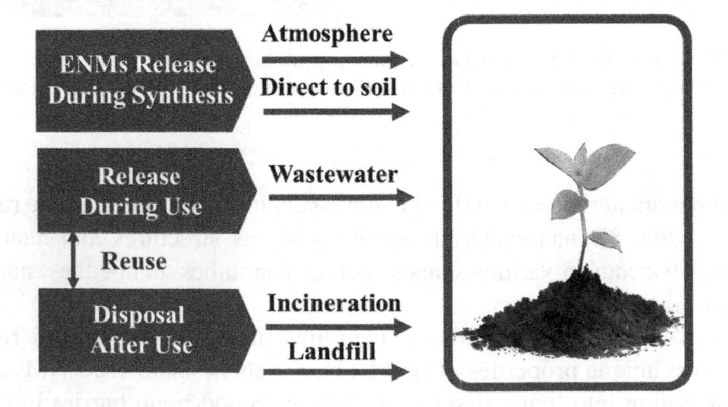

Fig. 10.2 Release routes of ENMs in soil. Few details are currently available on release routes of ENMs in various environments

nia, iron and zinc oxides are released into soil, water and air. While, SiO_2, titania (TiO_2), iron, and zinc oxides (ZnO), and alumina oxides (Al_2O_3) are the most abundant ENMs released to the soil and landfills. Most reach soil directly due to the nature of their use; for example, pigments containing TiO_2 used for pigments directly reaches into landfills (Sun et al. 2014a). These ENMs are released into the air from manufacturing, which subsequently reaches the soil. Additionally, upon incineration burning of ENM-s containing products in waste incineration plants (WIPs) also the ENMs will eventually reach the soil. ENMs such as TiO_2 contact soil when sewage treatment effluents containing ENMs are used as irrigation water. It is estimated that irrigation with nanomaterial-containing wastewater may result in an increase of 89 μg TiO_2 per kg of soil annually (Gottschalk et al. 2009). It was found that highest concentrations of ENMs measured in soils are those treated with municipal sludge. The same study calculates that 0.28–1.28 μg kg^{-1} y^{-1} of TiO_2 is released into soil in Europe. Similarly, 0.093 and 0.050 μg kg^{-1} y^{-1} of ZnO are released into soil in Europe and the US, respectively. While, other studies have, reported that highest amounts of TiO_2 are reaching all the compartments of the environment including soil, followed by ZnO (Gottschalk et al. 2009). The fate of ENMs upon reaching soil varies with soil type and properties of the nanomaterial. ENMs may dissolve in soil water, interact with charged particles in soil, or can be taken up by organisms resulting in their bioaccumulation. The latter depends on bioavailability of ENMs. ENMs may also reach water bodies underneath and sediments.

The fate of nanomaterials in soil and other environments varies based on their inherent properties as well as the properties of the environment (Klaine et al. 2008). Size, shape, chemical nature, and surface properties are key characteristics that influence ENM behavior and fate in soil. All ENMs undergo aging (weathering); with aging, particles may undergo chemical transformation, aggregation, and disaggregation (Bundschuh et al. 2018). Chemical transformations of ENMs in soil may include dissolution, sulfidation, adsorption and desorption.

10.5 Effect of Nanomaterials on Plant Health and Plant Microbiome

Although the use of nanomaterials in agriculture is still in its infancy, some commercial nano-based products are already available in the market (Sekhon 2014; Servin et al. 2015a). Studies showing the beneficial effects of ENMs have already been published (Khan et al. 2018; Faizan et al. 2018). This shows that the ENMs are added to soil intentionally in addition to the ENMs that are released in the soil through the routes discussed above. The presence of nanomaterials in the plant rhizosphere is a unique scenario which embraces an interplay of the plant microbiome, the plant, and the ENM. For example, as discussed above, plants secrete exudates that craft the microbial community within their respective rhizosphere. The rhizospheric microbial communities are also influenced, however, by ENMs present in the soil. Furthermore, the chemical nature of the ENMs may also change in the presence of root exudates and the extracellular substances from microbes. The presence of ENMs in soil may affect plant health either directly or by influencing the plant microbiome.

10.5.1 Direct Effects of ENMs on Plants

The uptake of ENMs by higher organisms and their interaction with various biomolecules is well established. It is known that many ENMs are taken up by plants and can be transported to leaves and other aerial parts, influenceing plant growth directly (Lin et al. 2009). Both growth-promoting and inhibitory activities of ENMs have been reported. These effects depends on a number of factors including the chemical nature of ENMs and their size, surface charge and dose (Husen and Siddiqi 2014; Faizan et al. 2018). Furthermore, the ability of ENMs to migrate from soil to the plant and their propensity for transport to various tissues also play an important role in plant health. Several reports are available on migration of nanomaterials to leaves and other aboveground tissues following absorption by roots. Movement of nanomaterials in plants can either be apoplastic (i.e., through extracellular spaces and xylem vessels) or symplastic (through plasmodesmata) (Pérez-de-Luque 2017). TiO_2 NPs and multi-walled carbon nanotubes (MWNCTs) exhibited limited mobility from soil to wheat and red clover leachates (Gogos et al. 2016).

Various molecular mechanisms have been proposed of nanomaterial toxicity (Khodakovskaya et al. 2011; Servin et al. 2015b); one of the most commonly reported is induction of stress which is detected by increased activity of superoxide dismutase and peroxide dismutase. When different carbon nanomaterials including fullerenes, reduced graphene oxide, and MWCNTs were added to the rhizosphere of rice at a dose of 50–500 mg/kg of soil for 30 days, the MWCNMs triggered the induction of four phytohormones including auxins, brassinosteroids, indoleacetic acid, and gibberellins (Hao et al. 2018). In addition, increased activities of superoxide dismutase and peroxide dismutase were observed. The study concluded that the CNMs resulted in toxicity to both rice plants and microbial communities. A 60 and 75% growth inhibition of zucchini (*Cucurbita pepo*) was reported in the presence of Ag NPs and MWCNTs, respectively (Stampoulis et al. 2009). Similarly, Ag NPs inhibited the germination of ryegrass and flax (*Linum usitatissimum*). At a concentration of 1.5 g L^{-1}, Ag NPs reduced the germination of barley (*Hordeum vulgare* L.) by 13% (Yehia and Joner 2012). Accumulation of CeO_2 in soybean roots and their translocation and accumulation in edible tissue is also reported (Hernandez-Viezcas et al. 2013). It has been demonstrated that exposure to ZnO NPs (200–300 mg/L^{-1}) results in a decrease in chlorophyll content of plants leading to deterioration of plant health (Wang et al. 2016). Poly(acrylic acid) nano ceria increased carbon assimilation rates by 67% in *Arabidopsis thaliana* plants (Wu et al. 2017). Nano ceria protected the plant from abiotic stress by scavenging free radicals. This was in contrast to the findings of other works, where nanomaterials induced stress in plants through ROS generation. CuO nanoparticles inhibited denitrification, nitrification, and soil respiration. This inhibition was observed, however, at a high concentration of 100 mg/kg dry soil.

In contrast, growth-promoting activities of ENMs have also been reported (Faizan et al. 2018; Zhu et al. 2019). For example, addition of nano TiO_2 alleviated Cd stress in soybean plants by promoting plant growth; this was achieved through increased chlorophyll or carotene content following TiO_2 treatment, consequently increasing the rate of photosynthesis (Singh and Lee 2016). Nanomaterials of graphene oxide have been shown to promote seed germination through increased water retention (He et al. 2018). Priming of aged rice with 5 and 10 ppm of Ag NPs improved seed germination and seedling vigor (Mahakham et al. 2017). Similarly, addition of ZnO promoted the growth of tomato plants by providing Zn as a micronutrient (Faizan et al. 2018). Addition of a water-soluble wood-based pyrolysis waste product termed nano-onions (wsCNOs) was found to enhance the overall growth rate of gram (*Cicer arietinum*) plants (Sonkar et al. 2012). Some ENMs serve as micronutrients for plants, but also as carriers for various nutrients (Jampílek and Kráľová 2017). ENMs act as an effective carrier for nutrients due to their small size and excellent penetration capabilities (Pérez-de-Luque 2017). In experiments on *Arabidopsis thaliana* it was observed that exposure to nano-cerium resulted in increased plant biomass and numbers of rosette leaves (Tumburu et al. 2017). These findings were supported by the microarray data presented in the same study. Change in gene expression of tobacco cells upon exposure to carbon nanotubes was studied; genes involved in cell division and water transport were upregulated at low concentration of CNT. Low exposure concentrations promoted cell growth (Khodakovskaya et al. 2012). Nano Fe_3O_4 was found to be beneficial for growth of *Triticum aestivum* L., and an increase in antioxidant enzyme activity was also observed (Iannone et al. 2016). Although many reports on this aspect are available, further systematic studies are required. A careful evaluation of environmentally realistic doses should be considered before reaching any conclusion regarding toxicity of ENMs to plants.

10.5.1.1 Influence of ENMs on Soil and Plant Microbiome

In addition to directly affecting growth of plants, ENMs also influence the microbial community in the soil and plant microbiomes consequently affecting the plant health (Fig. 10.3). For example, it has been reported that nano TiO_2 and ZnO influence soil microbial communities and a comparison of the two ENMs suggests that ZnO NPs induce more pronounced toxicity than does TiO_2 (Ge et al. 2011). Bacteria that carry out nitrogen fixation and methane oxidation were among the populations that decreased significantly with treatment of ENMs (Ge et al. 2011). While the population of members of *Sphingomonadaceae* increased, members of this family are well-known for decomposition of recalcitrant organic pollutants increased.

TiO_2 was assessed for its effect on the microbiome of wheat. It was observed that populations of certain prokaryotes changed, but growth of the plant and arbuscular mycorrhizal root colonization remained largely unaffected (Moll et al. 2017). It was

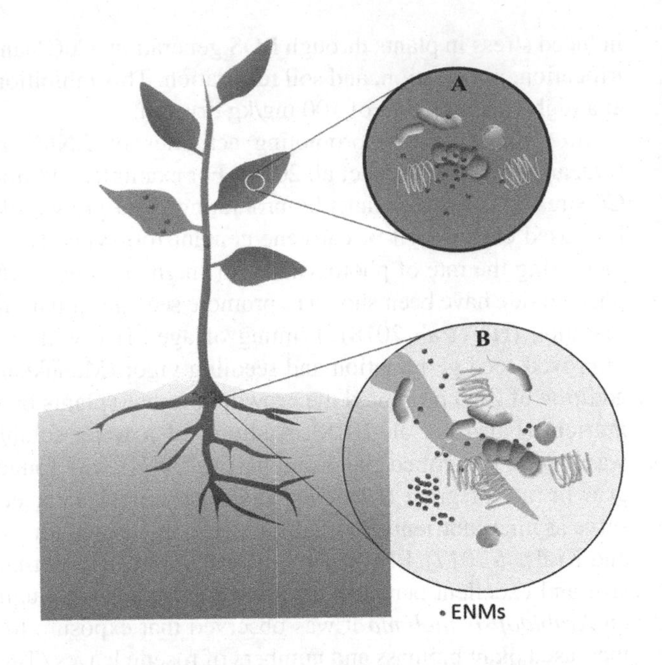

Fig. 10.3 ENMs present in soil may affect plant-microbe interactions, especially in rhizospheric microflora (**b**). These ENMs can also reach aerial parts of the plant and can affect the phyllospheric microflora (**a**)

suggested that the change in the prokaryotic community can be used as a marker for nano TiO_2 contamination in soil. In another study (Grün et al. 2019) it was observed that silver NPs affect the soil microbial community significantly (Grün et al. 2019). The same authors demonstrated that populations of β-proteobacteria and ammonia oxidizers decreased significantly upon exposure to silver nanoparticles; in contrast, populations of *Acidobacteria*, Actinobacteria, and Bacteroidetes increased significantly (Grun et al. 2018). Exposure to Ag NPs decreased soil microbial biomass, leucine aminopeptidase activity, and abundance of nitrogen-fixing microorganisms (Grun et al. 2018). When soil was treated with C60 fullerenes of 50 nm size, growth of fast-growing bacteria was suppressed three to four-fold (Johansen et al. 2008). When TiO_2 and polystyrene nanomaterials were added to the rhizosphere of lettuce seedlings numbers of rhizospheric soil bacteria decreased, which consequently inhibited root and shoot growth (Kibbey and Strevett 2019). Soils irrigated with waste effluent containing ENMs experienced an increased population of cyanobacteria and an unknown group of Archaea. The life cycle of *A. thaliana* was significantly shortened (Liu et al. 2018).

Carbon nanomaterials have been shown to affect the microbial community of the rice rhizosphere and incur toxicity (Hao et al. 2018). In a study on tomato plants, treatment of soil with carbon nanotubes did not significantly alter the soil microbial community (Khodakovskaya et al. 2013). Among the various carbon nanomaterial, reduced graphene oxide resulted in the most significant changes to the microbial community. The antimicrobial activity of ENMs is well-known and the mechanism of their antimicrobial activity has also been extensively studied and reported. Mechanisms include: (i) bacterial cell membrane disruption; (ii) perturbation of metabolic functions such as purine metabolism; (iii) protein denaturation; (iv) DNA damage; (v) inhibition of respiration through disruption of the respiratory chain; (vi) free radical formation and induction of oxidative stress; (vi) mutagenesis; and (vii) inhibition of DNA replication through DNA binding (Khan et al. 2016). ENMs are not always microbicidal in their action but simply be inhibitory to specific microbial enzymes and processes. The toxicity of nanomaterials also depends on their inherent properties including shape, size, chemical nature, surface charge, and hydrophobicity. Furthermore, some microorganisms may be more sensitive to a nanomaterial than others. For example, some nanomaterials are more effective against Gram-positive bacteria than Gram-negative bacteria. The growth rate of bacteria and the ability to produce extracellular polysaccharides (EPS) also influence sensitivity of bacteria to engineered nanomaterials.

10.6 Effect of ENMs on Soil Microbial Processes

Many nanomaterials are known to possess microbicidal properties and hence can inhibit the proliferation of microorganisms involved in crucial biogeochemical processes such as nitrogen fixation, ammonification, denitrification, phosphate solubilization, and other plant growth promoting (PGPR) activities, thus inhibiting geochemical processes (Fig. 10.4). Study of the literature shows that ENM toxicity is reported usually at high concentrations. To the contrary, it has been demonstrated that certain nanomaterials occurring at low concentrations promote some biogeo-

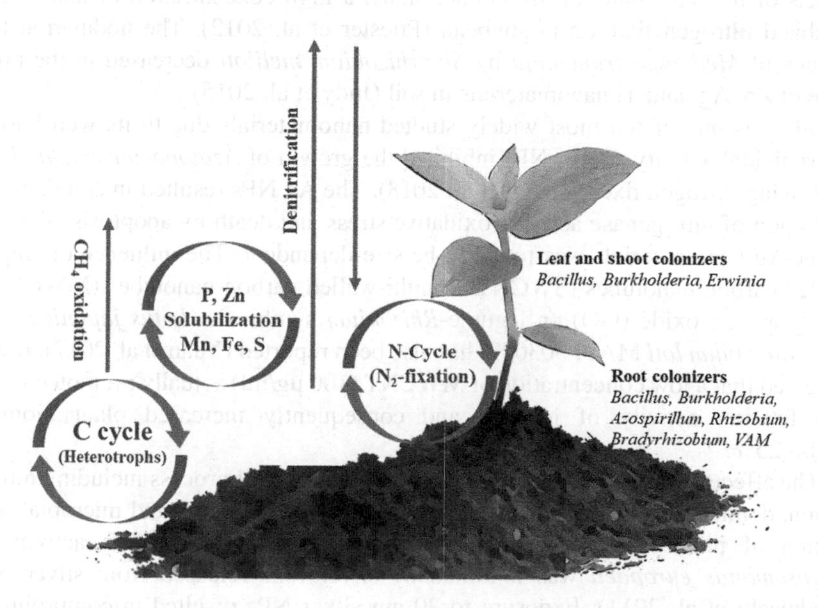

Fig. 10.4 Role of microorganisms in geochemical processes. Many of these microorganisms are free-living in soil while some live in close association with plants, playing an important role in nutrient cycling and consequently influencing plant growth and soil health

chemical processes, consequently promoting plant growth (Khan et al. 2018; Yuan et al. 2017).

Few reports are available on the influence of ENMs on reactions of one of the most important nutrients, i.e., carbon. Reduced graphene oxide (RGO) inhibit photosynthesis in pea plants, consequently affecting carbon fixation and biomass production (Chen et al. 2019). The RGO damaged the oxygen-evolving-complex on the donor site, consequently inhibiting the activity of photosystem II (PS II). This inhibition was attributed to oxidative stress induced by RGO. Various reports reveal increased biomass following treatment with nanomaterials at low concentration. For example, upon treatment with MWNCT (100 μg/kg soil), an increase in nitrogen fixation activity was observed, leading to an increase of biomass (Yuan et al. 2017). Microorganisms play an indispensable role in the nitrogen cycle from ammonification to nitrogen fixation to denitrification. Nitrogen fixation is one of the most important processes mediated by microorganisms, both free-living and symbiotic. The influence of ENMs on nitrogen fixation by both free-living and symbiotic microorganisms has been studied. In the presence of nano TiO_2 the growth rate of *Anabaena variabilis*, its nitrogen fixation rate, and rate of nitrogen storage were inhibited (Cherchi and Gu 2010). Time of exposure was found to be a more important factor than concentration of nanomaterial on microbial processes. The negative effects of copper nanoparticles on microbial carbon and nitrogen cycles has been reported (Simonin et al. 2018). In this study, however, the low ENM concentrations (0.1–1 mg/kg of soil) were not inhibitory to the process. Furthermore, denitrification was most sensitive to CuO-NPs. The presence of plants did not mitigate the effects of the nanomaterial. In another study, a high concentration of nano-CeO_2 inhibited nitrogen fixation in soybean (Priester et al. 2012). The nodulation frequency of *Medicago truncatula* by *Sinorhizobium meliloti* decreased in the presence of Zn, Ag, and Ti nanomaterials in soil (Judy et al. 2015).

Silver is one of the most widely studied nanomaterials due to its well-known microbicidal activity. Silver NPs inhibited the growth of *Azotobacter vinelandii*, a free-living nitrogen fixer (Zhang et al. 2018). The Ag NPs resulted in cell damage, inhibition of nitrogenase activity, oxidative stress and death by apoptosis. Toxicity of the Ag nanomaterials was found to be size-dependent. The influence of single-walled carbon nanotubes (SWCNTs), multi-walled carbon nanotubes (MWCNTs) and graphene oxide (GO) on legume-*Rhizobium* symbiosis (*Lotus japonicus* and *Mesorhizobium loti* MAFF303099) has also been reported (Yuan et al. 2017). It was observed that a low concentration of MWCNT (100 μg/ml) actually promoted nitrogen fixation activity of nodules and consequently increased plant biomass by 14–25%.

The effects of ENMs on various other PGPR microbial process including nitrification, denitrification, phosphate and potassium solubilization, and microbial protection of plants against diseases are also reported. Nitrification activity of *Nitrosomonas europaea* was inhibited by silver ions released from silver NPs (Radniecki et al. 2011). Exposure to 20 nm silver NPs resulted in compromised outer membranes and inhibition of ammonium oxidase activity. ZnO, SiO_2, TiO_2, and CeO_2 nanoparticles were tested for their activity against PGPR bacteria including *Azotobacter*, phosphate-, and potassium-solubilizing bacteria and their enzymatic activities (Chai et al. 2015). A rate of 1 mg/g ZnO and CeO_2 individually hindered thermogenic metabolism, reducing *Azotobacter* colony numbers and P- and K-solubilizing bacteria. These ENMs also inhibited activities of urease and catalase and decreased fluorescein diacetate hydrolysis activities. The activity of silica (SiO_2) and alumina (Al_2O_3) nanoparticles against plant growth promoting rhizobacteria, and other PGPR bacteria like *Pseudomonas fluorescens*, *Bacillus megaterium,* and *Bacillus brevis* have been studied. Nano Al_2O_3 particles were highly toxic to these organisms at 1000 mg/L (Karunakaran et al. 2014).

The effects of a metabolite of DDT, dichloro diphenyl dichloroethylene; *p,p'*-DDE, on plants was influenced by the presence of C60 fullerene nanomaterials (De La Torre-Roche et al. 2012). The level of the contaminant in soybean shoots decreased by 48% while in another case it increased. This study demonstrates that nanomaterials in soil may influence the uptake of various environmental pollutants.

10.7 Role of ENMs in Protecting Plants Against Pathogens

ENMs can influence plant-pathogen interactions. Nano-formulations of some pesticides, such as DMM [sodium dodecyl sulfate-modified photocatalytic TiO_2/Ag nanomaterial conjugated with dimethomorph] is already in use (Khot et al. 2012). Many studies have reported the inhibitory effect of ENMs to plant pathogens. The inhibitory effect can be due to the direct activity of ENMs or release of metal ions from ENMs, or to augmenting the activities of microorganisms that inhibit phytopathogens (Khan et al. 2018; Elmer and White 2018; Servin et al. 2015b). Silver is known for its antimicrobial activity; thus, its effect on various phytopathogens has been reported. The inhibitory effect of Ag NPs on phytopathogenic fungi that cause disease in ryegrass was evaluated (Jo et al. 2009). Ag NPs were found to reduce the growth of phytopathogenic fungi and were also found to reduce the disease in the plants. In another study, the ability of ENMs (Fe_2O_3, TiO_2, MWCNTs and C60) to inhibit tobacco mosaic virus and turnip mosaic virus was studied (Hao et al. 2018). It was observed that ENMs decreased the pathogenicity of these viruses by decreasing the coat proteins of the viruses by 15–60%. However, it is interesting to note that quite high doses were used for the study (50-200 mg/L) . A low dose (500 ng/mL) of ENMs (Ag, SiO_2, TiO_2, and ZnO) promoted the antifungal activity of *Pseudomonas protegens* CHA0 against *C. albicans* (Khan et al. 2018). Inhibition of *F. graminearum*, a plant pathogen, by ZnO nanoparticles in vitro and in vivo has been reported (Dimkpa et al. 2013). Treatment of wheat plants with ZnO NPs reduced *F. graminearum* infection, wherein a significant reduction in *F. graminearum* CFU was observed compared to control (Savi et al. 2015). Treatment with Zn nanoparticles did not harm the plant and the levels of zinc in wheat grains were within permissible limits. In another study the effect of six different carbon nanomaterials (SWCNT, MWCNTs, GO, RGO, C60, and activated carbon) against two phytopathogenic fungi (*Fusarium graminearum* and *Fusarium poae*) was evaluated

(Wang et al. 2014). Except for C_{60} and AC all nanomaterials inhibited the growth of the two fungi. Spores of these fungi were inactivated primarily through (i) deposition on the surface of the spores; (ii) inhibition of water uptake; and (iii) plasmolysis. In field trials the inhibition of *Colletotrichum* spp. (a phytopathogen which causes anthracnose) by Ag nanoparticles was demonstrated (Lamsal et al. 2011). Application of nanomaterials before the spread of disease resulted in significantly reduced infection of pepper plants by *Colletotrichum* spp. From the review of the literature discussed above, it is concluded that ENMs may help protect plants against phytopathogens. However, toxicity concerns remain to be evaluated carefully.

10.8 Conclusion

ENMs are released to the soil both intentionally and unintentionally without evaluating the risks involved. Among different ENMs that are released into the soil, the major ENMs are SiO_2,Titania (TiO_2) iron and zinc oxides (ZnO), and alumina (Al_2O_3). The unregulated and continuous use of ENMs has resulted in the release of tons of these ENMs to soil. Studies have estimated the annual changes in their concentrations in soil; concentrations depend on persistence combined with the cumulative additions of ENMs to soil. Little evaluation has been accomplished to date regarding the risks involved from such accumulation.

Upon reaching soil ENMs experience variable fates. Among these are interactions with soil and plants, and soil and plant microbiomes. Interaction with ENMs influences the soil microbiome, plant microbiome, and overall plant health. Microbiomes of bulk soil, and the plants vary from soil to soil and plant to plant and therefore cannot be generalized. Many studies have evaluated the potential toxicity of these ENMs on soil microorganisms and plants. Microorganisms play an important role in the biogeochemical cycling of nutrients and in affecting the overall health of plants as well. Unfortunately, many studies have evaluated only very high doses of ENMs such as mg/kg levels, while realistic concentrations are far lower. In many studies, almost no effect was observed at lower doses; some studies have shown growth promoting and other positive effects of ENMs on plants and the plant microbiome, resulting in improved plant health and productivity. Hence, the use of realistic doses should be considered as a criterion for conducting and publishing such studies.

It is becoming increasingly necessary to regulate the release of ENMs during production, use, and disposal following use, as continuous and increasing quantities of ENMs released may pollute soil to the point of severe consequences. Studies show the beneficial and plant growth promoting effect of ENMs, thus arguing for their use in agriculture. Few studies have investigated the simultaneous accumulation of ENMs in plants and the greater environmental consequences of their use. For example, if Ag promotes plant growth, it is also taken up by the plant. Continuously consuming plants having elevated levels of Ag or other ENMs may become a health concern. Therefore, the wise use of nanotechnology, keeping in view all applications and consequences, will ensure the sustainable use of nanotechnology in agriculture. The targeted use of those nanomaterials having a short life in soil is desirable. Biocompatible ENMs having a short lifespan in soil that can be easily recycled or removed from soil by natural processes, should be designed.

To conclude, ENMs are a boon for the soil ecosystem if concentrations are low and application is regulated. If uncontrolled release in the soil remains unabated, their presence in soil at high concentrations may result in serious environmental and public health hazards risking crop productivity, food security and the status of soil as a sustainable resource.

References

Aitken RJ, Chaudhry MQ, Boxall ABA, Hull M (2006) Manufacture and use of nanomaterials: current status in the UK and global trends. Occup Med 56(5):300–306. https://doi.org/10.1093/occmed/kql051

Berendsen RL, Pieterse CMJ, Bakker PAHM (2012) The rhizosphere microbiome and plant health. Trends Plant Sci 17(8):478–486. https://doi.org/10.1016/j.tplants.2012.04.001

Berendsen RL, Vismans G, Yu K, Song Y, de Jonge R, Burgman WP, Burmølle M, Herschend J, Bakker PAHM, Pieterse CMJ (2018) Disease-induced assemblage of a plant-beneficial bacterial consortium. ISME J 12(6):1496–1507. https://doi.org/10.1038/s41396-018-0093-1

Berg G, Rybakova D, Grube M, Köberl M (2015) The plant microbiome explored: implications for experimental botany. J Exp Bot 67(4):995–1002. https://doi.org/10.1093/jxb/erv466

Berube DM, Searson EM, Morton TS, Cummings CL (2010) Project on emerging nanotechnologies – consumer product inventory evaluated. Nanotechnol Law Bus 7(2):152–163

Bundschuh M, Filser J, Lüderwald S, McKee MS, Metreveli G, Schaumann GE, Schulz R, Wagner SJESE (2018) Nanoparticles in the environment: where do we come from, where do we go to? Environ Sci Europe 30(1):6. https://doi.org/10.1186/s12302-018-0132-6

Chai H, Yao J, Sun J, Zhang C, Liu W, Zhu M, Ceccanti B (2015) The effect of metal oxide nanoparticles on functional bacteria and metabolic profiles in agricultural soil. Bull Environ Contam Toxicol 94(4):490–495. https://doi.org/10.1007/s00128-015-1485-9

Chen L, Wang C, Yang S, Guan X, Zhang Q, Shi M, Yang S-T, Chen C, Chang X-L (2019) Chemical reduction of graphene enhances in vivo translocation and photosynthetic inhibition in pea plants. Environ Sci Nano 6(4):1077–1088. https://doi.org/10.1039/C8EN01426D

Cherchi C, Gu AZ (2010) Impact of titanium dioxide nanomaterials on nitrogen fixation rate and intracellular nitrogen storage in Anabaena variabilis. Environ Sci Technol 44(21):8302–8307. https://doi.org/10.1021/es101658p

De La Torre-Roche R, Hawthorne J, Deng Y, Xing B, Cai W, Newman LA, Wang C, Ma X, White JC (2012) Fullerene-enhanced accumulation of p,p'-DDE in agricultural crop species. Environ Sci Technol 46(17):9315–9323. https://doi.org/10.1021/es301982w

Dimkpa C, McLean J, Britt D, Anderson A (2013) Antifungal activity of ZnO nanoparticles and their interactive effect with a biocontrol bacterium on growth antagonism of the plant pathogen Fusarium graminearum. Biometals 26. https://doi.org/10.1007/s10534-013-9667-6

Elmer W, White JC (2018) The future of nanotechnology in plant pathology. Ann Rev Phytopathol 56(1):111–133. https://doi.org/10.1146/annurev-phyto-080417-050108

Faizan M, Faraz A, Yusuf M, Khan ST, Hayat SJP (2018) Zinc oxide nanoparticle-mediated changes in photosynthetic efficiency and antioxidant system of tomato. Plan Theory 56(2):678–686. https://doi.org/10.1007/s11099-017-0717-0

Fierer N (2017) Embracing the unknown: disentangling the complexities of the soil microbiome. Nat Rev Microbiol 15:579. https://doi.org/10.1038/nrmicro.2017.87

Fitzpatrick CR, Copeland J, Wang PW, Guttman DS, Kotanen PM, Johnson MTJ (2018) Assembly and ecological function of the root microbiome across angiosperm plant species. J Proc Natl Acad Sci 115(6):E1157–E1165. https://doi.org/10.1073/pnas.1717617115

Ge Y, Schimel JP, Holden PA (2011) Evidence for negative effects of TiO2 and ZnO nanoparticles on soil bacterial communities. Environ Sci Technol 45(4):1659–1664. https://doi.org/10.1021/es103040t

Gogos A, Moll J, Klingenfuss F, van der Heijden M, Irin F, Green MJ, Zenobi R, Bucheli TDJJN (2016) Vertical transport and plant uptake of nanoparticles in a soil mesocosm experiment. J Nanobiotechnol 14(1):40. https://doi.org/10.1186/s12951-016-0191-z

Gottschalk F, Sonderer T, Scholz RW, Nowack B (2009) Modeled environmental concentrations of engineered nanomaterials (TiO2, ZnO, Ag, CNT, Fullerenes) for different regions. Environ Sci Technol 43(24):9216–9222. https://doi.org/10.1021/es9015553

Grun AL, Straskraba S, Schulz S, Schloter M, Emmerling C (2018) Long-term effects of environmentally relevant concentrations of silver nanoparticles on microbial biomass, enzyme activity, and functional genes involved in the nitrogen cycle of loamy soil. J Environ Sci (China) 69:12–22. https://doi.org/10.1016/j.jes.2018.04.013

Grün A-L, Manz W, Kohl YL, Meier F, Straskraba S, Jost C, Drexel R, Emmerling C (2019) Impact of silver nanoparticles (AgNP) on soil microbial community depending on functionalization, concentration, exposure time, and soil texture. Environ Sci Eur 31(1):15. https://doi.org/10.1186/s12302-019-0196-y

Hao Y, Ma C, Zhang Z, Song Y, Cao W, Guo J, Zhou G, Rui Y, Liu L, Xing B (2018) Carbon nanomaterials alter plant physiology and soil bacterial community composition in a rice-soil-bacterial ecosystem. Environ Pollut 232:123–136. https://doi.org/10.1016/j.envpol.2017.09.024

He Y, Hu R, Zhong Y, Zhao X, Chen Q, Zhu HJNR (2018) Graphene oxide as a water transporter promoting germination of plants in soil. Nano Res 11(4):1928–1937. https://doi.org/10.1007/s12274-017-1810-1

Hernandez-Viezcas JA, Castillo-Michel H, Andrews JC, Cotte M, Rico C, Peralta-Videa JR, Ge Y, Priester JH, Holden PA, Gardea-Torresdey JL (2013) In situ synchrotron X-ray fluorescence mapping and speciation of CeO_2 and ZnO nanoparticles in soil cultivated soybean (Glycine max). ACS Nano 7(2):1415–1423. https://doi.org/10.1021/nn305196q

Hunter P (2016) Plant microbiomes and sustainable agriculture: deciphering the plant microbiome and its role in nutrient supply and plant immunity has great potential to reduce the use of fertilizers and biocides in agriculture. EMBO Rep 17(12):1696–1699. https://doi.org/10.15252/embr.201643476

Husen A, Siddiqi KS (2014) Carbon and fullerene nanomaterials in plant system. J Nanobiotechnol 12(1):16. https://doi.org/10.1186/1477-3155-12-16

Iannone MF, Groppa MD, de Sousa ML, Fernández van Raap MB, Benavides MP (2016) Impact of magnetite iron oxide nanoparticles on wheat (Triticum aestivum L.) development: evaluation of oxidative damage. Environ Exp Bot. 131:77–88. https://doi.org/10.1016/j.envexpbot.2016.07.004

Jacoby R, Peukert M, Succurro A, Koprivova A, Kopriva S (2017) The role of soil microorganisms in plant mineral nutrition-current knowledge and future directions. Front Plant Sci 8:1617–1617. https://doi.org/10.3389/fpls.2017.01617

Jampílek J, Kráľová K (2017) Nanomaterials for delivery of nutrients and growth-promoting compounds to plants. In: Prasad R, Kumar M, Kumar V (eds) Nanotechnology: an agricultural paradigm. Springer, Singapore, pp 177–226. https://doi.org/10.1007/978-981-10-4573-8_9

Janssen PH (2006) Identifying the dominant soil bacterial taxa in libraries of 16S rRNA and 16S rRNA genes. Appl Environ Microbiol 72(3):1719–1728. https://doi.org/10.1128/AEM.72.3.1719-1728.2006

Jo Y-K, Kim B, Jung G (2009) Antifungal activity of Silver ions and nanoparticles on phytopathogenic fungi. Plant Dis 93. https://doi.org/10.1094/PDIS-93-10-1037

Johansen A, Pedersen AL, Jensen KA, Karlson U, Hansen BM, Scott-Fordsmand JJ, Winding A (2008) Effects of C60 fullerene nanoparticles on soil bacteria and protozoans. Environ Toxicol Chem Int J 27(9):1895–1903. https://doi.org/10.1897/07-375.1

Judy JD, McNear DH, Chen C, Lewis RW, Tsyusko OV, Bertsch PM, Rao W, Stegemeier J, Lowry GV, McGrath SP, Durenkamp M, Unrine JM (2015) Nanomaterials in biosolids inhibit nodulation, shift microbial community composition, and result in increased metal uptake relative to bulk/dissolved metals. Environ Sci Technol 49(14):8751–8758. https://doi.org/10.1021/acs.est.5b01208

Karunakaran G, Suriyaprabha R, Manivasakan P, Rajendran V, Kannan N (2014) Influence of Nano and bulk SiO_2 and Al_2O_3 particles on PGPR and soil nutrient contents. Curr Nanosci 10(4):604–612

Keller AA, McFerran S, Lazareva A, Suh S (2013) Global life cycle releases of engineered nanomaterials. J Nanopart Res 15(6):1692. https://doi.org/10.1007/s11051-013-1692-4

Khan ST, Musarrat J, Al-Khedhairy AA (2016) Countering drug resistance, infectious diseases, and sepsis using metal and metal oxides nanoparticles: current status. Colloids Surf B Biointerfaces 146:70–83. https://doi.org/10.1016/j.colsurfb.2016.05.046

Khan ST, Ahmad J, Ahamed M, Jousset A (2018) Sub-lethal doses of widespread nanoparticles promote antifungal activity in Pseudomonas protegens CHA0. Sci Total Environ 627:658–662. https://doi.org/10.1016/j.scitotenv.2018.01.257

Khodakovskaya MV, de Silva K, Nedosekin DA, Dervishi E, Biris AS, Shashkov EV, Galanzha EI, Zharov VP (2011) Complex genetic, photothermal, and photoacoustic analysis of nanoparticle-plant interactions. Proc Natl Acad Sci U S A 108(3):1028–1033. https://doi.org/10.1073/pnas.1008856108

Khodakovskaya MV, de Silva K, Biris AS, Dervishi E, Villagarcia H (2012) Carbon nanotubes induce growth enhancement of tobacco cells. ACS Nano 6(3):2128–2135. https://doi.org/10.1021/nn204643g

Khodakovskaya MV, Kim BS, Kim JN, Alimohammadi M, Dervishi E, Mustafa T, Cernigla CE (2013) Carbon nanotubes as plant growth regulators: effects on tomato growth, reproductive system, and soil microbial community. Small (Weinheim an der Bergstrasse, Germany) 9(1):115–123. https://doi.org/10.1002/smll.201201225

Khot LR, Sankaran S, Maja JM, Ehsani R, Schuster EW (2012) Applications of nanomaterials in agricultural production and crop protection: a review. Crop Prot 35:64–70. https://doi.org/10.1016/j.cropro.2012.01.007

Kibbey TCG, Strevett KA (2019) The effect of nanoparticles on soil and rhizosphere bacteria and plant growth in lettuce seedlings. Chemosphere 221:703–707. https://doi.org/10.1016/j.chemosphere.2019.01.091

Klaine SJ, Alvarez PJJ, Batley GE, Fernandes TF, Handy RD, Lyon DY, Mahendra S, McLaughlin MJ, Lead JR (2008) Nanomaterials in the environment: Behavior, fate, bioavailability, and effects. Environ Toxicol Chem 27(9):1825–1851. https://doi.org/10.1897/08-090.1

Kreyling WG, Semmler-Behnke M, Chaudhry Q (2010) A complementary definition of nanomaterial. Nano Today 5(3):165–168. https://doi.org/10.1016/j.nantod.2010.03.004

Lamsal K, Kim SW, Jung JH, Kim YS, Kim KS, Lee YS (2011) Application of Silver nanoparticles for the control of Colletotrichum species in vitro and pepper anthracnose disease in field. Mycobiology 39. https://doi.org/10.5941/MYCO.2011.39.3.194

Lin S, Reppert J, Hu Q, Hudson JS, Reid ML, Ratnikova TA, Rao AM, Luo H, Ke PC (2009) Uptake, translocation, and transmission of carbon nanomaterials in Rice. Plan Theory 5(10):1128–1132. https://doi.org/10.1002/smll.200801556

Liu J, Williams PC, Geisler-Lee J, Goodson BM, Fakharifar M, Peiravi M, Chen D, Lightfoot DA, Gemeinhardt ME (2018) Impact of wastewater effluent containing aged nanoparticles and other components on biological activities of the soil microbiome, Arabidopsis plants, and earthworms. Environ Res 164:197–203. https://doi.org/10.1016/j.envres.2018.02.006

Lugtenberg B, Kamilova F (2009) Plant-growth-promoting rhizobacteria. Annu Rev Microbiol 63:541–556. https://doi.org/10.1146/annurev.micro.62.081307.162918

Mahakham W, Sarmah AK, Maensiri S, Theerakulpisut P (2017) Nanopriming technology for enhancing germination and starch metabolism of aged rice seeds using phytosynthesized silver nanoparticles. Sci Rep 7(1):8263. https://doi.org/10.1038/s41598-017-08669-5
Moll J, Klingenfuss F, Widmer F, Gogos A, Bucheli TD, Hartmann M, van der Heijden MGA (2017) Effects of titanium dioxide nanoparticles on soil microbial communities and wheat biomass. Soil Biol Biochem 111:85–93. https://doi.org/10.1016/j.soilbio.2017.03.019
Novikov LS, Voronina EN (2017) Potential space applications of nanomaterials. In: Protection of materials and structures from the space environment. Springer, Cham, pp 139–147
Panke-Buisse K, Poole AC, Goodrich JK, Ley RE, Kao-Kniffin J (2014) Selection on soil microbiomes reveals reproducible impacts on plant function. ISME J 9:980. https://doi.org/10.1038/ismej.2014.196. https://www.nature.com/articles/ismej2014196#supplementary-information
Pérez-de-Luque A (2017) Interaction of nanomaterials with plants: what do we need for real applications in agriculture? Front Environ Sci 5(12). https://doi.org/10.3389/fenvs.2017.00012
Perez-Jaramillo JE, Mendes R, Raaijmakers JM (2016) Impact of plant domestication on rhizosphere microbiome assembly and functions. Plant Mol Biol 90(6):635–644. https://doi.org/10.1007/s11103-015-0337-7
Pérez-Jaramillo JE, Carrión VJ, Bosse M, Ferrão LFV, de Hollander M, Garcia AAF, Ramírez CA, Mendes R, Raaijmakers JM (2017) Linking rhizosphere microbiome composition of wild and domesticated Phaseolus vulgaris to genotypic and root phenotypic traits. ISME J 11:2244. https://doi.org/10.1038/ismej.2017.85. https://www.nature.com/articles/ismej201785#supplementary-information
Piccinno F, Gottschalk F, Seeger S, Nowack B (2011) Industrial production quantities and uses of ten engineered nanomaterials in Europe and the world. J Nanoparti Res:14. https://doi.org/10.1007/s11051-012-1109-9
Prakash O, Sharma R, Rahi P, Karthikeyan N (2015) Role of microorganisms in plant nutrition and health. In: Rakshit A, Singh HB, Sen A (eds) Nutrient use efficiency: from basics to advances. Springer, New Delhi, pp 125–161. https://doi.org/10.1007/978-81-322-2169-2_9
Priester JH, Ge Y, Mielke RE, Horst AM, Moritz SC, Espinosa K, Gelb J, Walker SL, Nisbet RM, An Y-J, Schimel JP, Palmer RG, Hernandez-Viezcas JA, Zhao L, Gardea-Torresdey JL, Holden PA (2012) Soybean susceptibility to manufactured nanomaterials with evidence for food quality and soil fertility interruption. J Proc Natl Acad Sci 109(37):E2451–E2456. https://doi.org/10.1073/pnas.1205431109
Radniecki TS, Stankus DP, Neigh A, Nason JA, Semprini L (2011) Influence of liberated silver from silver nanoparticles on nitrification inhibition of Nitrosomonas europaea. Chemosphere 85(1):43–49. https://doi.org/10.1016/j.chemosphere.2011.06.039
Santhanam R, Luu VT, Weinhold A, Goldberg J, Oh Y, Baldwin IT (2015) Native root-associated bacteria rescue a plant from a sudden-wilt disease that emerged during continuous cropping. J Proc Natl Acad Sci 112(36):E5013–E5020. https://doi.org/10.1073/pnas.1505765112
Savi GD, Piacentini KC, de Souza SR, Costa ME, Santos CM, Scussel VM (2015) Efficacy of zinc compounds in controlling Fusarium head blight and deoxynivalenol formation in wheat (Triticum aestivum L.). Int J Food Microbiol 205:98–104. https://doi.org/10.1016/j.ijfoodmicro.2015.04.001
Sekhon BS (2014) Nanotechnology in agri-food production: an overview. Nanotechnol Sci Appl 7:31–53. https://doi.org/10.2147/NSA.S39406
Sergaki C, Lagunas B, Lidbury I, Gifford ML, Schäfer P (2018) Challenges and approaches in microbiome research: from fundamental to applied. Front Plant Sci 9:1205–1205. https://doi.org/10.3389/fpls.2018.01205
Servin A, Elmer W, Mukherjee A, De la Torre-Roche R, Hamdi H, White JC, Bindraban P, Dimkpa C (2015a) A review of the use of engineered nanomaterials to suppress plant disease and enhance crop yield. J Nanoparti Res 17(2):92. https://doi.org/10.1007/s11051-015-2907-7
Servin A, Elmer W, Mukherjee A, De La Torre Roche R, Hamdi H, White JC, Bindraban P, Dimkpa C (2015b) A review of the use of engineered nanomaterials to suppress plant disease and enhance crop yield. J Nanoparti Res 17. https://doi.org/10.1007/s11051-015-2907-7

Simonin M, Cantarel AAM, Crouzet A, Gervaix J, Martins JMF, Richaume A (2018) Negative effects of copper oxide nanoparticles on carbon and nitrogen cycle microbial activities in contrasting agricultural soils and in presence of. Plan Theory 9(3102). https://doi.org/10.3389/fmicb.2018.03102

Singh J, Lee B-K (2016) Influence of nano-TiO_2 particles on the bioaccumulation of Cd in soybean plants (Glycine max): a possible mechanism for the removal of Cd from the contaminated soil. J Environ Manag 170:88–96. https://doi.org/10.1016/j.jenvman.2016.01.015

Sonkar SK, Roy M, Babar DG, Sarkar S (2012) Water soluble carbon nano-onions from wood wool as growth promoters for gram plants. Nanoscale 4(24):7670–7675. https://doi.org/10.1039/C2NR32408C

Stampoulis D, Sinha S, C White J (2009) Assay-dependent phytotoxicity of nanoparticles to plants. Environ Sci Technol 43. https://doi.org/10.1021/es901695c

Sun TY, Gottschalk F, Hungerbühler K, Nowack B (2014a) Comprehensive probabilistic modelling of environmental emissions of engineered nanomaterials. Environ Pollut 185:69–76. https://doi.org/10.1016/j.envpol.2013.10.004

Sun Z, Liao T, Dou Y, Hwang SM, Park M-S, Jiang L, Kim JH, Dou SX (2014b) Generalized self-assembly of scalable two-dimensional transition metal oxide nanosheets. Nat Commun 5:3813. https://doi.org/10.1038/ncomms4813. https://www.nature.com/articles/ncomms4813#supplementary-information

Tumburu L, Andersen CP, Rygiewicz PT, Reichman JR (2017) Molecular and physiological responses to titanium dioxide and cerium oxide nanoparticles in Arabidopsis. Environ Toxicol Chem 36(1):71–82. https://doi.org/10.1002/etc.3500

Wang X, Liu X, Chen J, Han H, Yuan Z (2014) Evaluation and mechanism of antifungal effects of carbon nanomaterials in controlling plant fungal pathogen. Carbon 68:798–806. https://doi.org/10.1016/j.carbon.2013.11.072

Wang X, Yang X, Chen S, Li Q, Wang W, Hou C, Gao X, Wang L, Wang S (2016) Zinc Oxide Nanoparticles Affect Biomass Accumulation and Photosynthesis in Arabidopsis. Front Plant Sci 6(1243). https://doi.org/10.3389/fpls.2015.01243

Wu H, Tito N, Giraldo JP (2017) Anionic cerium oxide nanoparticles protect plant photosynthesis from abiotic stress by scavenging reactive oxygen species. ACS Nano 11(11):11283–11297. https://doi.org/10.1021/acsnano.7b05723

Yang H, Huang X, Thompson JR, Flower RJ (2014) Soil pollution: urban brownfields. Science 344(6185):691–692. https://doi.org/10.1126/science.344.6185.691-b

Yehia E-T, Joner E (2012) Impact of Fe and Ag nanoparticles on seed germination and differences in bioavailability during exposure in aqueous suspension and soil. Environ Toxicol 27. https://doi.org/10.1002/tox.20610

Yuan Z, Zhang Z, Wang X, Li L, Cai K, Han H (2017) Novel impacts of functionalized multi-walled carbon nanotubes in plants: promotion of nodulation and nitrogenase activity in the rhizobium-legume system. Nanoscale 9(28):9921–9937. https://doi.org/10.1039/C7NR01948C

Zhang L, Wu L, Si Y, Shu K (2018) Size-dependent cytotoxicity of silver nanoparticles to Azotobacter vinelandii: growth inhibition, cell injury, oxidative stress and internalization. PLoS One 13(12):e0209020–e0209020. https://doi.org/10.1371/journal.pone.0209020

Zhu Y, Xu F, Liu Q, Chen M, Liu X, Wang Y, Sun Y, Zhang L (2019) Nanomaterials and plants: positive effects, toxicity and the remediation of metal and metalloid pollution in soil. Sci Total Environ 662:414–421. https://doi.org/10.1016/j.scitotenv.2019.01.234

Chapter 11
Nanoparticles: A New Threat to Crop Plants and Soil Rhizobia?

Hassan Rasouli, Jelena Popović-Djordjević, R. Z. Sayyed, Simin Zarayneh, Majid Jafari, and Bahman Fazeli-Nasab

Abstract Nanoparticles (NPs) are extremely small units occurring at the scale of nanometers (nm) which have been synthesized from both chemical and natural sources. The applicability of these particles has expanded over the past decade so that thousands of useful applications are now attributed to these remarkable particles. The impact of nanotechnology on medicine and other branches of material science suggests that researchers can craft particles for improving and developing agricultural products. The potential benefits of different types of NPs for enhancing the sustainable growth of plants have evaluated under *in vitro* and greenhouse conditions; results show that nanoparticles cause both positive and adverse effects to plants. In some cases, NPs trigger the growth of aerial parts of plants; for other species, no benefits are observed, and in others, growth of target plants decrease or are partially inhibited. Introduction of nanoparticles to agricultural systems, after consideration of possible safety concerns and possible side effects to crop plants and soil ecosystems, may be helpful to farmers for enhancing crop growth, and for conserving arable lands and managing them sustainably. This chapter aims to present

H. Rasouli (✉)
Medical Biology Research Center (MBRC), Kermanshah University of Medical Science, Kermanshah, Iran
e-mail: hrasouli@kums.ac.ir

J. Popović-Djordjević
Faculty of Agriculture, Department of Food Technology and Biochemistry, University of Belgrade, Belgrade, Serbia

R. Z. Sayyed
Department of Microbiology, PSGVP Mandal's Arts, Science & Commerce College, Shahada, Maharashtra, India

S. Zarayneh
Department of Biology, Science and Research Branch, Islamic Azad University, Tehran, Iran

M. Jafari
Faculty of Chemistry, University of Mazandaran, Babolsar, Iran

B. Fazeli-Nasab
Research Department of Agronomy and Plant Breeding, Agricultural Research Institute, University of Zabol, Zabol, Iran

S. Hayat et al. (eds.), *Sustainable Agriculture Reviews 41*, Sustainable Agriculture Reviews 41, https://doi.org/10.1007/978-3-030-33996-8_11

and briefly discuss several nanoparticles and report potential side-effects to plants and soil microorganisms.

Keywords Agriculture · Rhizobacteria · Nanoparticles · Toxicity

11.1 Introduction

Nanotechnology is an exciting field of research for discovering, inventing, and developing methods and approaches for beneficial uses of nanoparticles (Rasouli 2018). Researchers have successfully introduced nanotechnology for applications in medicine, drug delivery, painting, textiles, building, agriculture, biotechnology and so on (Zarayneh et al. 2018). The term 'nanoparticle' (NP) describes a group of minute chemical compounds at the scale of nanometers (nm). These are produced by synthetic methods and also occur naturally (Mohanraj and Chen 2006).

Impending global climate change and increased global population have exacerbated demands for food supplies; however, estimates indicate that available arable lands cannot bear the burden of these pressures in the coming years (Rosenzweig et al. 2001; Fanzo et al. 2018). Therefore, attempts to enhance the production of food supplies using novel methods has received attention from crop biotechnologists (Fanzo et al. 2018; Nuccio et al. 2018). The introduction of supplementary NPs to agricultural systems have opened many possibilities for fortifying the growth of crop plants (Prasad et al. 2017). Various forms of NPs, including powders and emulsions, are commercially available in agricultural markets in which each product is designed for a unique application within a target plant species (Prasad et al. 2017). In addition, utilization of NPs for improving the formulations of herbicides, pesticides, and fungicides is being investigated; products are now formulated with NPs to improve the effectiveness of chemical agents for fighting weeds, forbs, pests, and pathogens (Pérezde and Rubiales 2009; Abigail and Chidambaram 2017).

Nano-based chemical agents are now entering global markets for integrated management scenarios in agricultural systems (Abigail and Chidambaram 2017). Nano-based substances are easily absorbed by target plants based on their specific application; the NP amendment could help plant populations experience sustainable growth (Abigail and Chidambaram 2017, Prasad et al. 2017). It has been proposed, for example, that nano-herbicides can reduce the total amount of applied chemical agents for crop management in comparison to conventional herbicides (Abigail and Chidambaram 2017).

Despite the potential benefits of NPs for agricultural products, the issue of possible toxicity to target plants or soil microorganisms concerns many researchers (Mazumdar and Ahmed 2011). Application of nano-based products for enhancing the growth of plants is widespread, however, the possible toxicity of NPs for agricultural systems remains unclear.

Agriculture and sustainable food production are two determinant factors for establishing the stability of nations; without these elements, nations must confront poverty, economic recession and possible political instability (Borlaug 2000, Tilman et al. 2011). Due to the crucial roles of crops in worldwide food security, it is essential to assess the possibility of NP toxicity if being used to support agriculture. This chapter briefly aims to highlight the potential side-effects of NPs for agriculturally cultivated plants.

11.2 Toxicity of NPs to Plants

Many variables present in the environment may adversely affect plant growth and decrease productivity (Soliman et al. 2018), including anthropogenic chemical agents, physical stressors, microbes, insects, and so on. Among these factors, chemical agents are classified as the most dangerous variables facing plants, posing challenges to the soil, and consequently, to the sustainable growth of plants (Ashraf et al. 2018; Blum 2018). In addition to concerns about chemical agents, potential NP toxicity is a new challenge for agronomists. The adverse effects of these compounds on plants have received increased attention across a number of disciplines in recent years. Studies have shown that among nanoparticles, metallic particles are potentially harmful agents to natural ecosystems. Metallic NPs or their oxides have the potential to genetically or morphologically alter the physiologic status of plant tissues (Rastogi et al. 2017). Evidence suggests that these classes of NPs upregulate genes involved in defense mechanisms, consequently leading to significant oxidative stress in target plants (Rastogi et al. 2017).

Nano-toxicological studies have reported that prolonged exposure of plants to NPs imparts significant side effects to tissues and vital organs. Mazumdar and Ahmed (2011) have shown that silver nanoparticles (AgNPs) measuring 25 nm at concentrations of 50, 500 and 1000 mg/L damaged the cell walls and root vacuoles of *Oryza sativa* plants. Similarly, Rajput et al. (2018a, b) reported that prolonged exposure to NPs could adversely affect soil biological indices. Observed effects include changes to the diversity of soil microorganisms and changes in the distribution of microbial communities. The authors also suggest that NPs cause significant abnormalities in plants (*e.g.* reduced shoot and root growth, and flowers lacking fertile seeds) (Rajput et al. 2018a, b). Koelmel et al. (2013) reported that gold nanoparticles (AuNPs) at a concentration of 0.14 mg/L have the potential to cause toxicity to above-ground tissues of *Oryza sativa* plants. In this context, Taylor et al. (2014) suggested that plant tissues will absorb gold particles in an ionic form, but exposure to gold particles is followed by up-regulating genes for regulating oxidative stress and down-regulating of specific metal transporters for decreasing the total quantity of Au particle uptake (Taylor et al. 2014).

Toxicity of NPS to plants can be linked to the Accumulation of these NPs in topsoil; however, systematic studies and continuous monitoring is required to detect toxic effects of NPs. Rajput et al. (2018a) published a critical review regarding the phytotoxic effects of copper nanoparticles (CuO) to crop plants. According to their discussion, plants experienced negative effects during prolonged exposure to CuO NPs they had the potential to inhibit seed germination and shoot/root elongation in crop plants (Rajput et al. 2018a). Rastogi et al. (2017) critically reviewed the possible side effects of CuO on plants and found that Cu and CuO negatively affect antioxidant systems of plants. The particles displayed moderate to high inhibitory effects to plant growth and seedling development in a dose-dependent manner. Specifically, interference with the uptake of micronutrients, inhibition of root development, triggering of lipid peroxidation, DNA damage, changing phytohormone levels, and altered photosynthetic pathways were observed (Rastogi et al. 2017).

Some plants have expressed obscure behaviors when the tissue was exposed to different concentrations of nanoparticles. Some plants remain unaffected by NPs while others experience significant changes in physiology (Rastogi et al. 2017; Rajput et al. 2018a, b). Yang et al. (2017) reported that plants showed variable responses to NPs. In some cases, activities of plant enzymes decreased while in others root growth or shoot elongation was significantly impaired. Following exposure to different classes of NPs, some plants experienced significant changes in hormonal levels. The chemical essence of NPs (*e.g.* size, surface area, and dose) apparently influenced the potential to increase or decrease concentrations of phytohormones in local compartments. The authors expressed concern regarding the accumulation of nanoparticles in plant tissues, concluding that this may affect food quality in the human diet (Yang et al. 2017). Sukhanova et al. (2018) reported that physicochemical properties of nanoparticles including electric charge, shape, size, and lethal concentrations are key factors for determining possible toxic properties (Sukhanova et al. 2018).

In addition to metallic NPs, other classes of nanoparticles such as quantum dots (QDs) have shown toxic effects in model plants (Santos et al. 2010). Some have reported that QDs have the potential to decrease cell viability and seed germination (Nair et al. 2011). The potential of QDs for changing morphological traits such as root formation is well established (Alimohammadi et al. 2011). However, there is little information about the toxic properties of these NPs within natural ecosystems; therefore, QD toxicity studies under field conditions must be performed to determine whether QDs can disrupt or enhance the sustainable growth of plant species. Table 11.1 presents a number of NPs and their adverse effects on common agricultural plants.

As shown in Table 11.1, NPs results in differing degrees of toxicity when various concentrations are exposed to plants. The behavior of crop plants when in contact with NPs varies from plant to plant so that prediction of the interaction between NPs and plant is difficult. Much excitement has been generated from the fact that some NPs such as SiO_2 and TiO_2 have the potential to improve plant growth (Yang et al. 2017); however, their behavior in ecosystems is uncertain because various factors including light, temperature, other chemical agents, and inherent soil properties might modify the ability of these particles to enhance or, conversely, mitigate their effects (Chen 2018). Researchers must, therefore, examine the behavior of NPs under field conditions to classify them into safe or hazardous categories. This

Table 11.1 Possible toxic effects of NPs to agriculturally cultivated plants

Particle	Plant	Dose	Effects	References
Ag	*Triticum aestivum* L.	10 mg/L	Particles had no effects on wheat DNA but releasing Ag ions caused primary abnormality for root tips.	Vannini et al. (2014)
			Particles could affect the expression of some proteins involved in cell defense and primary metabolism	
	Sorghum bicolor	5–40 μg/mL	Plant growth inhibited but was not significant.	Lee et al. (2012)
	Lycopersicone sculentum	50–5000 μg/mL	Root elongation reduced and physiological status of tomato cells modified.	Song et al. (2013)
			Seed germination decreased.	
	Oryza sativa L.cv. KDML 105	10 or 100 mg/L	Leaf deformation.	Thuesombat et al. (2014)
			Seed germination decreased.	
			Seedling growth reduced.	
			Small particles (~20 nm) showed less negative effects in comparison to larger particles (~150 nm)	
	Cucumis sativus	<200 mg/L	Vegetative growth delayed.	Cui et al. (2014)
			Less negative effects observed for the germination stage.	
ZnO	*Oryza sativa* L.	25, 50, 100 mg/L	Seedling growth inhibited.	Chen et al. (2018)
			Oxidative damage observed.	
			Chlorophyll content decreased.	
	Allium cepa	5 and 50 μg/mL	Oxidative stress and ROS generation observed.	Sun et al. (2019)
			Root growth decreased.	
			Release of Zn^{2+} from ZnO was responsible for cytotoxicity and genotoxicity effects.	
	Triticum aestivum L.	400 and 500 mg/L	Reduced root growth.	Prakash and Chung (2016)
			Decreased shoot growth.	

(continued)

Table 11.1 (continued)

Particle	Plant	Dose	Effects	References
			Oxidative damage observed and lipid peroxidation in roots increased.	
			Redox imbalance and lignification observed.	
	Cucumis sativus	20, 225, 450, and 900 mg/kg soil	In calcareous soil, biomass decreased at higher concentrations.	García-Gómez et al. (2018)
	Zea mays		In acidic soil, germination decreased.	
			Oxidative stress observed.	
	Beta vulgaris		In calcareous soil, biomass decreased at higher concentrations.	
			Photosynthetic pigments increased.	
			In acidic soil, germination decreased.	
	Solanum lycopersicum		In acidic soil, germination decreased.	
			Oxidative stress observed.	
	Pisumsativum		In acidic soil, germination decreased.	
	Lactuca sativa		In acidic soil, germination decreased.	
	Phaseolus vulgaris		In acidic soil, germination decreased.	
			Oxidative stress observed.	
	Raphanussativus		In acidic soil, germination decreased.	
TiO_2	*Triticum aestivum* L. *Viciafaba*	10 g/kg of soil	Wheat biomass decreased.	Du et al. (2011)
		5, 25, 50 mg/L	Plant growth and PSII maximum quantum yields remained unchanged.	Foltête et al. (2011)
			Oxidative stress in shoots did not occur.	
			At 50 mg/L glutathione reductase activity decreased in roots.	
			At 5 and 25 mg/L ascorbate peroxidase activity reduced in roots.	

(continued)

Table 11.1 (continued)

Particle	Plant	Dose	Effects	References
CuO	*Oryza sativa* L.	100 mg/L	Particles reduced root elongation and changed root morphology.	Peng et al. (2015)
			Cell viability decreased.	
			ROS caused oxidative stress	
	Hordeum vulgare L.	0.5–1.5 mL/ medium	The maximum quantum yield of PSII was not affected.	Shaw et al. (2014)
			Oxidative stress observed.	
			Photosynthesis performance decreased.	
	Raphanus sativus	100, 500, 1000 mg/mL	Induced DNA damage.	Atha et al. (2012)
			Oxidative stress observed.	
	Vigna radiata	500 mg/L	Shoot length and biomass reduced.	Nair et al. (2014)
			Chlorophyll content decreased.	
			Growth of primary and lateral roots inhibited.	
			ROS generation and lipid peroxidation in roots increased.	
			Root lignification increased.	
	Cicer arietinum	1000–2000 μg/ mL	Root development inhibited.	Adhikari et al. (2012)
			Root necrosis occurred.	
	Latuca sativa	20 mg/L	The total length of shoot and root decreased.	Hong et al. (2015)
			Nutrient availability and enzymatic activity decreased.	
NiO	*Allium cepa*L.	10–500 mg/L	Higher concentrations of particles decreased mitotic indices.	Manna and Bandyopadhyay (2017)
			Chromosomal aberration and break observed (genotoxic effect)	
			Cell physiology modified.	
			Lipid peroxidation and oxidative stress increased.	
	Solanum lycopersicum	2 mg/mL	Oxidative stress and mitochondrial dysfunction observed.	Faisal et al. (2013)

(continued)

Table 11.1 (continued)

Particle	Plant	Dose	Effects	References
			Particles triggered cell death through increasing ROS molecules.	
Al_2O_3	*Triticum aestivum* L.	5, 25, 50 mg/L	Root elongation decreased.	Yanık and Vardar (2015)
			Peroxidase activity increased.	
			DNA fragmentation induced.	
			Morphological changes increased.	
	Zea mays	1000 mg/L	Seed germination delayed.	Karunakaran et al. (2016)
			Root elongation decreased.	
QDs[a]	*Medicago sativa*	100 Mm CdSe/ZnS QD added to suspension culture	Oxidative stress and ROS generation were observed.	Santos et al. (2010)
			Cell growth significantly reduced.	
SiO_2	*Gossypium hirsutum*	10, 100, 500 and 2000 mg/L	Root and shoot biomass decreased.	Rui et al. (2014)
			Cotton height decreased.	
			SOD activity and IAA level modified.	
			Cytotoxic effect was in a dose-dependent manner.	
CeO_2	*Glycine max*	500, 1000, 2000, 4000 mg/L	Genotoxic effects observed.	Santos et al. (2010)
	Oryza sativa L.	62.50, 125, 250, 500 mg/L	Increased lipid peroxidation in shoots	Rico et al. (2013)
	Hordeum vulgare L.	500 mg/kg soil	Oxidative stress induced.	Rico et al. (2015)
			Plant height increased.	
			Chlorophyll content increased.	
			Potassium leakage increased.	
			Biomass and dry weight increased.	

[a]*QDs* quantum dots, *ROS* reactive oxygen species, *IAA* indole 3-acetic acid, *SOD* superoxide dismutase, *PSII* photosystem II

strategy can help agronomists determine the boundaries for developing environmentally–friendly NPs to use in agricultural systems for increasing productivity of strategic plants.

Rana and Kalaichelvan (2013) reported that several side-effects are commonly linked with NPs, including negative impacts on soil microbial communities, delayed hatching times in aquatic organisms, induction of excited electrons after interaction with UV radiation, production of ROS molecules and oxidative stress, cellular disruption in vital organs, membrane peroxidation, interference with translocation of nutrients in plant cells, altering the normal function of plant roots and stomata, and impaired transpiration and respiration (Rana and Kalaichelvan 2013). Additionally, the occurrence of toxic effects of NPs concurrently with abiotic stresses such as salinity and drought might amplify cellular damage caused by oxidative stress. Therefore, regular monitoring of agricultural soil following the residual nanoparticles is a recommended strategy to identify susceptible plants and soils, along with limiting the application of a specific nanomaterial, if necessary. According to Table 11.1, enhancing oxidative stress, delaying root and shoot elongation, leaf deformation, inhibiting seed germination and seedling growth, modifying photosynthetic pathways, and decreasing biomass are significant practical problems when plants subjected to different concentrations of these nanoparticles.

11.3 Toxicity of NPs to Soil Rhizobia

The potential toxic effects of NPs to soil microorganisms have been investigated. Information on toxic effects of NPs is available from various *in vitro* assays (Dinesh et al. 2012). Chen (2018) reported that the environmental behavior of metal NPs is directly dependent on the complexity of soils, where our understanding of the behavior of metallic particles remains unclear. Other studies have suggested that processes such as shifts in oxidation and reduction status may affect the chemical structure of metallic NPs, consequently leading to release of ions under certain conditions (Jena and Raj 2007; Lok et al. 2007).

Some metallic NPs (*i.e.* AgNPs) impart adverse effects to soil bacterial communities and biological processes such as nitrification. Juan et al. (2017) reported that AgNPs affect the distribution of different types of soil bacterial species including *Acidobacteria, Actinobacteria, Cyanobacteria,* and *Nitrospirae* in a dose-dependent manner. The powerful antibacterial properties of AgNPs modify the bacterial species distribution in soils. In such conditions, a suitable environment for growing *Proteobacteria* and *Planctomycetes* phyla is provided, and as a consequence, populations of useful bacteria involved in soil biological process may decrease (Juan et al. 2017). The authors also report that the impact of AgNPs on *Nitrosomonas europaea*, a well-known ammonia-oxidizing bacterial strain, is directly linked to bacterial cell wall damage and oxidative stress (Juan et al. 2017).

Possible side effects of C60 fullerene (CF) NPs to soil microorganisms have been evaluated. Evidence suggests that CF-NPs are not entirely toxic to soil bacteria (Johansen et al. 2008).

The toxicity of NPs and their inhibitory potential against plant growth promoting rhizobacteria (PGPRs) is yet another important issue. PGPRs are diverse classes of soil bacteria with the potential to have a symbiosis with plant root systems to provide nutrition, immunity and disease resistance to plants (Verma et al. 2019). These microorganisms also have critical roles in maintaining soil structure stability and quality. PGPRs are unique microorganisms in the soil to help plants to escape from salinity, drought and other catastrophic stresses in the soil and surrounding environment. Since NPs have antibacterial properties, the time has come to explore their detrimental effects on soil bacteria.

The extinction of PGPR species in the agricultural soil environment is equal to a significant decrease in the productivity of crop plants. Lewis (2016) has reported that engineered silver NPs and ions could inhibit the growth of three PGPRs namely *Bacillus amyloliquefaciens* GB03, *Sinorhizobium meliloti* 2011, and *Pseudomonas putida* UW4. Mishra and Kumar (2009) have proved the potentially toxic effects of NPs against phytostimulatory soil PGPRs.

The possible toxic effects of metallic NPs to plant growth promoting rhizobacteria (PGPRs) have been surveyed; evidence suggests that AgNPs are inhibitory to *Bacillus amyloliquefaciens* GB03, *Sinorhizobium meliloti* 2011, and *Pseudomonas putida* UW4 strains (Lewis 2016). Mishra and Kumar (2009) evaluated the effects of metallic NPs (*e.g.* Au, Ag, Al) on PGPRs; these NPs had the potential for killing phytostimulatory soil bacteria. Dinesh et al. (2012) reported that metal NPs and metal oxides are a possible hazard to soil bacteria in comparison to fullerene particles. NPs displayed considerable antibacterial properties which enabled them to strongly interact with beneficial microbes under controlled conditions (Dinesh et al. 2012).

Other studies have reported that the chemical and physical properties of soils including pH, electrical conductivity index, mineral content, ionic capacity, dissolved organic matter and humidity play a pivotal role in the mobility, degradation or potential risks of NPs to plants and soil microorganisms (Chen 2018). Taken together, these results suggest that the toxic effects of NPs to soil microbes must be conducted under environmentally realistic conditions to better understand their behavior and effects.

11.4 Conclusion

The available literature states that NPs impart both positive and negative effects to crop plants and microbial population of agricultural soils. The findings discussed in this chapter suggest that among NPs, metals and metal-oxide particles are the most dangerous to plants, causing a variety of abnormalities and dysfunction in both agriculturally cultivated plants and soil bacterial communities. The major limitation with toxicity studies to date is that most have been performed under *in vitro* conditions or in limited soil-based media. The behavior of NPs within the natural environment is likely to be entirely different than that determined under controlled conditions. Current evidence suggests that metallic nanoparticles are a potential hazard to crop plants because of their ability to induce oxidative stress and cause morphological changes and physiological abnormalities. The complexity of the soil is making studies difficult to follow up the fingerprint of NPs in agricultural ecosystems because the numerous environmental factors and chemical properties of NPs increased the number of uncertainties about the interaction of these particles with living organisms.

Safety data about NPs must be compiled and carefully followed; further critical investigations are required to understand the toxic effects of NPs to agricultural systems. Without providing large-scale field-based data about NPs, concern about their adverse effects may not be true. Considering the potential hazards may support chemists and biotechnologist to coin new safe and effective particles for enhancing the sustainable growth of agricultural crop plants in the near future.

Acknowledgments This chapter is dedicated to Prof. Nicolas Taylor, University of Western Australia, for his kindness, and for his endless support.

References

Abigail EA, Chidambaram R (2017) Nanotechnology in herbicide resistance. In: Seehra MS (ed) Nanostructured materials: fabrication to applications. IntechOpen, Rijeka, pp 207–212

Adhikari T, Kundu S, Biswas AK, Tarafdar JC, Rao AS (2012) Effect of copper oxide nanoparticle on seed germination of selected crops. J Agric Sci Technol A2(6A):815

Alimohammadi M, Xu Y, Wang D, Biris AS, Khodakovskaya MV (2011) Physiological responses induced in tomato plants by a two-component nanostructural system composed of carbon nanotubes conjugated with quantum dots and it's in vivo multimodal detection. Nanotechnology 22(29):295101

Ashraf MA, Akbar A, Askari SH, Iqbal M, Rasheed R, Hussain I (2018) Recent advances in abiotic stress tolerance of plants through chemical priming: an overview. In: Advances in seed priming. Springer, Singapore, pp 51–79

Atha DH, Wang H, Petersen EJ, Cleveland D, Holbrook RD, Jaruga P, Dizdaroglu M, Xing B, Nelson BC (2012) Copper oxide nanoparticle mediated DNA damage in terrestrial plant models. Environ Sci Technol 46(3):1819–1827

Blum A (2018) Plant breeding for stress environments. CRC Press, Boca Raton

Borlaug NE (2000) Ending world hunger. The promise of biotechnology and the threat of antiscience zealotry. Plant Physiol 124(2):487–490

Chen H (2018) Metal based nanoparticles in agricultural system: behavior, transport, and interaction with plants. Chem Spec Bioavailab 30(1):123–134

Chen J, Dou R, Yang Z, You T, Gao X, Wang L (2018) Phytotoxicity and bioaccumulation of zinc oxide nanoparticles in rice (Oryza sativa L.). Plant Physiol Biochem 130:604–612

Cui D, Zhang P, Ma Y-h, He X, Li Y-y, Zhao Y-c, Zhang Z-y (2014) Phytotoxicity of silver nanoparticles to cucumber (Cucumis sativus) and wheat (Triticum aestivum). J Zheijang Univ Sci A 15(8):662–670

Dinesh R, Anandaraj M, Srinivasan V, Hamza S (2012) Engineered nanoparticles in the soil and their potential implications to microbial activity. Geoderma 173:19–27

Du W, Sun Y, Ji R, Zhu J, Wu J, Guo H (2011) TiO 2 and ZnO nanoparticles negatively affect wheat growth and soil enzyme activities in agricultural soil. J Environ Monit 13(4):822–828

Faisal M, Saquib Q, Alatar AA, Al-Khedhairy AA, Hegazy AK, Musarrat J (2013) Phytotoxic hazards of NiO-nanoparticles in tomato: a study on mechanism of cell death. J Hazard Mater 250:318–332

Fanzo J, Davis C, McLaren R, Choufani J (2018) The effect of climate change across food systems: implications for nutrition outcomes. Glob Food Sec 18:12–19

Foltête A-S, Masfaraud J-F, Bigorgne E, Nahmani J, Chaurand P, Botta C, Labille J, Rose J, Férard J-F, Cotelle S (2011) Environmental impact of sunscreen nanomaterials: ecotoxicity and genotoxicity of altered TiO_2 nanocomposites on Vicia faba. Environ Pollut 159(10):2515–2522

García-Gómez C, Obrador A, González D, Babín M, Fernández MD (2018) Comparative study of the phytotoxicity of ZnO nanoparticles and Zn accumulation in nine crops grown in a calcareous soil and an acidic soil. Sci Total Environ 644:770–780

Hong J, Rico CM, Zhao L, Adeleye AS, Keller AA, Peralta-Videa JR, Gardea-Torresdey JL (2015) Toxic effects of copper-based nanoparticles or compounds to lettuce (Lactuca sativa) and alfalfa (Medicago sativa). Environ Sci: Processes Impacts 17(1):177–185

Jena BK, Raj CR (2007) Synthesis of flower-like gold nanoparticles and their electrocatalytic activity towards the oxidation of methanol and the reduction of oxygen. Langmuir 23(7):4064–4070

Johansen A, Pedersen AL, Jensen KA, Karlson U, Hansen BM, Scott-Fordsmand JJ, Winding A (2008) Effects of C60 fullerene nanoparticles on soil bacteria and protozoans. Environ Toxicol Chem Int J 27(9):1895–1903

Juan W, Kunhui S, Zhang L, Youbin S (2017) Effects of silver nanoparticles on soil microbial communities and bacterial nitrification in suburban vegetable soils. Pedosphere 27(3):482–490

Karunakaran G, Suriyaprabha R, Rajendran V, Kannan N (2016) Influence of ZrO_2, SiO_2, Al_2O_3 and TiO_2 nanoparticles on maize seed germination under different growth conditions. IET Nanobiotechnol 10(4):171–177

Koelmel J, Leland T, Wang H, Amarasiriwardena D, Xing B (2013) Investigation of gold nanoparticles uptake and their tissue level distribution in rice plants by laser ablation-inductively coupled-mass spectrometry. Environ Pollut 174:222–228

Lee W-M, Kwak JI, An Y-J (2012) Effect of silver nanoparticles in crop plants Phaseolus radiatus and Sorghum bicolor: media effect on phytotoxicity. Chemosphere 86(5):491–499

Lewis RW (2016) Toxicity of engineered nanomaterials to plant growth promoting Rhizobacteria. PhD

Lok C-N, Ho C-M, Chen R, He Q-Y, Yu W-Y, Sun H, Tam PK-H, Chiu J-F, Che C-M (2007) Silver nanoparticles: partial oxidation and antibacterial activities. JBIC J Biol Inorg Chem 12(4):527–534

Manna I, Bandyopadhyay M (2017) Engineered nickel oxide nanoparticle causes substantial physicochemical perturbation in plants. Front Chem 5:92

Mazumdar H, Ahmed G (2011) Phytotoxicity effect of silver nanoparticles on Oryza sativa. IJ ChemTech Res 3(3):1494–1500

Mishra VK, Kumar A (2009) Impact of metal nanoparticles on the plant growth promoting rhizobacteria. Dig J Nanomater Biostruct 4:587–592

Mohanraj V, Chen Y (2006) Nanoparticles-a review. Trop J Pharm Res 5(1):561–573

Nair R, Poulose AC, Nagaoka Y, Yoshida Y, Maekawa T, Kumar DS (2011) Uptake of FITC labeled silica nanoparticles and quantum dots by rice seedlings: effects on seed germination and their potential as biolabels for plants. J Fluoresc 21(6):2057

Nair PMG, Kim S-H, Chung IM (2014) Copper oxide nanoparticle toxicity in mung bean (Vigna radiata L.) seedlings: physiological and molecular level responses of in vitro grown plants. Acta Physiol Plant 36(11):2947–2958

Nuccio ML, Paul M, Bate NJ, Cohn J, Cutler SR (2018) Where are the drought tolerant crops? An assessment of more than two decades of plant biotechnology effort in crop improvement. Plant Sci 273:110–119

Peng C, Duan D, Xu C, Chen Y, Sun L, Zhang H, Yuan X, Zheng L, Yang Y, Yang J (2015) Translocation and biotransformation of CuO nanoparticles in rice (Oryza sativa L.) plants. Environ Pollut 197:99–107

Pérez-de-Luque A, Rubiales D (2009) Nanotechnology for parasitic plant control. Pest Manag Sci Former Pesticide Sci 65(5):540–545

Prakash MG, Chung IM (2016) Determination of zinc oxide nanoparticles toxicity in root growth in wheat (Triticum aestivum L.) seedlings. Acta Biol Hung 67(3):286–296

Prasad R, Bhattacharyya A, Nguyen QD (2017) Nanotechnology in sustainable agriculture: recent developments, challenges, and perspectives. Front Microbiol 8:1014

Rajput VD, Minkina T, Sushkova S, Tsitsuashvili V, Mandzhieva S, Gorovtsov A, Nevidomskyaya D, Gromakova N (2018a) Effect of nanoparticles on crops and soil microbial communities. J Soils Sediments 18(6):2179–2187

Rajput V, Minkina T, Suskova S, Mandzhieva S, Tsitsuashvili V, Chapligin V, Fedorenko A (2018b) Effects of copper nanoparticles (CuO NPs) on crop plants: a mini review. BioNanoScience 8(1):36–42

Rana S, Kalaichelvan P (2013) Ecotoxicity of nanoparticles. *ISRN toxicology2013*

Rasouli H (2018) Devil's hand conceals behind the obscure side of AgNPs: a letter to the editor. Int J Biol Macromol 125:510–513

Rastogi A, Zivcak M, Sytar O, Kalaji HM, He X, Mbarki S, Brestic M (2017) Impact of metal and metal oxide nanoparticles on plant: a critical review. Front Chem 5:78

Rico CM, Hong J, Morales MI, Zhao L, Barrios AC, Zhang J-Y, Peralta-Videa JR, Gardea-Torresdey JL (2013) Effect of cerium oxide nanoparticles on rice: a study involving the antioxidant defense system and in vivo fluorescence imaging. Environ Sci Technol 47(11):5635–5642

Rico CM, Barrios AC, Tan W, Rubenecia R, Lee SC, Varela-Ramirez A, Peralta-Videa JR, Gardea-Torresdey JL (2015) Physiological and biochemical response of soil-grown barley (Hordeum vulgare L.) to cerium oxide nanoparticles. Environ Sci Pollut Res 22(14):10551–10558

Rosenzweig C, Iglesias A, Yang X-B, Epstein PR, Chivian E (2001) Climate change and extreme weather events; implications for food production, plant diseases, and pests. Glob Change Hum Health 2(2):90–104

Rui Y, Gui X, Li X, Liu S, Han Y (2014) Uptake, transport, distribution and bio-effects of SiO_2 nanoparticles in Bt-transgenic cotton. J Nanobiotechnol 12(1):50

Santos AR, Miguel AS, Tomaz L, Malhó R, Maycock C, Patto MCV, Fevereiro P, Oliva A (2010) The impact of CdSe/ZnS quantum dots in cells of Medicago sativa in suspension culture. J Nanobiotechnol 8(1):24

Shaw AK, Ghosh S, Kalaji HM, Bosa K, Brestic M, Zivcak M, Hossain Z (2014) Nano-CuO stress-induced modulation of antioxidative defense and photosynthetic performance of Syrian barley (Hordeum vulgare L.). Environ Exp Bot 102:37–47

Soliman S, El-Keblawy A, Mosa KA, Helmy M, Wani SH (2018) Understanding the phytohormones biosynthetic pathways for developing engineered environmental stress-tolerant crops. In: Biotechnologies of crop improvement, vol 2. Springer, Cham, pp 417–450

Song U, Jun H, Waldman B, Roh J, Kim Y, Yi J, Lee EJ (2013) Functional analyses of nanoparticle toxicity: a comparative study of the effects of TiO_2 and Ag on tomatoes (Lycopersicon esculentum). Ecotoxicol Environ Saf 93:60–67

Sukhanova A, Bozrova S, Sokolov P, Berestovoy M, Karaulov A, Nabiev I (2018) Dependence of nanoparticle toxicity on their physical and chemical properties. Nanoscale Res Lett 13(1):44

Sun Z, Xiong T, Zhang T, Wang N, Chen D, Li S (2019) Influences of zinc oxide nanoparticles on Allium cepa root cells and the primary cause of phytotoxicity. Ecotoxicology 28(2):175–188

Taylor AF, Rylott EL, Anderson CW, Bruce NC (2014) Investigating the toxicity, uptake, nanoparticle formation and genetic response of plants to gold. PLoS One 9(4):e93793

Thuesombat P, Hannongbua S, Akasit S, Chadchawan S (2014) Effect of silver nanoparticles on rice (Oryza sativa L. cv. KDML 105) seed germination and seedling growth. Ecotoxicol Environ Saf 104:302–309

Tilman D, Balzer C, Hill J, Befort BL (2011) Global food demand and the sustainable intensification of agriculture. Proc Natl Acad Sci 108(50):20260–20264

Vannini C, Domingo G, Onelli E, De Mattia F, Bruni I, Marsoni M, Bracale M (2014) Phytotoxic and genotoxic effects of silver nanoparticles exposure on germinating wheat seedlings. J Plant Physiol 171(13):1142–1148

Verma M, Jitendra M, Naveen KA (2019) Plant growth-promoting rhizobacteria: diversity and applications. In: Environmental biotechnology: for sustainable future. Springer, Singapore, pp 129–173

Yang J, Cao W, Rui Y (2017) Interactions between nanoparticles and plants: phytotoxicity and defense mechanisms. J Plant Interact 12(1):158–169

Yanık F, Vardar F (2015) Toxic effects of aluminum oxide (Al_2O_3) nanoparticles on root growth and development in Triticum aestivum. Water Air Soil Pollut 226(9):296

Zarayneh S, Sepahi AA, Jonoobi M, Rasouli H (2018) Comparative antibacterial effects of cellulose nanofiber, chitosan nanofiber, chitosan/cellulose combination and chitosan alone against bacterial contamination of Iranian banknotes. Int J Biol Macromol 118:1045–1054

Index

S. Hayat et al. (eds.), *Sustainable Agriculture Reviews 41*, Sustainable Agriculture Reviews 41, https://doi.org/10.1007/978-3-030-33996-8

Printed in the United States
By Bookmasters